THE HISTORY OF BIOLOGY: AN INTRODUCTION

by F. S. Bodenheimer

FRONTISPIECE

BUST OF ARISTOTLE, BY AN UNKNOWN ARTIST AFTER A PORTRAIT IN THE 4TH CENTURY, B.C., AND THUS APPARENTLY AUTHENTIC.

Original in the Louvre, Paris.

THE HISTORY OF BIOLOGY: AN INTRODUCTION

by

DENHEIMER

& SONS LTD.
DON

DEDICATED TO THREE GREAT HISTORIANS OF BIOLOGY

IN GRATITUDE

CHARLES SINGER

F. J. COLE

WALTER PAGEL

First published 1958

Printed in Great Britain by
HARDING & CURTIS LTD.,
Somerset Hall, Bath

Introduction

ANOTHER History of Biology? Yes and no. This book does not intend to compete with any other book of similar title. Its main purpose is to assist the teacher when he gives lectures on the History of Science. In most cases he can scarcely give the most elementary facts, but almost never will he be able to tell about his own studies, or the historical development of problems which he has studied. The preliminary or parallel reading of this book will make him free to talk about what he wants and to give lectures really having the dignity of higher learning.

Hence in Part One we have shown in very brief outlines the relations and associations of history of science with other branches of culture, such as sociology, economic developments, religion and the sciences of the values (humanities), etc., because everyone interested in the history of science must know how deep are these connections today, and how much science is being stimulated from these other provinces of culture. We have also discussed a small number of methodoligical problems of history of biology, such as homology and analogy, readyness of the time, the philosophical preparations of the background of biological solutions, the very slow developments of the banal truth of today, misunderstandings even where the historical problem is readily available, and last but not least, the uniqueness of Aristotle in the history of biology.

The second part is very short and offers a condensed factual history until second part of the nineteenth century, where only a very few landmarks like Aristotle, Leonardo da Vinci, Spallanzani and Darwin are treated somewhat more fully to permit some discussion. We could equally have chosen Beer, Pasteur, Bernard, Liebig, and similar outstanding men for an enlarged discussion, but to do so in a more extended way would have led by necessity into a veritable chronological history of biology. This was not our intention. A very short, sometimes almost tabular description should help to create a skeleton around which the student, self-taught or otherwise, shall create the flesh.

Part Three is formed by a Source Reader. No two scientists will agree upon any such selection, but the short paragraphs from

130 authors reaching to the beginning of this century are the careful selection of a lifetime of reading in the history of biology. To the teacher who is supposed to be already versed in biology and its history many cross-references will impress themselves upon his mind. Such cross-references give points for interesting discussions with the students.

ACKNOWLEDGEMENTS

To many colleagues the author owes his sincere thanks for many friendly and stimulating talks during the many years of his life. In connection with the present book he wishes to mention especially: D. MacKie—London, S. Kahn—Jerusalem, R. Klebanski—Montreal, L. Kopf—Jerusalem, J. Needham—Cambridge, W. Pagel — London, C. Singer — Cornwall, W. Wilkie — London. Especial appreciation is due to Miss Joyce Allan—Farnham, who helped much in the technical preparations of the book and in reading the proofs. The errors of this book—and it is improbable that none have remained—are of course entirely the author's fault.

The Librarians of the Museum of Natural History—South Kensington, of the British Museum—London, the Bibliothèque Nationale—Paris, the Bibliothèque du Musée National de l'Histoire Naturelle—Paris and of the National Library—Jerusalem may also be assured of the author's gratitude. J. Underwood of the Wellcome Museum—London kindly gave us a number of photographs.

And the following publishers kindly permitted the reprint of quotations (all to be found in the text) from their books. We have always attempted to give the translations of an English or American writer, and deleted many translations which we had prepared ourselves, in order to guarantee a more readable style. The publishers are: The Royal Society of Medicine—London, The Royal Aeronautical Society—London, the Loeb Classical Library—London, the Clarendon Press—Oxford, Harvard University Press—Cambridge, Massachusetts, Jonathan Cape Limited—London, Routledge and Keegan Paul Limited—London, William Heinemann Limited—London, Chronica Botanica Company—Waltham, Massachusetts, Charles T. Brandford Company—Newton Centre, Massachusetts.

CONTENTS

Bath (1140). 31. Frederick II of Palermo (1240). 32. Albertus Magnus (1260). 33. Al Qasvini (1260). 34. Gershon ben Shlomoh (1280). 35. Al Damiri (1400). 36. Da Vinci (1500). 37. Oviedo (1535). 38. Valerius Cordus (1541). 39. Fuchs (1542). 40. Vesalius (1543). 41. Gesner (1555). 42. Olaus Magnus (1555). 43. Belon (1557). 44. Paracelsus (1567). 45. Coiter (1575). 46. Serres (1600). 47. Aldrovandi (1602). 48. Baco of Verulam (1620). 49. Harvey (1628). 50. Pelsaert (1629). 51. Galilei (1630). 52. Descartes (1637). 53. Van Helmont (1640). 54. Boyle (1660). 55. Malpighi (1661). 56. Redi (1668). 57. Van Graaf (1672). 58. Grew (1672). 59. Van Leeuwenhoek (1676). 60. Kircher (1680). 61. Borelli (1680). 62. Swammerdam (1680). 63. Ray (1684). 64. Camerarius (1694). 65. Tournefort (1700). 66. Linné (1720). 67. Hales (1733). 68. Trembley (1744). 69. Bonnet (1745). 70. Lamettrie (1747). 71. Réaumur (1748). 72. Buffon (1749). 73. Von Haller (1749). 74. Wolff (1769). 75. Cook (1773). 76. Spallanzani (1773). 77. Priestley (1774). 78. Blumenbach (1775). 79. Lavoisier (1777). 80. Ingenhousz (1779). 81. White (1789). 82. Von Goethe (1792). 83. Sprengel 1793). 84. Jenner (1798). 85. Malthus (1798). 86. De Saussure (1804). 87. Knight (1806). 88. Bell (1807). 89, 90. Wilson and Audebon (1808, (1827) 91. Lamarck (1809). 92. Cuvier (1812). 93. Von Chamisso (1821). 94. Bichat (1822). 95. Mueller (1826). 96. Von Baer (1828). 97. Woehler (1828). 98. Beaumont (1833). 99. Lyell (1833). 100. Brown (1833). 101. Dutrochet (1837). 102. Unger (1837). 103. Schleiden (1838). 104. Schwann (1839). 105. Von Liebig (1840). 106. Boussingault (1843). 107. Mayer (1845). 108. Darwin (1845). 109. Hofmeister (1851). 110. Von Naegeli (1855). 111. Wallace (1858). 112. Berthelot (1860). 113. Leuckart (1860). 114. Bernard (1865). 115. Mendel (1868). 116. Strassburger (1877). 117. Thomson (1877). 118. Manson (1878). 119. Von Beneden (1883). 120. Pasteur (1885). 121. Ehrlich (1885). 122. Roux (1888). 123. Loeb (1890). 124. Metschnikoff (1892). 125. Von Behring (1893). 126. Schimper (1828). 127. De Vries (1900). 128. Grijn (1901). 129. McClung (1902). 130. Pavlov (1902). 131. Harrison (1907). 132. Howard (1911). 133. Morgan (1917).

Why Study The History of Science?

MANIFOLD are the reasons which induce the individual to take an interest in the History of Science.

There are some people who would like to read for their amusement a more or less anecdotal history of their speciality. We should not say that this is bad in itself, but it is definitely an inadequate motive for making the History of Science an object of serious study. Others would like to find a more serious outline of the development of theories and factual discoveries of their own branch of science. And again, there are some students who are on the look out for a very detailed historical description and analysis of the development of certain problems, such as heredity, experimental embryology, the history of our knowledge of the role of sperm and ovum in fertilisation, the fertilisation of flowers, etc., etc. Few laymen realize the interest which an objective history of this kind is apt to awake.

Another group, mainly led by humanists, believes, and not without very good reason, that the history of scientific thought is the history of human culture and civilization proper. This school is guided by the grand old man of the History of Science, George Sarton, who aims at the creation of a new Humanism based upon the History of Science (in its broadest aspects). Still others, again often humanists, are interested in the culture and history of a certain period, a certain civilization or country. They feel that the History of Science in that period, civilization, or country helps them to understand the general trend of thought. The reverse is, of course, equally true: that the general cultural and philosophical background is important for the understanding of the History of Science. Outstanding recent contributions of that type are J. Needham's *History of Philosophy and Science in China*, or H. Butterfield's *History of Modern Science*. Another group is especially interested in the influence of environmental factors, such as religion and the social or political environment, on the development of science; and there are also those who are interested in the reasons why certain ideas remain for a long time without influence and suddenly gain general recognition with a lightning speed. Darwin's *Origin of Species* is a classical illustration for the latter, Mendel's *Laws of Hybridization* one for the former group. The *History of Biology* (2nd Edition, 1950) of the

Senior amongst the Historians of Biology, Charles Singer, does not aim at a mainly historical point of view. It mainly wishes to give an historical introduction into the modern problems of biology, offering thus another important aspect of the History of Science.

These are only some of the major reasons which have recently, rapidly and on a large scale, increased the interest in the History of Science. All these groups, led by the most varying reasons to take an interest in the History of Science, need a primer orienting them in the development of their own science. It may be frankly stated that—apart from the unique phenomenon of George Sarton—no single person is able to embrace the whole field of the History of Science. The London Graduate Institute for the History and Philosophy of Science has made the requirement, and the author has become convinced that it is right, that only the student and the teacher who has actively worked in the field of one branch of science, who has actually tried to solve one of its actual problems, is fit to take an interest in, to study and to teach, that one branch of science in which he has worked (in its largest definition). We share this view and deal here, therefore, only with the history of biology, as in other fields we could only give second-hand knowledge and second-hand errors.

We have mentioned that an initial orientation is desirable. This is true as well for University courses in the History of Biology, as for those who desire a first introduction for the purpose of more detailed and more intensive self study. The advanced lecturer, especially, should be able to refer the student to such a first introduction, so that he may be able to discuss more freely the development of theories, of problems or other special topics of his own research, from which discussion the students will gain stimulation and guidance for research of their own. This present book intends to be such a basic preliminary to the hearing of University lectures, or to self study.

PART I

Parerga and Paralipomena instead of a General Introduction

1. A few words on Epistemology and on the Principles of Biology.
2. Methods and Aims of the History of Science.
3. External influences upon the History of Science.
4. The History of Science in its Humanistic Interrelations.
5. The Philosophical Background of Scientific Innovations.
6. Erroneous Interpretations of good Observations may lead to important Discoveries.
7. The long Way of Errors of some Biological Problems:
   (*a*) The History of Generation.
   (*b*) Some Facts about the Revolution of the Renaissance.
   (*c*) The long History of the Interpretation of Insect Parasitism, an illustration of the Accumulation of Knowledge.
8. Errors of Tradition in the History of Biology.
9. Analogy and Homology in the History of Biology.
10. Problems of Priority.
11. An Introduction, for the Beginner, into the Bibliography of the History of Biology.
12. References.

1

A FEW WORDS ON EPISTEMOLOGY AND ON THE PRINCIPLES OF BIOLOGY

IN an earlier analysis of the concept of biotic organization in synecology, namely that life-communities are well integrated super-organisms (Bodenheimer, 1953), we came to the conclusion that general *apriori* approaches to a physico-chemical unity of the universe, such as that proposed by A. J. Lotka (*Elements of Physical Biology*, 1925), have no reference to any general *biological* problem. The ruling theory is that life-communities are in a highly integrated equilibrium and represent actually true homologies to the structures and functions of an organism. A full statement of this view is found, for example, in the recent manual of Allee, and others (*Principles of Animal Ecology*, 1949). Bodenheimer (*Problems of Animal Ecology*, 1938) strongly refutes this point of view insofar as it claims to be proven by a synthesis of observation and induction. That similar life-communities occur in similar environments is merely the outcome of natural selection, which results empirically in similar life-associations.

The epistemological approach, generally speaking, teaches us that man is much more inclined to accept preconceived intuitive ideas than a critical analysis would agree with. We shall cite only a few relevant quotations:

Kant, the founder of modern epistemology, states in his *Critique of Pure Reason:* " Experience is by no means the only field to which our understanding can be confined. Experience tells us what is, but not that it must be necessarily what it is and not otherwise. It therefore never gives us any really general truths; and our reason, which is particularly anxious for that class of knowledge, is roused by it rather than satisfied. General truths, which at the same time bear the character of an inward necessity, must be independent of experience—clear and certain in themselves."

B. Russell states, in his *Human Knowledge*: " It is customary to speak of induction as what is needed to make the truth of scientific laws probable. I do not think that induction, pure and simple, is fundamental. . . . Every finite set of observations is compatible with a number of mutually inconsistent laws, all of which have exactly the same inductive evidence in their favour. Therefore pure induction is invalid, and is, moreover, not what we in fact believe. . . . The law is one which had suggested itself more or less independently of the evidence, and had seemed to us in some way likely to be true."

We add the resigned conclusion of A. Standon (*Science is a Sacred Cow*, 1952), " that biology is one vast mass of analysis, very different indeed from the cold logical thinking of the physicist. . . . In its central content biology is not accurate thinking, but accurate observation and imaginative thinking, with great sweeping generalisations. . . . The ' Unity of Life ' is a catch phrase biologists are addicted to, although this can hardly be regarded as confirmed by experiment, because it is impossible to say what it means, if indeed it has any meaning at all. . . . Neither science nor scientists have any notion what goes on in the zone of higher knowledge."

Russell even defends the idea that the professional authority is a factor of great weight in human knowledge. Most competent analyses agree that induction alone can never lead methodologically to the idea of a superorganistic organisation of life-communities. It is clear that here we have before us an intuitive hypothesis, which may be correct or wrong. A number of facts contradict it, many more may be either analogies or true homologies. As long as the biologist realizes this situation and does not claim, as most ecologists do today, that this superorganistic structure is definitely proven (by induction!), we may regard it as an hypothesis, useful as long as it leads to further research. The only mortal sin is to anticipate theories as established, when the inductive material may be sufficient to stimulate an inspiring vision, but does not—or does not yet—form a solid base for truth or even for a high degree of scientific probability. Science advances only when it passes successfully between this Scylla and Charybdis into the open sea of balanced judgement between observation and intuition (Bodenheimer, 1952).

Woodger once remarked that biology has not yet found its Galilei. To this dogmatic statement I wish to oppose the suggestion that such an answer is due essentially to what appears to be the frog perspective of many modern biological thinkers. The Galilean revolution was needed in physics because Aristotle's physics was built upon the concept of a geocentric universe. But it is usually forgotten that Aristotle was primarily a biologist, and his methodology, epistemology and bio. logical principles did grow from his deep delving biological experience. It is true we find many errors in detail, especially in his physiology, to mention only the four elements and the four humours, the innate temperature and the physiology of the heart and of digestion; but, on the other hand his taxonomic, morphological, embryological, ecological and other principles are so well conceived, that they still are the equivalent of the physical Galilean revolution. His wide and deep experience enabled Aristotle also to use those of his principles which cannot any longer claim general approval, like his teleology—which, in any case, was much more subtle than is often assumed—in a fashion elastic enough to avoid serious errors. He never was afraid of making observation the prime arbiter of his conclusions (see, for example, *De Gen. Anim.* 760b 28ff.). We have little hesitation in declaring his animal system superior even to that of Linnaeus. The taxonomic principles of his *Analytica Posteriora* are the only methodological analysis of this branch of natural history until the very recent present. The following are some of his important principles, which often are regarded as discoveries of the twentieth century:

(1) The principle of the correlation of the organs within the organism, i.e., that any surplus on the development of one organ induces—in consequence of the limited amount of building material available—the reduction of another organ, such as the reduction in the number of teeth in horn-carrying mammals. Yet, Aristotle knows also that another type of correlation exists where he was unable to discover any functional correlation, e.g., the correlation of cotyledons with lack of front teeth in the upper jaw, and their absence where all front teeth are present but horns are missing (*De Gen. Anim.* 746a 9).

(2) The principle of structural and functional adaptation to habitat and habits. This principle is excellently demonstrated for the birds (*De Part. Anim.* 693a 10ff.).

(3) The principle of parsimony, according to which nature never expends any unnecessary energy.

(4) The principle of homoiomerous and anhomoiomerous, i.e., of homogenous and heterogenous parts of the body; we would say today: tissues (bone, muscle, etc.) and organs (eye, hand, etc.). Yet here structure and function sometimes are intermingled, e.g., when the heart is declared to be homoiomerous by reason of its structure, and anhomoiomerous by reason of its being a functional organ.

(5) In the classification of animals we find a number of important principles. Methodologically his refusal of the mechanical application of the dichotomous principle is of paramount

importance. Making a division between animals with wings and without wings would separate ants of the same species widely from one another. He builds his biology upon the principle of unity of plan of the *Sanguinea* (vertebrates), as well as of every other big group amongst the evertebrates. We may add here another principle, that of the ladder of nature, describing the entire world of living beings as a continuous chain or ladder or organisation, instead of an anarchic diversity.

(6) Most important is the principle of homology and analogy, which enabled the foundation of comparative anatomy. A definition of homology is found in *De Part. Anim.* (644b 8ff.): " It is generally similarity in the shape of particular organs or of the whole body, but has determined the formation of the larger groups. It is in virtue of such a similarity that birds, Cephalopoda and Testacea have been made to form each a separate class. For within the limits of each such class, the parts do not differ in that they have no nearer resemblance than that of analogy—such as exists between the bone of man and the spine of fish—but differ merely in respect of such corporeal conditions as largeness, smallness, softness, hardness, smoothness, roughness, and other similar oppositions, or, in one word, in respect of degree." In the *History of Animals*, nail and hoof, the hand of man and a crab's claw, the scale of a fish and the feather of a bird are given as analogous features, and the alimentary canals of all animals as well. While many mistakes are still made in detail, the principle is clearly established: architectural homology and functional analogy, perhaps one of the most important discoveries of general morphology.

(7) To the principle of the unity of epigenetical development and " invisible " preformation should be added the basic principle that the general structures appear in development before the special or differentiated ones (*De Gen. Anim.* 736b 2ff.); also, that tissues originate before the organs (*De Part. Anim.* 646b 5ff.).

(8) Concerning heredity, we may say that Aristotle's knowledge was not more extended than the ignorance of principles on which Charles Darwin built up his classical theories of variation and natural selection. Yet from the methodological point of view, his concepts were sounder. Aristotle attacked

the Hippocratic theory of pangenesis, revived by Darwin, namely that small particles from all parts of the body migrate into the semen and from it or within it build up the corresponding parts of the embryo. His arguments may be summarized as follows (Bodenheimer, 1952 p. 17):

(*a*) It is not possible to explain certain immaterial characters, such as the timbre of a voice or the particular body attitudes of the individual, in this manner.

(*b*) Young parents without grey hair and beard produce children the hair of which becomes grey in senescence. Also, in plants the offspring often resembles its parents in characters which the latter did not yet manifest at their reproductive season.

(*c*) Children sometimes are not similar to their parents but to farther distant ancestors from whom they cannot possibly have obtained material particles by direct transmission.

(*d*) The pangenetic particles are supposed to be derived from organs, not from tissues, which makes this hypothesis much too complicated to be accepted.

(*e*) As the body is really only the wrapping of the embryo, this hypothesis is not more acceptable than that the clothes and shoes send particles into the semen, an obviously absurd assumption.

(*f*) Children of crippled or mutilated parents do not always copy their mutilations, which would be an unavoidable necessity according to the Hippocratic theory. It took Weismann to revive this view 2100 years later.

(*g*) Both semina, the maternal as well as the paternal, in principle participate in the characters of the offspring, according to the virility of each. Also, the sex is only determined after the mixing of both these semina, i.e., at the beginning of embryonic development. This is also true where environmental factors, such as the direction of the compass to which the copulating animals turn, determine the sex.

It is unnecessary to underline how many sound principles are condensed in these few propositions.

It is proper to return to the remarks of J. Needham (*History of Embryology*, 1934, p. 40 f.) regarding the personality of Aristotle:

"A metaphysician as well as a scientific worker, he was able to use the concept of purposiveness as a heuristic aid, but he never *rested* upon it. The trouble was that he introduced it into the discussion at all. It is an interesting speculation to consider what would have happened if the first great biologist had not brought final causes into his teaching . . . Final causes led (the epigones) irresistibly to the theological blank alleys . . . Perhaps A. would not have made so many great discoveries if he had been more of a Democritus. For teleology is like other varieties of common sense, useful from time to time . . .; every biologist acts in the same way at the present time. The important thing is not to give the last word to teleology. A. did know how to change rapidly from metaphysician into physicist and back again, how to bow politely to the final cause and press on with the dissection. . ."

It is true that the knowledge of the master in physiology was imperfect, because cellular and micro-hereditary analysis, etc., were branches of biology not understood at the Aristotelian period, and not only for lack of the microscope. But was the knowledge of Darwin in many of these fields more competent? It is probable that the biological revolution which we are approaching is much more concerned with the "grands molécules organiques" of Maupertuis, Leibnitz and Buffon, which are more and more being recognized as building up the cell—in the same way in which the cells build up the organism—yet are so much nearer to the origin of life. Quite a few leading biologists: W. Thompson, E. S. Russell, and J. Needham, to name a few, agree with us that of all biological principles those of Aristotle have most often retained their youth and freshness till our day. It seems to be the best way for a revival of biological thought to include the reading (with competent commentaries) of parts of the *Analytica Posteriora* (and related books) as well as of the *Parts of Animals* (with selections from *On Generation* and the *History of Animals*) in every biological curriculum. It would be well if special translations in an easily understandable language would be prepared for that purpose. This seems to be the soundest return to biological thought, which has sunken into deep decline by its having lost the memory of Aristotle; and at the same time it would be the best preparation for the coming reform or revolution in biological principles. To this end it is also needed that those responsible for the curricula realize better than they do today, that not the teaching of facts is the main aim of the university, but education towards an improvement of the ability of thinking in the students.

Mathematicians and physicists have recently tried to give us an introspective view into the ' workshop ' of the genesis of the conception of new scientific ideas (see Taton, 1955). We have little of such introspection available in biology. Yet a careful reading of Darwin's *Beagle* report clearly shows that his first revelations of evolution and natural selection came to him as impressive intuitions. The deep impact of these first intuitions was followed by decades of careful observations, reading and critical thought. Different is the story of that work of genius, the discovery of the rings of organic chemistry by Kékulé, who dreamed on the roof of a London bus that jumping cats, biting each into the tail of its predecessor, formed a closed ring. This was the model for the modern ring-theory of organic chemistry.

The sarcode had already been discovered by Dujardin; protoplasm, by Purkinje; the presence of a nucleus in all plant cells, by Robert Brown (1829). But the comprehensive cellular theory of plants, still imbued with highly unsatisfactory ideas concerning the origin of these cells, was given in 1838 by Schleiden. The following year saw a much more satisfactory general theory of all organisms by Schwann. About the history of the latter representation we read (J. R. Baker, 1946, p. 154.f): " It was in October 1838 that the ex-lawyer M. J. Schleiden and the anatomist Theodor Schwann dined together in Berlin. They were a strangely assorted pair. The volatile Schleiden, having shot himself in the forehead and recovered, can have had little in common with the placid Schwann apart from their intense interest in the minute anatomy of organisms. Schleiden described to Schwann the nucleus of plant cells, and Schwann at once recognized it as corresponding to something with which he was familiar in cells of the spinal cord of vertebrates. The two men repaired forthwith to Schwann's laboratory in the Anatomical Institute of the University. Schwann showed his friend the cells of the spinal cord, and Schleiden at once recognized the nuclei as corresponding to those with which he was familiar in plants. Due recognition must be given the researches of those who had preceded them, but this occasion may nevertheless be justly regarded as marking the first general formulation of the cell theory."

There is no longer any doubt that, not ' scientific method ', but intuition is the cradle of every great biological new idea, as well as it is in mathematics and physics. The scientific method has its well-merited place in that induction which advocates the careful checking of every new theory.

2

METHODS AND AIMS OF THE HISTORY OF SCIENCE

THE history of science, as part of the historical sciences, is bound to work with historical methods, apart from the obligation of being familiar to no mean degree with the object—science itself. This task calls for a combination of knowledge and other efforts which is beyond the forces of the average individual student, especially when we consider that, in addition to this amazing extent of knowledge, no historian of science can successfully tackle any major problem without a gift of intuition, with a certain amount of phantasy, wherever he aims at creative synthesis.

(1) As long as the great majority of the original texts of history of science are unpublished manuscripts, they are not available to the general student. In addition, even where texts in Arabic, Hebrew, Greek, Chinese, Indian, Assyrian, Egyptian, and other languages are published, they usually are not available in reliable translations, and hence they are closed to all those who are not familiar with these languages. It is to be regretted that most linguists have shown so far very little inclination to study the texts of interest to the historian of science, which have remained until our day a stepchild of literary research. The translation of these original texts is urgently needed, though usually impossible without the aid of the philologist. The first translation of such a text will give an idea of the general attitude, the main sources, and many interesting facts contained in that text. Yet very often the texts, even printed ones, are not up to the standard of modern philological science. A number of manuscripts must be compared in order to reconstitute the original text as far as possible, which can be done only with the help of a philologist. In addition, even the reconstituted texts still offer many problems, where the help of the philologist again is indispensable. Thus, the biological interpretation of certain terms of Aristotle is ambiguous and obscure today. His term ' entelecheia ' is interpreted by some modern biologists

(Russell, Needham) in the sense of modern experimental embryology, while a certain philologist (Mauthner) takes its meaning for a misunderstood application of the dative, with many other intermediate interpretations. Only historical, philological and biological teamwork will finally succeed in making precise the ambiguity of such terms. Senn (1933), who has devoted many years of good research to Theophrastus, stresses the changing interpretation of certain words, such as *symbainein* and *symbebekos* in Aristotle and Theophrastus from chance observation to purposeful observation. Here again the philologist's help is needed. By far too little attention has hitherto been given to the changing meaning of the same word in various periods. Thus, *experimentum* still referred even in Bacon of Verulam chiefly to observation. And the meaning of evolution (= preformation) in the 17th century certainly differed from the 'transformation' of our day.

(2) From the philosophical point of view, we must never forget the saying of Kant (*Kritik der Reinen Vernunft.* Leipzig, 1930. p. 862): "Nobody can try to build a science without being guided by an idea." It is by far too often regarded as a criterion of true science, that it should be unprejudiced. In a certain sense this is a banality, yet in another one it is nonsense. Senn (1933) has tried to demonstrate that Theophrastus in his later research was infinitely superior to Aristotle: while the master was only guided by leading ideas and arranged his facts according to a preconceived system of order and causes, Theophrastus renounced any guiding or preconceived ideas and restricted himself to the mere description of observations and facts from which ultimately he derived his generalisations. Yet we know no case of any such procedure in scientific research. No scientist ever has reached generalisations when he was unguided by heuristic theories. And with regard to the final value of such heuristic theories nobody has ever said a truer word about their temporary value than Aristotle himself in his discussion of the generation of bees. After he has extensively discussed his observations and his (often incorrect) conclusions, he concludes (*De Generatione Animalium* 760b 28ff.): "Such appears to be the truth about the generation of bees, judging from theory and from what are believed to be the facts about them; the facts, however, have not yet been sufficiently grasped; if ever they are, then credit must be given rather to observation than to theories, and to theories only if what they affirm agrees with the observed facts."

This is the proper attitude of true and sincere research, namely, to strive after the solution of problems without personal, material, or political intentions, and as free from such influences as from the prejudices of one's own mind. The sincerity and truth of science is at the opposite pole from self-complacency and from bombastic phraseology. Such true seekers after truth are common amongst the men of science of all ages. A good illustration is found in the first words of Shemtob Falaquera's (13th century) *Introduction into Science:* " If you want to gain knowledge, first prepare your soul. Withdraw from all extremes and move on the middle path. The beginning of wisdom is the fear of God." Another of Falaquera's books, *The Seeker after Truth,* is entirely devoted to the story of a youth who goes out to inquire, from the masters of success and of learning of all professions, after the truth which they have experienced. We have to respect this note of sincerity, wherever we find it in history. On the other hand, there has been no lack of self-complacent, shallow students of restricted mind—or of clever, agile students of greedy mind—either in the past or in the present. The criterion of truth and sincerity is especially important in the Middle Ages, when the conception of originality differed so much from our own that it is often the decisive factor in determining the value of a book.

(3) The mastering of proper historical methods, from those of bibliography and documents to a critical attitude towards an analysis of dates, is of no less importance. Often dates given range widely, even of almost contemporaneous events. Even the date of printing of one of the early editions of Ptolemy's *Geography* deviates 15 years from the actual date of publication (1462 for 1477). Such errors occur naturally, apart from intended falsifications or mystifications regarding the origin of a book. The latter aim was successfully achieved for some centuries, for example, with the *Book of Agriculture* of Ibn Wahyah (9th cent.), who pretended that his work was the translation of an ancient Chaldaean or Nabathaean book.

Of great importance is the proper perspective of time. We may compare, from Darmstaedter's *Handbook of the Discoveries* (1908), the various periods and their percentage of total history with the number of pages devoted to the discoveries of that period. The low number of discoveries in early periods is, of course, not based alone on the lower productivity of science, nor upon the earlier narrow localisation of scientific activity, nor upon the low population densities of earlier ages, but largely upon inadequate documentation for the past, as compared with those for the present.

Centuries.	Duration in years.	Percentage of total time.	Percentage of total discoveries.
3500 B.C.—1600 A.D.	500	94.14	10
XVIIth cent. ...	100	1.85	5
XVIIIth cent. ...	100	1.85	11
XIXth cent. ...	100	1.85	67
1901—1908 ...	8	0.14	7

Yet what are these 5400 years of scientific tradition in comparison with the 500,000 years during which man slowly developed language, fire, tools, clothing, habitation, hunting, primitive agriculture, and husbandry, and so forth, from which we have no written documentation at all? Such comparisons help us to put our present period into a proper perspective: as a passing moment in the steady flow of science and of its history. (G. Sarton.)

(4) There are various ways of writing the history of science : a biography, the monograph dealing with a problem or a period, the full history of a science. The main types are those exhibiting the descriptive, the judging, and the understanding methods so classified by Lütjeharms, (1936, p. 14).

The first, or descriptive, method compiles, accumulates and tabulates all the material facts : of biography, description, and theory in a strictly chronological order. Such a compilation of facts in suitable historical sequence, while not an aim in itself, is most useful, especially if accompanied by a reliable index. It is always the first step and a necessary step, yet it is never the final word in the history of science. A useful book of that type is C. Pickering's *Chronological History of Plants* (Boston 1879), which enumerates all plants of economic, medical, or cultural interest, together with the facts about the beginnings of their use and discovery.

The second or judging method is the one preferred by positivism. The historical material is selected, mainly from the point of view of its survival in modern theories, and is often arranged according to the history of the various branches, theories or problems of a science. J. Sachs' *Geschichte der Botanik* (München, 1875) is typical, revealing all the advantages and drawbacks of this method.

The third or understanding method is that which we have repeatedly defended here. Troeltsch (1916) has described it: " To understand a period means to measure it by its own problems and ideals." All good modern histories of science aim to apply this method, with varying success.

We now come to the conclusion of this introductive survey with the following deductions (Sarton, 1936, p. 5) :

Definition: Science is systematized positive knowledge, or what has been taken as such at different ages and in different places.

Theorem: The acquisition and systematization of positive knowledge are the only human activities which are truly cumulative and progressive.

Corollary: The history of science is the only history which can illustrate the progress of mankind. In fact, progress has no definite and unquestionable meaning in other fields than in the field of science.

In short, the purpose of the history of science is to establish the genesis and the development of scientific facts and ideas, taking into account all intellectual exchanges and all influences brought into play by the very progress of civilization. It is indeed a history of human civilization, considered from its highest point of view. The centre of interest is the evolution of science and of thought, but general history always forms the background (Sarton, 1948, p. 33).

The basic criticism and analysis of scientific work—its methods, bases, and aims—springs from two main sources: from philosophy and from history. The philosophical approach is mainly through epistemology. The analysis of the ways and means of human understanding and knowledge is indispensable in order to gain the true perspective and limitations of scientific work. B. Russell in his *Human Knowledge* (1948) has admirably shown in what manner our individual theory of epistemology influences the judgment of present achievements. Yet not less important is the historical way for the understanding of the progress of knowledge, for learning the essentials of its development—the difficulties, indigenous as well as external, to be overcome; the errors from which every period has to extricate itself for the progressive purification of its trends of thought and ideas. The complicated paths of pro- and retrogressions in the development of every theory ask for an historical analysis. Both the philosophy and the history of science have in common their lifting the watertight compartments of the particular sciences, of national and racial boundaries above their isolation, to demonstrate the essential interrelations of the various branches of science, of various countries and civilisations, of the various periods of the past. Thus we gain a higher base for grasping the unity and the continuity of the human mind.

We conclude with two conclusions of George Sarton (1948, p. 58, 57), that grand old man amongst the historians of science, who has done more than anybody else to raise the history of science from the

Fig. 1. George Sarton (1872-1956). (From the dust-cover of Horus).

status of a mere speciality and to make it a keystone of modern humanism:

> " History of science, i.e., the history of human thought and civilization in its broadest form, is the indispensable basis of any philosophy. History is but a method, not an aim."
>
> " We must try to humanize science, to show its various relations with other human activities—its relation to our own nature. It will not lower science; on the contrary, science remains the centre of human evolution and its highest goal; to humanize it is not to make it less important, but more significant, more impressive, more amiable. The new humanism, as I venture to call the intellectual movement that has been defined here, will also disentangle us from many local and national prejudices, also from many of the common prejudices of our own time . . . Our age is not necessarily the best or the wisest, and anyhow it is not the last! We have to prepare for the next one, and I hope a better one . . . To build up this future, to make it beautiful, as the glorious times of synthetic knowledge like those of Phidias and Leonardo da Vinci, it is necessary to prepare a new synthesis, a more intimate collaboration between scientist, philosopher and historian. This would give birth to so much beauty that the collaboration of the artist would also be secured: an age of synthesis is always an age of art. This new humanism is something in the making—not a dream. We see it growing, but no one can tell how big it will grow."

The History of Science confirms the essential unity of East and West. Nobody deserves greater credit for the detailed documentation of this than J. Needham (1953). He has shown, for the first thirteen centuries of the Christian era, a continuous but lagging transmission of technological and ideological knowledge from East to West, followed by a similar but reversed trend. Arabs, Jews, and Jesuits were important transmitters, while Chinese scientists frequently visited Western courts up to Persia and Bagdad. Many analogies and even homologies are just convergences, such as the conception of the ' ladder of souls ' (Aristotle, Hsün Tzu). Needham is, however (with G. Childe), of the opinion that the earliest basic inventions, like fire, wheel, plough, weaving, domestication of plants and animals, bronze metallurgy, etc., " can scarcely be envisaged as of dual origin", the original centre being apparently Mesopotamia. While this theory is, of course, very different from, and much more soundly based than, that of the monophyletic origin of all culture, it still seems to be far-reaching. Fire, plough, and domestication are decidedly polyphyletic in origin. Or, if fire was a monophyletic invention (against all probability), it goes back at least to the *Pithecanthropus* of Java. Many of the latter homologies have certainly taken their origin independently in the East and in the West. Apart from direct transmission, we have stimulus diffusion, as exemplified by the windmill. The windmill is probably a Persian invention which remained based upon a horizontal wheel, until it emerged, rather late, in Western Europe on a vertical wheel. We are again doubtful, however, whether the description of western schizophrenic antitheses, in contrast to the Chinese synthesis of mind, which Needham exemplifies by the micro-macro-cosmos conception, is a correct one. We had in certain periods and personalities a complete synthesis of the Universe, and I am practically sure that a better knowledge of Chinese literature will reveal that there also existed periods and personalities inclined to ‘ schizophrenic ’ antitheses. Such a condition will only tend to make the Unity of East and West more real and more complete.

The vice of contemporary science is over-specialisation and atomisation of the problems and of the students. Yet, as Singer (1950) points out since most specialised problems are limited in extent, they will be solved within a reasonable time and cease to attract new students. The student of protozoan diseases, for example, will be more and more attracted by the problems of general cytology, pharmacology, heredity, etc., and thus lift a narrow field of specialisation from the isolation back into the general trends of science in the future. “ Specialisation shows the solution of problems, but as soon as these are solved, the results are added to general science and the specialisation loses its justification”. This hopeful outlook is, however, often inhibited by the effect of mere inertia.

An interesting aspect of the problem is that, when great men discover a new theory or a new method of research, they themselves often speedily exhaust their potential value almost to the limit to be explored within the framework of the science of their time. We may refer to the exhaustive exploration of C. Darwin into the problems of the transformation of species. His maturity, his lack of prejudice, and his fairness, which led him to inquire into all possible objections at every step , make his *Origin of Species* one of the classics of biology, in spite of the fact that he was ignorant of even the most basic facts of heredity. Thus Roux exhausted the main possibilities of his new method of experimental embryology at his time with his own experiments and thoughts, but his method survives and is intensively applied to new problems arising from the progress of biology. And the same is true for Carrell's discovery of tissue culture: he himself explored by his new method all the general problems to which the new method promised a new approach and solution, but it flourishes today as an auxiliary method for the solution of a number of special problems.

The problem of the importance of the individual hero exists in the history of science just as prominently as it does in general history. We still are far from a real understanding of the role which the individual thinker plays or may play in directing the future trend of science, or to what degree the individual reaches maturity and ascendance only or mainly upon the background of the collective trend of his period. One thing is clear however: no great idea has ever been conceived *and* successively spread by a mediocre person or by a merely learned man, but always and only by true explorers of science and of mind by personalities aware of wide horizons with eyes open to problems which had not been recognised as such by earlier students and minds ready to work towards the solutions of the new problems—personalities usually with a broadly human and humanistic outlook. One is often surprised at the greatness of conception so often found in apparently parchment-dry specialists who opened up one specialised field or another. Who would have expected to find the following sentences in that monument of hyperpedantry *Die Hebraeischen Übersetzungen des Mittelalters* (Berlin 1893, p. xxiv), of Steinschneider, which compiles with deep understanding of cultural interrelations the bibliography of Hebrew mediaeval science and literature, a 'terra' almost 'incognita' before and after him: "Saying farewell to this book, which occupied me for the greater half of a long life, is to some extent a good-bye to life itself. And every end makes us realise, how tiny the individual is in confrontation with the universe. Yet just this thought stimulates us to explore more fully the interrelations between the individual and the universe, teaching us that the finite merges into the infinite, without being annihilated by it."

One of the basic mistakes which the historian of science may commit is to measure the thought, the achievements, the theories, and the systems of a period of the past with the gauge of modern science. This analogy, often used by the adherents of positivism, falsifies the perspective. In order to understand any period, any early scientist, or any theory of the past, one has first to learn the entire development which preceded them and to forget all later developments. Every period, every scientist, every theory has to be understood within the frame of his contemporaneous knowledge and thought, against the background of the political, cultural, social and technical events and achievements of his time (Sergescu, 1951). It is indispensable to read the publications under study as a contemporary would have read them when they were first published. This is by no means an easy task. A positivist will deny that this is possible or desirable at all. He is merely interested in finding out that part of ancient theories which still forms part of modern theory (Metzger, 1935). Yet actually the historian cannot regard a theory or a system of thought as isolated phenomena leading to modern development. Every period is a spiritual entity with its own aims, its own problems, its own traditions, its own methods. In order to understand a Cartesian physicist, we must become temporarily a Cartesian ourselves. Only thus can we avoid wrong and shallow analogies, wrong and inadequate gauges. This obviously requires a great effort on the part of the historian of science. Elgood (1951, p. 584 ff.) airs this question for Arabic science and medicine: " What is the reward for the great labour involved in becoming a specialist fit to study Arabic Medicine? " asks Elgood. " Medicine, Arabic, Greek Medicine are only the minor prerequisites. Glossaries are almost valueless."

How should it be pursued? For many years the best way is to make more facts available. Many portions of the Mss. can safely be neglected, as today already the sections on drugs:

" Already one is beginning to feel that too much time has been spent on fruitless attempts to identify botanical names. Names must have varied from place to place, and it is idle to expect consistency. Confusion must always have existed owing to the wide variations in one and the same species produced by the extremes of climate within which Arabian medicine held sway. I feel sure that the Arab and Persian writers themselves must often have made the grossest errors through ignorance and lack of care in checking their descriptions " (p. 593).

Next therapeutics, general medicine, may also be neglected, whilst clinical observations will always be of interest.

" But far more important than these is the philosophy and psychology that underlie all these works. Here is to be found a return to the first principles which are still in dispute. What is disease? Fashion influences this fundamental conception. The reality of diseases as independent entities has been supported, challenged, reasserted within living memories . . . In this respect the thoughts of the great Arabian and Persian scientists will continue to be worthy of study. Who would dream of studying moral philosophy without reading Aristotle and Plato? Arab and Persian views are worth as much as our own on these great fundamental questions . . ." (p. 594).

Man is more than flesh and blood, and beyond the experimental circle the Arab and Greek speaks with an authority equal to that of the modern student.

This last point is of paramount importance. Not only that the scientist of every period speaks with an authority equal to that of the modern student, but within the possibilities of knowledge and method the great men of every period did think with the same scientific effort and success as we do within those of our own time. We need only to mention people like Aristotle, Galen, Ibn Rushd, Galilei, Boerhave, Réaumur, Newton, and many others to make our point clear.

A mere accumulation of knowledge is not yet science. Only the unification or systematisation of these facts makes science of them. None of the scientific systems of the past and of the present is a closed one. Only the systems of logic are closed and not subject to change (Kant). Scientific systems are open, and thereby enable their further development, a development which is conditioned by external as well as by internal factors. The former will be discussed later. The internal factors are often difficult to understand. Very often we see that all the scientific facts and tools were ready for some time, but that the progressive conclusions, so patent to our retrospection, were not drawn. This is often caused by the dominance of other problems in contemporary science, which attracted the attention of the students. Often environmental influences, from spheres beyond the internal drive of a special science or its internal trends of development, forced the use of long-established facts into new and fruitful paths of conclusions and problems. Thus, the mere readiness of the facts, the mere possibility to draw important conclusions, is never identical with their necessity. But it is always a prerequisite of new development that the scientific tools—facts as well as methods—be ready. This point is very important in an age like ours, which so often tends to overlook the fact that applied science can proceed only if pure science continues to develop such new tools, without being able to judge when and where they may be applied practically. The great discovery of Banting concerning insulin as the base of diabetes would have been retarded for long if a mere fact-searching doctor's thesis many decades before had not stated that in the pancreas of fishes in the islands of Langerhans, which are the insulin-producing part of this organ, occur as a massive part of it, thus enabling him to start the proper experimentation. No great progress in applied science is possible if pure science remains at a standstill. Its function is to prepare the tools for its own progress, incidentally preparing the path for that of the applied sciences.

We have mentioned before the importance of chance, and of the open eyes of the scientist ready to make use of this chance. We have no better illustration of this principle than the discovery of penicillin in our time. Hundreds and thousands of bacteriologists have had the opportunity to observe that in cultures of bacteria an empty space surrounds small colonies of moulds growing in the same culture. Yet only Fleming was astonished enough to ask for the reason of this common phenomenon, a question which led to the discovery of the antibiotics. We can scarcely find a better illustration of Aristotle's statement that the beginning of science is in ' wonder '.

The solutions of the fundamental problems of life are often regarded as the aim of biology. Yet science has never and nowhere succeeded in solving any of its fundamental problems. All solutions of the last questions are rooted in belief, not in knowledge. We all are united in the desire to explore the world to the very limits of our knowledge and mind. Thus we all strive to learn everything within our powers with regard to the physico-chemical background of the phenomena of life. The materialists and the mechanists assume that our present incompetence is only temporary and that the progress of physics and chemistry will solve and ' explain ' one day all the outstanding lacunas. But this is mere belief, to which any other belief, as far as it does not clash with the facts known to us, may be opposed with the same justification. Not solution of the last problems, but a progressive unification and systematisation of facts is the aim of science, or at least the end in our grasp. Everything beyond is belief. The great physiologist Dubois Raymond stated as his belief with the same right: *Ignoramus* and *ignorabimus* about the last problems of life. When Schroedinger (1941) gave a physical paraphrase of life, he did not add anything to its explanation. This situation explains the survival and revival of almost every basic and age-old antithesis in the scientific outlook. These antitheses will persist, even if for a short period the victory may be claimed by one or the other side.

The apriorism of our mind, based upon belief instead of knowledge, is the cause of this persistence. Mechanism versus vitalism was a problem in ancient Greece, as it is for our days. The 17th-century discussion between the preformists, who assumed that every man of the past as well as of the future was actually contained within Eve, and the epigenetists, who assumed that every embryo is formed anew from a generalised matter, is by no means closed. While modern geneticists have resumed the preformistic view in a different form, the workers in experimental embryology are inclined to an epigenetic one. A hundred years ago it became a certainty that the cell is the lowest organisation of living matter, and twenty-five years ago still no biologist doubted that this cellular theory is one of the great and final milestones of biology. Today, this view has to give way to another one, promoted 200 years ago by Maupertuis, Leibnitz, and others, namely that the cell is composed of a huge quantity of smaller entities of the size of the virus, the genes, the plasmogenes, the mitochondria and other big protein moleules of the most varied shapes and properties which together compose the living cell in the same way as the cells compose the body.

3

EXTERNAL INFLUENCES ON THE HISTORY OF SCIENCE

INTERNAL drive as well as environmental impact together determine, in any given period, the trend of development of science and civilisation. We shall deal here briefly with the main environmental factors.

1. *The geographical factor.* In the past century, the geographical environment has been often stressed to the extreme as the dominating factor forming social, political, and cultural life. Ratzel and Huntington are outstanding fighters for this view, while C. Semple (1911, 1932) tries to maintain a moderate judgement. There is

not the slightest doubt that geographical conditions as a whole determine largely social organisation and social processes, and this in proportion to the primitiveness of the human organisation. The isolated families of tribes inhabiting primaeval forests are just as conditioned by their means of existence, as primitive agriculture, brought to a higher level by the great irrigation civilisations of the Indus, Euphrates and Tigris, as well as of the Nile, tends to social aggregation. Once man has reached a higher social level, his dependency upon topography becomes less conspicuous, yet still persists to some degree. The primitive Delaware Indians used the wood of the forests only for cooking and primitive heating purposes. The modern inhabitant of Pennsylvania has the choice between wood, coal, oil, and electricity for heating; but on the other hand, almost every one of his activities is bound up with the use of one or more of these materials.

The same geographic environment has a very different meaning for peoples of different character, especially where industrial utilisation is concerned. The mountains of Switzerland and of Andorra, and the wide grass plains of Russia exemplify this point. The unfavourable ports of Sidon and Tyre and their unfavourable hinterland certainly did not make the Phoenicians into great seafarers and colonisers: their innate energy and their commercial ingenuity were the determining factors. None of Huntington's strong theories about the exclusive culture-promoting values of temperature or humidity has withstood a reasonable analysis. People are not pessimistic because they inhabit a small and limited territory, and dwellers on the great plains are not happy because they may roam over wide areas, as Ratzel wishes. Are the Danes really more pessimistic, and the Russians more happy, than other people? Her geographical analysis has not prevented Miss Semple (1933) from including the saharo-sindian deserts and the irano-turanian steppes in the Mediterranean region. Strygowski and Bodenheimer have shown that no cultural analysis of the Eastern Mediterranean is possible without taking these biogeographical aspects into account. We must beware of blind admiration for theories, like those promoted by Huntington or MacCowan, about the geographic determination of the origin of the three western monotheistic religions. The soft, rounded hills of Samaria and Galilee, the rugged hills of Judaea, and the wide spaces of the desert did not determine Christianity, Judaism and Islam. As if there were not plenty of homologous topographies all over the world which have never produced similar, or even any, religions! The location of London, Paris, and Rome, the hearts of historical and modern empires, were not chosen because of their geographical importance for these empires; rather, geographical causes strongly influenced the choice of their sites as suitable for the establishment of small towns, under very different conditions. Coastal peoples are not forced by their environment to become important sea-faring nations, just as nobody is ever forced in history to become important. But if such people take to seafishery, and if they possess an innate energy, they may develop it in the direction of the sea.

Without denying the great importance of the geographical environment even for the highest social structures, we must deny it the dominating role as a truly deterministic force in the shaping of human civilisation. If this is true for the geographical factors in their entirety, it must be stated still more strongly for any isolated geographical factor, such as temperature. Actually we do not possess yet *one* adequate analysis of the impact of geography upon the progress of civilisation and science. We look forward eagerly to the first analysis of this type.

2. *The social factor.* More attention is paid today to the social implications of science, which are basically twofold. On the one hand, applied science and technology have changed our daily life fundamentally, from nutrition and housing to warfare, and promise to continue to do so in the close future ; and on the other hand the social impact of a period influences the development of its scientific thought in a much deeper measure than was realised heretofore. Not the least important aspect of the social influences is that they break down the barriers among the particular sciences. Thus, the roots of Darwin's slowly developing thoughts during the cruise of the *Beagle* are found in a study of geology (Lyell) which appeared 35 years before the *Origin of Species*; and Darwin's concept of the struggle for life speedily influenced human sociology as well as astronomy. Mendel's laws of plant hybridisation were rediscovered about 40 years after their first publication, when the growing interest in plant breeding as a means of furthering agriculture and horticulture became more and more urgent. The Soviet Union has developed new trends in agricultural science, such as vernalisation, driven by the urge to enlarge the area of cultivated crops in its northern areas, where the annual sum of daylight is too small to permit the untreated varieties of wheat, etc., to mature; as well as in its southern arid areas, where the natural period of vegetation is

determined by very short periods of rainfall, preventing development of the crops into mature seeds, which have not been submitted previously to vernalisation. There has been no more urgent stimulus for the development of higher mathematics than the need for better ballistic analysis in modern artillery in the 17th century, as well as in the time of Napoleon Bonaparte, as Walter, Sergescu, and others, have shown very clearly. And the rapid development of mathematics urged upon us by the development of aerodynamics in our day is another most striking illustration.

Rosenfeld (1948), Lilley (1949), and others, have analysed well the social impact of science as illustrated in the development of thermodynamics, one of the central problems of the last hundred years in physics. All the basic concepts were ready before 1800, some 40 years before the subject was actually established: the concept of quantity of heat and of calorimetry (Black, Laplace, Lavoisier) in chemistry; that of kinetic energy as the *vis viva* in the theoretical mechanics of even the 17th century; and that of work and the measurement of heat evolved from the friction involved in the boring of a cannon by the practical engineer Rumford. Yet these concepts were developed in three different fields of science, which were largely unconnected. " The story of the development of thermodynamics is very largely the story of how these concepts, available in separate fields of science and technology, were brought together into a synthesis " (Lilley, 1949, p. 391). The practical engineer's idea of work had been developed within the frame of the industrial revolution. Chemistry in its 18th-century period of rapid development also had some contact with the industrial revolution, but the much older science of theoretical mechanics followed its more internal lines and had few relations with modern industrialisation. The penetration of the problems arising from the technical requirements of the industrial revolution led to the foundations of thermodynamics. Energy changes from heat to mechanical energy were utilized in the steam-engine. The improvement of this steam machine was of crucial importance for the industrial economy, suggesting the ' right ' questions to a few inspired minds. Yet, what circumstances made energy so important in the eyes of the scientist in 1840, when it had not seemed important at all about 1800? Until the *vis viva* (kinetic energy) had some mechanical connection, it was a pure mathematical abstraction. When, after 1800, the production of heat by friction was discovered (Rumford), relating heat produced to work done, it was still not yet applied to the steam engine. There followed the electromagnetic discoveries connecting mechanical movement with electrical and chemical phenomena and with heat, which led Faraday and others to ideas about the ' correlation of physical forces' which vaguely suggested the first thermodynamic law of the conservation of energy. But only outsiders to theoretical physics—like Joule, who had started his career as a physicist from an engineer's point of view, and like the physician R. Mayer, who was interested in human physiology—could conceive the needed synthesis.

The former (1838, 1843) tried to build battery-driven electric motors more efficient than the steam-engine, which attempts led him to the discovery of the quantitative equivalence of heat and mechanical energy, involving the principle of the conservation of energy. And Mayer (1841) developed Lavoisier's theory of heat production in the organism as combustion (published 60 years earlier), by merely deductive considerations, into the statement that the mechanical work of a moving animal must be quantitatively equivalent to the heat produced by the same food in the body of the resting animal, and he later generalised this idea into the law of the conservation of energy. But in the 60 years since Lavoisier nobody had ever tried to interpret the actual data in this direction! The reason is that the problems of official chemistry tended in other directions. While Lavoisier regarded the organism as a fire, a producer of heat like the furnaces he used in his chemical experiments, Mayer regarded it as a heat engine, in which the heat is subsidiary to the mechanical effect produced, and he remarked that his discovery gave hope for an improvement of the prime motors of the steam-engine. However, Joule and Mayer were still ignored for some time by the scientific world, which failed to understand the meaning of the new conceptions. Regnault's (1847) practical tables of the numerical characteristics of steam did more than the scientific foundations to arouse a rapidly growing interest in the theoretical and practical importance of the first law of thermodynamics, with imposing developments, mainly within one decade. Once its industrial importance was established, social (i.e., economic in this case) pressure forced its rapid development, based upon concepts and principles which for over 60 years had lain ready, but which were now driven towards synthesis primarily by the social drives.

These few examples illustrate only a small part of the many impacts of the social environment upon the progress of science, which together form only one side of the medal. The other side, the impact of science upon its social environment is now highly appreciated in its importance and implications. Both sides still need much further analysis.

3. *The cultural factor.* Many are the relations between general culture and science. Some have recently been discussed by Cortesao (1949), of which we may mention: 'Tensions affecting international understanding', 'history of science as the third dimension of human civilisation', and 'the impact of the creation of the scientific method upon the development of culture'. We here intend only to cast a glance upon one special aspect, namely the unity of science as the symbol of the unity of mankind and its spirit. Needham (1949) and Sarton (1948) have taken great pains to lay bare the manifold threads of cultural and scientific interrelations which have joined, since at least late antiquity, the civilisations and scientific as well as technical achievements of China, India, and Persia with the western world. We hear about Chinese physicians copying Galenus and the Arab physicians at Bagdad in the 10th century, and about Chinese astronomers in Persia in the 13th century. We still do not realise the debt we owe to early Indian literature at the early stages of the literatures of the Middle East and of Europe. From the same source we obtained the discovery of the zero (independently discovered by the Mayas) and of the 'Arabic' ciphers. While technical inventions, such as printing, artillery, etc., spread slowly during the first fourteen centuries of our era from China to Europe, the more theoretical aspects did not pass either way in the Middle Ages beyond the Arab world. This is an important and still unexplained phenomenon.

G. Sarton (1948, p. 163 ff.) stresses the unity of East and West, as follows: "The seeds of science, including experimental methods and mathematics, in fact, the seeds of all the forms of science, came from the East, and during the Middle Ages they were largely developed by Eastern people. Thus, in a large sense, experimental science is a child not only of the West, but also of the East: the East was its mother, the West its father.

"And I am fully convinced that the West still needs the East today, as much as the East needs the West. As soon as the Eastern peoples have unlearned their scholastic and argumentative methods, as we did in the 16th century, as soon as they are truly inspired with the experimental spirit, there is no telling what they may be able to do . . . We must not make the same mistake as the Greeks, who thought for centuries that their spirit was the only one, who ignored altogether the Semitic spirit and considered foreign people barbarians . . . Remember the rhythm between East and West. Many times already has our inspiration come from the East. Why should that never happen again? The chances are that great ideas will still reach us from the East and we must be ready to welcome them . . .

"The unity of mankind includes East and West. They are like two moods of the same man. They represent two fundamental and complementary phases of human experience. Scientific truth is the same East and West, and so are beauty and charity. Man is the same everywhere with a little more emphasis on this or that."

4. *Religion as a factor.* Religion has often been subjected to very unfavourable criticism by historians of science. The retarding influence of dogmatic churches has been stressed again and again. Yet this is only part of the story. First, if by inquisition the Catholic church has retarded the progress of science, if during long centuries it has not shown any active interest in its promotion, it participated fully in the flowering of science at other periods. It is not the Catholic faith which is responsible, but the period. And the Catholic church has wisely survived and passed through all the last 1950 years, participating in all the spiritual revolutions, accepting the truth whenever it could grasp its truth, neglecting the progress of science only when the period was not interested in its development. Religion, in general, was a dominating influence whenever souls were in ferment, sometimes stimulating, sometimes retarding. To state that monotheism has barred the progress of science while polytheistic or pantheistic religions have favoured it (Putnam 1949), is sheer nonsense. Apart from the general inadvisability of such generalisations in principle, there exists no type of religion which in all its phases either favoured or barred science and the search for truth.

In view of the widespread prejudice against religion as a factor in human progress, it may be appropriate to cite here one instance, at least, in which religion stimulated, supported and furthered, very much indeed, the development of scientific interest and thought. Merton (1938) has demonstrated, in a methodically well conceived study of England in the 17th century, that puritanism was one of the important factors forming a mighty impulse for scientific development. The glorification of God by exploring the miracles of His creation, and the promotion of help for our fellowmen by creating better conditions of work and of living, the attitude of a social utilitarianism—to both of these powerful motives and trends science appeared as the main resource by which to gain this common aim: the glorification of God by knowledge. In other religious sects, which have sought to reclaim their followers, body and soul, a similar turn of devotion has often mightily stimulated scientific research and interest.

Yet it would be unfair to ascribe this influence of puritanism, or of certain other sects, to protestantism as a whole, either in contrast to other religions or to other protestant sects or even to puritanism through all the phases of its development. The aforementioned trend, of worshipping God by the discovery of the greatness of His works and creations, was widespread in the 17th and 18th centuries in western Europe. The physicotheology of Durham, Bonnet, the Abbé Pluche, and many others, was not less prevalent at that period in the Protestant Netherlands than in Catholic France. The whole life of a Jan Swammerdam, whose *magnum opus* quite properly is named *Biblia Naturae*, and of many others, was entirely and fanatically dedicated only to this aim.

Neither should we forget that, since the days of Mesopotamia and Egypt, all science was largely, if not for certain ages exclusively, in the hands of the priests. Without the monasteries of Byzantium the science of antiquity would scarcely have survived to our days and in the monasteries of the Middle Ages, scholasticism forged the weapons of the spirit which prepared for the Renaissance. The interrelations between science and religion are most complex and varying, and they still await a historian without prejudice for the full analysis of their intricacies.

5. *Art as a factor.* Art cannot be overlooked as an important impact upon the development of science. Art, as a primary experience, is intimately connected with every period of upheaval and enthusiasm. Harmony of art indicates a harmony of the soul of that period. All great periods of high scientific fruition coincided with those of artistic fruition. In a measure which is not yet sufficiently understood and stressed, the artistic temperament is surprisingly suited for scientific creation. In the age of Leonardo da Vinci, the great creative personalities of Italy were more attracted to art than to the dry humanistic studies of books. They had the open mind, the broad vision, the deep and penetrating spirit necessary to gain an entirely new and fascinating view and access to science as a whole, as a spiritual experience, wherein no branch of nature

remained neglected; every star, every plant, every organ of man, every machine called forth new responses of the soul of the artist vibrating in unison. As with religious and social influences, such important stimulations of art to science are not permanent factors. They manifested themselves at times when the soul was uplifted, and it is useless to ask which preceded which. It is the uplift of the time which has opened the souls of its great men wide to new experiences of the fundamental aspects of life and of science.

This brings us to our conclusion : the aim of science as seen from the historian's point of view is, not in the least of its significations, the transmission of the eternal flame of the spiritual soul, from generation to generation. The nutritive matter, and with it the colour, size, and shape of the flame, are continuously changing, yet the flame remains alive and burning. This burning flame of the uplifted soul is the purest symbol of humanity and of the true humanism. To its maintenance the historian of science is called, as is nobody else in the same degree.

4

HISTORY OF SCIENCE IN ITS HUMANISTIC INTERRELATIONS

THE history of human thought is ultimately of immensely greater importance than that of human wars and of dynasties. Thought precedes action, though both are usually not united in the same persons. The discovery and spread of new thought, from its first spark to its successful manifestation and victory, is a most intricate and complex process, often—or even usually—accompanied by human tragedies. The real martyr is not of the type of Galileo, who conceived and confessed a great truth in all its entity and implications, yet who compromised in renouncing his truth formally, in order to continue a quiet life. We do not intend to judge him from a moral point of view, but, in any case, his soul remained in full possession of the treasure of his new knowledge. The true martyrs are the early precursors who grasp the first sight of new horizons without being able to build a bridge from their new discovery to the ruling authoritarian view of their time.

Sergescu (1951, p. 22 ff.) illustrates this by the problem of the infinite: " We do not imagine properly the tragedy lived by many scientists of the Middle Ages." Aristotle, this dominating authority of mediaeval science, refuted the idea of the infinitely small, i.e., the atoms, already accepted in antiquity by Archimedes! They had to search for a compromise to create a bridge between their observations, their deductions, and their beliefs, and Aristotelian science. Here is one of the great achievements of the much blamed scholasticism: it forged the spiritual tool for finer analysis. These mediaeval discussions, with their abstracted logic, often seem empty to us. Yet every historian of science now agrees that without the logical analysis of the scholastics the spiritual revival of the Renaissance would have been impossible. Thus Gilles of Rome (1247-1316) helped solve one of his problems by accepting three different kinds of size: first, a size which is abstracted from the matter in which it is realised; second, a size supported by a non-specified matter; third, a size made by a determined matter. Whilst the two former classes permit division into the infinite, the third one cannot be divided—in agreement with Aristotle—beyond a ' natural minimum '. This analysis reached its full antithesis in the theses of Albert of Saxonia (d. 1390) and Gregory of Rimini (d. 1358). There is no doubt that, if at that time the work of Archimedes had been available, the scholastic discussion would have led to the revolution in mathematics which was realised only 200 years later.

Many people overlook the subtlety and penetration of the scholastic analyses, by stressing only their lacunas and draw backs. We shower with ridicule the thirty years' polemic of the 13th century about the number of the teeth of the horse. This discussion was conducted with great intensity, with much ink and much temperament, but neither side ever thought of looking into a horse's mouth, and thus deciding the question. The tendency to accept prejudiced opinions is deeply inborn in the human mind, especially when they are based upon authoritative sources and arguments. We should confess that our own period has not advanced much in this respect. We live now almost a hundred years after Charles Darwin's *Origin of Species* (1859), yet it was only thirty years ago that a few abortive experiments were made to verify the struggle for existence, the protective value of camouflage in the animal world, etc. Fifty years ago A. Hensen propagated the quantitative exploration of the populations of the marine plankton and nekton, stating that only quantitative population studies would enable us to obtain any idea at all about the extent of the struggle for existence. He was strongly attacked by none other than E. Haeckel, who regarded this claim almost as a sacrilege, since the struggle for existence was indubitable and was not in need of any experimental or quantitative confirmation. This is ample proof that our mental attitude is still almost the same as that of the scholastics. An immense potential of progress in this respect is still before us.

The history of science as the history of human thought belongs to the humanistic sciences. No study of the development of observations, no laws of historical development, will ever permit us to forecast the development of an idea or of a science. From our retrospective point of view it may seem that a series of new facts or theories may point to one conclusion only. We can never be sure that this one logical conclusion will be drawn by some specific individual or by a definite period. One often gains the impression that the uphill work of new discoveries has tired and exhausted the forces of the pioneer. His spirit seems too tired to finish the work and to draw the apparently patent and natural conclusions. In other cases, it is just the novelty of observations or theories which may prevent them from arousing the interest of his contemporaries—even those of stature—and from being accepted or even refuted. Even if facts and conclusions are drawn to perfection, they need not find the proper attention. The theory of evolution or transformism (as distinct from that of its mechanism) was perfectly stated and well based by Lamarck in 1809, but it was neglected and forgotten until Darwin presented it 50 years later. And Mendel's classical experiments on the hybridization in plants (1865), which now form one of the keystones of modern genetics, did not find any response among the biologists of his time. His laws had to be rediscovered independently in 1900 by de Vries, Tschermak and Correns, before they were accepted.

This phenomenon is often described as follows: the discoveries of the precursors are made prematurely, while the rapid acceptance and spread of a theory takes place when the minds of the contemporaries are ripe for its conception. This is, of course, no causal or rational explanation. It is true that, where the precursors anticipate later ideas, their conceptions may be true but crude, still too crude to conquer the scientific spirit of their time: though their anticipations are often great from the retrospective point of view, they did not fit into the general system of scientific thought of their time. Yet this is not true of the two examples given above. Lamarck's exposition of the theory of the origin of species by transformation, as well as Mendel's laws of hybridization, were both competent and should have been accepted, or at least they should have found a wide response, at any time. The refutation, and still worse, the ignoring of new theories by the authorities of the day has often delayed and suppressed the progress of science. But no proper gauge has yet been found to determine the degree to which a period is ready to respond to new stimuli. We can only say that, once a fruitful and promising theory has been well based, it will crop up again and again, until it appears at the proper moment to conquer the minds of its contemporaries.

A good illustration of this point is provided by Leonardo da Vinci, that unique phenomenon in the history of science. All his discoveries, all his new approach to science, remained closed to the development of science. And when they were published, chiefly within the last hundred years, his observations, his conclusions, and his new approach had practically all been already rediscovered independently and become established long since. By the non-publication of his manuscripts, the progress of science had been delayed, but not barred for ever.

Chance plays a great role also in the history of science. Voltaire's insolent and improper jokes, that the fossils on mountain peaks had been left there by earlier pilgrims, killed de Maillet's correct and well-presented theory of their marine origin on the spot in a period of the past. Cuvier's ridicule of Lamarck's description of how the giraffe gained its long neck and its long legs by continued efforts to reach the leaves of high trees, contributed much to suppress the great truth involved in his *Philosophie Zoologique*.

Thus, authority is often a retarding element in the history of science. The illustrations are too numerous to call for specification. Yet we must point out that—harsh as its impact upon the individual may be—its effect is not detrimental only. The authority of a generation drives its contemporaries towards a deeper inquiry into a special theory or the wide application of a new method, thus contributing much to the stabilisation of science. Thereby it helps to establish, to explore and to exhaust the truths of one period, helps their working out into systems, and helps a later generation to realise its limitations and failures. Thus, the delaying effect of authority is only temporary.

While facts and observations are accumulated through the ages, in science more than in any other branch of human activity, their interpretation and the interest which they arouse change continually. This shift of interest to ever-new problems from generation to generation is one of the basic facts in the development of science. It is a mistake to assume that the development of science takes place in a straight and uninterrupted line. Haeckel in his stupendous *General Morphology of Organisms* states that certain lacunas remained in his proposed phylogeny, but that later generations will be lucky enough to fill these lacunas. Yet the following generations showed little interest in the details of phylogeny, since other problems, such as cytology, genetics, ecology, etc., captured and fascinated their minds, just as Haeckel's mind was fascinated by the problems of morphological and taxonomic phylogeny. This continual shift of every science from one set of problems to another, from generation to generation, is remedy against the lasting inhibition exercised by authority.

If the reader has followed me attentively, and if I have not failed in my exposition, he will by now be aware that the history of science, properly understood, is *one* key to all that is, and has been, creative in the history of mankind, and hence something which should concern each one of us vitally. It is, of course, not the knowledge, but the understanding which is so vitally important. A few words may be added in this direction.

The history of mankind, spiritual as well as material, has passed through dumb and stable, through slightly progressive, and through gloriously creative centuries. The higher the intensity of creation, the lower the degree of self-complacency. The real spiritual, scientific, moral, religious or social struggle of any age is less in man's reason, still far less in man's body, but in the souls of the individuals. Devotion to ideas and to general notions may dominate and throw the individual into the struggle for life in which many have burned and perished. Yet those who perished did not lose their lives in vain, neither Socrates, nor Giordano Bruno, nor the martyrs of all faiths or of science. Their courage and devotion enlighten the course of human history like beacons.

We often find the apodictic statement that the natural sciences and the humanistic ones are entirely different, insofar as the former are ruled by unalterable laws of nature, while the latter are influenced by values and sentiments, and, hence, are subjected to other kinds of laws, to the extent that laws can be advocated at all. Yet, within the last hundred years, from Comte, Büchner and Marx to Spengler and Toynbee, unalterable laws have also been advocated for human history. The latter is often regarded as 'sociological zoology', subjected to physical laws like those which apply to all other animals. This fundamental problem is very important for the History of Science, as the trend of discoveries is definitely subjected to the same laws as all other human history. The great role which intuition plays in its progress and development is ample proof of its belonging to the world of values. It belongs, of course, not less to that of the scientific method, as critical induction has been called since Descartes. But, as we have pointed out before, every real progress in science came from intuition, and not from induction. We wish to repeat, however, that intuition unchecked by induction is inadmissable in science.

But every *a priori* theory of determinism, whether it be teleological or mechanistic, or aesthetic, or scientific (I. Berlin, 1954), is wrong in human history, as well as in that of science. The strict separation of science as a kingdom of natural law, in contrast to the freedom—apart from the physical laws ruling all matter—of individuals and of groups, with the moral responsibility which this involves, is entirely wrong. No History of Science can be separated from that freedom and moral responsibility, any more than the history of politics, religion, etc., can be separated from it. This is clear when we read, in that inspiring lecture of I. Berlin (1955, p. 53): " The invocation to historians to suppress even that minimal degree of moral or psychological evaluation which is necessarily involved in viewing human beings as creatures with purposes and motives (and not merely as causal factors in the procession of events), seems to me to rest upon a confusion of the aims and methods of the humane studies with those of natural science. It is one of the greatest and most destructive fallacies of the last hundred years." While we agree in substance concerning the destructive fallacy as regards humane history, we must protest against the exception made for science. We must protest for two reasons: first, because the History of Science is made by human beings and not by the presupposed ' unalterable laws of nature ' themselves; and second, because of the importance which intuition and values have in the development of science.

Science as ' Shibboleth ', science as the ' Sacred Cow ', is one of the characteristic features of our time. And parallel to this raising of science to a godlike authority goes the emptying from our souls of the ' Erlebnis ' of the true and eternal values of the soul. Already Louis Pasteur, when he analysed the causes of the moral debacle of his fatherland in 1871, ascribed it to the lowering of the spirit as illustrated by higher education. Instead of men, of personalities who are scientists, our universities and our technical high schools produce competent technicians, on the production line, people competent for the technical work which is their task, but lacking the broad horizon, the wide understanding of problems, capable of synthesizing various branches of science. Payot described, at the end of the last century, and in a really prophetic way, the degeneration of the professors from spiritual leaders of their generation to competent teachers of facts in professional schools. In a deep analysis, Standon (1952) describes the danger of regarding the modern scientist as a superior being. He describes how the mere pleasure of solving technical problems put to them by politicians and commercialists has dulled their conscience against the most primitive demands of humanism. Or, Norbert Wiener, in his *The Human Use of Human Beings*, the first truly humanistic book from the world of technology, sees darkly into the future of the American economy and life. By the rigid restriction of the young scientists to very particular problems, life robs them of that initiative, of that broadness of mind and of horizon, which is required of the leaders of every economy. And even the university has forgotten this, its primary task, and produces specialists and technicians instead of true scientists. No higher approach, no development of thought and thinking, not even the slightest hint of the basic assumptions of every scientific work is given to the student. It is satisfied with a superficial, rationalistic, and mechanistic approach to the problems of life and of science. And the result is precisely the competent specialist, an empty personality, who eagerly runs away from any quiet dialogue with his own mind. And since the youth of our generation, like that of all generations, is thirsting for a cause to give it its soul, all kinds of false gods are chosen and followed blindly, and just by the best part of our youth: totalitarian fascism, communism or religious hyperorthodoxy, a narrow chauvinism, etc. Their eyes closed, they give their hearts and their souls to such causes. All the self-sacrifice, all the devotion of youth is spent on them, while the great eternal values of humanity are suppressed or falsified.

5

THE PHILOSOPHICAL BACKGROUND OF SCIENTIFIC INNOVATIONS

OUR knowledge of the mechanism of scientific discovery is still limited. We know that it very rarely, if ever, will appear without long subconscious and/or conscious preparations in the mind of the student. The apple of Newton and the ' Eureka ' of Archimedes are probably legends; if they are true, they represent just the last step to such extended preparations of the mind. Yet there are a few cases in which we can follow at least a partial chain of such preparation. These teach us in which direction the research on the mental development preceding the discovery should be extended. One such discovery is the achievement of William Harvey in demonstrating the circulation of the blood.

The anatomical analysis was followed by the famous ' calcul '. The left ventricle of the human heart may contain up to two ounces of blood in the diastole, of which one-quarter to one-eighth is ejected into the aorta during the systole. This ejected blood cannot return into the ventricle because of the valves at the aortal root. By multiplying the half-ounce of blood expelled by the number of pulses, we come to 1000 half-ounces in half-an-hour, more than is contained in the entire body and very much more than could be supplied by the ingesta or be contained in the veins at any one moment. Butchers know well that the entire blood is drained away in less than 15 minutes after cutting the throat of an ox. Thus the quantity of blood in the body is actually fixed and small.

From among the experiments made by Harvey, we may mention a few. When the heart of a snake is laid bare, and the *vena cava* is pressed, that part of it which is between the pressure and the heart will soon become and remain empty; but when the aorta is pressed the heart will expand to a considerable size. The effect of a very strong ligature upon the veins and arteries of an arm upward and downward from the ligature was studied. The discovery of the venal valves by Aquapendente was interpreted as procuring a steady, one-directional motion of the venal blood towards the heart. These valves become visible as small knots in the light ligature made during phlebotomy. When the vein is pressed above the knot, towards the heart, in that direction, the blood contained in it can be pressed to above the next valve. But any attempt to press it back into the now empty distance between the first and second knot is prevented by the presence of the upper valve.

Harvey's description of the circulation of the blood starts with anatomy and experimentation, goes over to the calculations, and ends with the philosophical background. Yet the actual progress was the reverse. The philosophical background gave the suggestion which led to the calculations for which proof was sought and found by dissection and experiment. The really convincing step in Harvey's argumentation, the mark of his genius, was the calculation. His experiments had precursors, and certainly do not show him as the excellent experimentalist he is so often said to have been. In anatomy, he failed to demonstrate the capillaries, the actual connection between the arterial and the venal system, but he promoted knowledge considerably by his statement, that he was unable to discover any trace of the Galenic communication between the right and left ventricles of the heart.

The neat and clear way in which Harvey presented his arguments is evident *ex posteriori*. But we still want to know what induced him to start his calculations. Here W. Pagel (1951) has recently opened up new horizons by scouting into the philosophical backgrounds of Harvey. Harvey pronounces himself everywhere a strong Aristotelian. Now, the circular movement is the most perfect motion; it alone is continuous, as is proven by the circular motion of the eight spheres. Following the Aristotelian adherence to the macro-microcosmos theory, man is built on the analogy of the cosmos. Within this cosmos, a ' circulation ' of water takes place, bringing it from the skies down to earth and from there back to the skies. A similar circulation was looked for by Harvey in man: "I began to think whether there might be a motion, as it were in a circle."

Similar general conceptions were not uncommon throughout the 16th and 17th centuries. Thus, we read in the *Dialoghi d'Amore* of Leone Ebreo (posthumous, 1525, Venetia. Dial. II): " The heart is very similar to the eighth sphere . . . The human heart moves by the same reason continuously and equally in a circle, never rests, and provides by its continuous and equal motion life to the entire body . . . The pulse is the beginning (of development)."

Giordano Bruno wrote in 1590 (*De rerum principiis*. See W. Pagel, 1950, 1951) about the ' circular ' pattern of the motion of the blood: " The spiritual life-force is effused from the heart into the whole of the body and (flows back) from the latter to the heart, as it were from the centre to the periphery and from the periphery to the centre, following the pattern of a circle . . . The material part of all these spirits is a fluid which cannot move on its own account, but by means of its innate spirit. Hence there is no circular or sphaerical motion outside the body, for the blood, which in the animal moves in order to distribute its motor, the spirit, lies immovable outside the body, is inert and decays, no longer deserving the name of blood . . . In us the blood and other fluids are being moved continually and very rapidly in a circle, flow and flow back, are diffused from the centre into the extreme periphery and from there return to the centre." Thus, for Bruno the perfection of the circle is a fundamental symbol of all life and action in the cosmos.

This reasoning finds a splendid interpretation by Andreas Caesalpino (1571 *Quaestiones Peripateticae* V: 4, e. a. See W. Pagel, 1953), who transports cosmic analogies into physiological thinking.

In 1571 he claimed " a continuous motion from the heart into all parts of the body ". In 1583 he said: " We see that the food is brought by the veins to the heart, the place where the innate warmth is produced, reaches there its perfection and is spread by the arteries all over the body." And in 1593: " There is a perpetual motion of the blood from the veins to the heart and from the heart to the arteries."

Caesalpino already mentions in this connection the direction of the venal flow, known to the barbers from phlebotomy, and the importance of the heart valves for the direction of the blood flow. He was also the first to use the term *circulatio*. He remained vague and indulged in guessing about the morphology of the blood's circulation, but he stressed its physiological interpretation, which to him was a distillation of the blood. The blood is 'boiled' to perfection in the heart, rises from there as vapour in the lungs, as in the roof of a distillation flask, where it is cooled off. The cooled blood, now full of its nutritious spirits given to it in the heart, returns via the heart to the aorta to nourish all the bodily organs. Through the *vena cava* it returns to the heart, where it is perfected again by ' boiling '. This stress upon the physiological circulation, the rejuvenation of the blood by distillation, and the neglect of the morphological details of circulation, prevented his proper recognition of the truth. But it is important to underline that this double meaning (in a slightly changed sense) of the term circulation is still fully maintained by Harvey. In *De Motu Cordis et Sanguinis* (1628, ch. VIII) we read: " I began to think whether there might not be a motion as it were in a circle. Now, this I afterwards found to be true; . . . which motion we may be allowed to call circular, in the same way as Aristotle says that the air and the rain emulate the circular motion of the superior bodies; for the moist earth, warmed by the sun evaporates; the vapours drawn upwards are condensed, and descending in the form of rain moisten the earth again; and by this arrangement are generations of living things produced; and in like manner, too, are tempests and meteors engendered by the circular motion, and by the approach and recession of the sun. And so, in all likelihood, does it come to pass in the body, through the motion of the blood: the various parts are nourished, cherished, quickened by the warmer more perfect vaporous, spiritous, and, as I may say, alimentive blood; which, on the contrary, in contact with these parts becomes cooled, coagulated, and, so to speak, effete: whence it returns to its souvereign, the heart, as if to its source, or to the inmost home of the body, there to recover its state of excellence or perfection. Here it resumes its due fluidity and receives an infusion of natural heat—powerful, fervid, a kind of treasury of life, and is impregnated with spirits, and, it might be said, with balsam; and thence it is again dispersed; and all this depends on the motion and the action of the heart. The heart, consequently, is the beginning of life; the sun of the microcosmos, even as the sun in his turn might well be designated as the heart of the world; for it is the heart . . . which . . . is indeed the foundation of life, the source of all action."

For Caesalpino as well as for Harvey, the *circulatio* was essentially also a *destillatio*. Both assume that the blood is not changed in its substance in the lungs. While Caesalpino, however, ascribes to them the important role of cooling in this distillation, Harvey regards the lungs as not different from other organs, namely, as simply extracting from the rejuvenated blood the nourishing material due to it.

I think that the foregoing has strongly demonstrated the importance of the philosophical myths of an age for the conception and progressive precision of scientific theories. This aspect, namely the general philosophical background of the period, has hitherto almost entirely been neglected. It is to be hoped that Pagel's successful attempt with respect to Harvey will be followed by other similar discoveries.

C. Gillispie (1956) has recently tried to inquire into the philosophical background of the Lamarckian theory of evolution. He describes as a real return to the theory of the fire as the mover of everything of the ancient Greek. This theory begins in the inanimate world and continues into the animate one as the progressive flux of evolution. Everything flows! Hence there exist no fixed species. This pyretic theory is strongly antagonistic to the qualitative and static theory of chemistry of Lavoisier. The latter was the great scientific revolution of those days. Yet Lamarck had the feeling of a revolutionary idealistic romanticism against the precise, mathematical attitude of the modern chemistry of his days. We have similar romantic revolutionary attitudes in Rousseau *versus* Kant, in Goethe *versus* Cuvier. With Lamarck this antagonism extended into politics. He felt himself pursued by the aristocratic spiritual world of Lavoisier. Actually, neither he nor his work was refuted, but his philosophical attitude was ignored and refuted by his contemporaries. Thus we begin to understand an attitude which was enigmatic to the modern historians of science who ignored this background which is, however, not yet sufficiently elucidated.

We may remind that in many other cases the philosophical background is very clear, thus the Hegelian philosophy of Oken and Kielmeyer.

6

ERRONEOUS INTERPRETATIONS OF GOOD OBSERVATIONS MAY LEAD TO IMPORTANT DISCOVERIES

LONG is the list of wrong interpretations of observed facts which have led to important discoveries in the history of science, to be discussed in this and the following sections.

Galvani and Volta. Already Swammerdam (*Biblia Naturae*, plate XLIX, fig. 5-9) had demonstrated contraction of an isolated nerve-muscle preparation of the frog by pulling the nerve with a metallic wire. A. von Haller had written in 1760 on muscular contraction after electric stimulation; Fontana (1781) had confirmed it and related it to the phenomena of electric fishes. Galeazzi had begun to experiment with the Leyden jar, etc.

J. G. Sulzer of Zürich, when studying in 1750 the physiology of taste, noticed an unpleasant, pungent sensation when placing his tongue between pieces of lead and silver when their edges were touching. Luigi Galvani, professor of anatomy in Bologna, studied the effect of electric discharges upon the nervous system of the frog's legs. In 1786, he made the following observations:

" I had dissected a frog and had prepared it (fig. 2 of plate v) . . . and had placed it upon a table on which there was an electric machine, while I set about doing certain other things. The frog was entirely separated from the conductor of the machine, and indeed

To test the possible influence of lightning on the frog's legs, Galvani experimented outdoors. Wire AA is a glass-insulated, iron wire leading to the nerve of the specimen suspended in the bottle C. The feet are in touch with another wire leading down the well into the water. Fig. 8 shows the use of an uninsulated wire. Galvani found that the legs moved under certain conditions even when the air was serene.

Fig. 2. The electrical frog-experiments of L. Galvani. (Of his pl. V, fig. 2).

was at no small distance away from it. While one of those who were assisting me touched lightly and by chance the point of his scalpel to the internal crural nerves of the frog, suddenly all the muscles of its limbs were seen to be so contracted that they seemed to have fallen into tonic convulsions. Another of my assistants, who was making ready to take up certain experiments in electricity with me, seemed to notice that this happened only at the moment when a spark came from the conductor of the machine. He was struck with the novelty of the phenomenon, and immediately spoke to me about it, for I was at the moment occupied with other things and mentally preoccupied. I was at once tempted to repeat the experiment, so as to make clear whatever might be obscure in it. For this purpose I took up the scalpel and moved its point close to one or the other of the crural nerves of the frog, while at the same time one of my assistants elicited sparks from the electric machine. The phenomenon happened exactly as before. Strong contractions took place in every muscle of the limb, and at the very moment when the sparks appeared, the animal was seized as it were with tetanus."

This first observation led to intensive experimentation, from which Galvani concluded that animals have a proper electricity of their own, the Animal Electricity; that animal electricity is concentrated in the nerves and mainly secreted in the brain; that the inner part of the nerve is specialised for conducting electricity, the outer part of it for isolating it and permitting its accumulation; that the receivers of animal electricity are the muscles, which like a Leyden jar are negative on their outside, positive on their inside; that the mechanism of motion consists in the discharge of the electric fluid from the inside of the muscle via the nerve to its outside, and that the discharge of the muscular ' Leyden jar ' furnishes an electrical stimulus to the irritable muscle fibres, which therefore contract.

The production of muscular contraction by touching a nerve and a muscle with an arc of two metal wires of zinc and copper respectively, became the amusement of the elegant world all over Europe, which everywhere set out hunting and killing frogs to perform this demonstration. In later experiments, Galvani showed that contractions could be produced even entirely without the use of metals: when a usual preparation of the frog was held by one foot and swung so rapidly that the vertebral column and the sciatic nerve touched the muscle of the other leg, the muscle would contract vigorously. It was many years later that it was shown (with the aid of the galvanometer) that this induction was caused by the injury-current of cut muscles.

During eight years of intensive experimentation and controversy all over Europe, the Pavia physicist Alessandro Volta proved that in all these experiments the electricity was produced by the touching in a humid medium, of two metals, and that the spectacular muscular contraction served merely as an indicator. This led to the construction of the Volta pile, the prototype of an electric cell, which was the first electrical machine which, after rapid development, furnished man with a new and abundant source of energy. The interested reader will find many more details in Dibner (Galvani—Volta, 1952). We leave this development at the point where Volta replaced Animal Electricity by Metallic Electricity, when the discussion changed from a physiological into a physical phase. This does not mean, that electricity ceased to interest the physiologist: rather, apart from the intrinsic problems, electricity furnished him with fine and sensitive indicators for many aspects of his experimentation.

7

THE LONG WAY OF ERRORS OF SOME BIOLOGICAL PROBLEMS

(*a*) *The History of Generation.*

THE discovery of the spermatozoa and that of the nature of the egg (especially in mammals) and their part in generation are among the major events in the history of biology. Many of the stages of the long and tedious road of true and erroneous observations ; of preconceived, well and badly interpreted theories, and speculations based upon them which led to the threshold of our present day ideas on the nature and mechanism of generation—have been analysed by J. R. Cole (1930) in a masterly essay, whose reading is warmly recommended to everybody interested in the history of biology. Here we must be content to throw light only upon a few landmarks of this remarkable and tortuous development, broken at every step by important or negligible blind side-tracks. We follow slowly how, after centuries of true or, more often, wrong hypotheses and incomplete or faulty interpretation of observations, that level of knowledge has been reached which today is held to be the truth of generation. We will do well to ask ourselves how far our present conceptions are based upon *a priori* assumptions, often made only in the subconscious. The masterly observations of Leeuwenhoek on a thousand and one objects still retain our fascinated interest, while the voluminous natural philosophy of Ch. Bonnet, one of the leading biologists of his day, are read by historians of science only, apart from the breath-taking story of his extended observations of the parthenogenesis of an aphid, or some records of experiments on regeneration in *Hydra* made in conjunction with A. Trembley. It is incorrect to blame only the want of optical instruments for this long way of errors. Some of the most important observations were made with simple biconvex lenses by a Leeuwenhoek or by a Spallanzani. Yet the philosophical prejudices of their ages prevented even those outstanding explorers from drawing the direct and proper conclusions from their own observations, as is also conspicuous in Malpighi, the founder of epigenesis by observation, but a preformist by prejudice.

The problem in Antiquity. The wisdom of Aristotle describes the male semen as the formative principle, the *causa efficiens*, of generation, while the female semen (read: the menstrual blood) is the nutritive principle for growth and development of the embryo. He concludes from his observations of the development of the chick that the organs do not appear all together, but " it is plain even to the senses (that no preformation exists), as some of the parts are clearly visible as already existing in the embryo, while others are not; that is not because of their being too small; that they are not visible is clear, for the lung is of greater size than the heart, and yet appears later than the heart in the original development . . . (The latter organs) come into being only *after* the others!" Aristotle's theory of heredity is a pangenesis, wherein the whole organism takes part in the generative act. Minute representative particles from all parts of the body migrate into the semina. Male and female particles mingle and they both exercise their influence, according to their relative strengths, transmitting characteristics of structure, of function, and of behaviour in the developing young.

Theories of the male Semen in modern Times. After centuries of mere speculation, the first male spermatozoa (a name given by C. v. Baer in 1826, on the erroneous assumption that they are parasitic infusorians) were held by Chr. Huygens (1678) to be a kind of animal arising from corruptions, but having a different origin "like those animalcules one discovers with the microscope in the semen of animals, which seem to belong to it, and are present in such great quantity as to compose almost the whole of it. They are formed of a transparent substance, their movements are very brisk, and their shape is similar to that of frogs before their limbs are formed."

But Huygens had already read the first letter by Leeuwenhoek, of November 1677 (publ. Phil. trans. 1679) which we recommend to the reader before he continues. (III: 596). In 1683, Leeuwenhoek assumes that " the eggs are impregnated by the seminal animalcula and that *one* of them should go into a certain point of the egge. But if no one animal should find this point, then the egge is unfruitful, and this may be the reason why there are so many thousand more animals in *semine masculino* than eggs in the female." To his critics, he answers by referring to the great wastage of seeds in the vegetable kingdom.

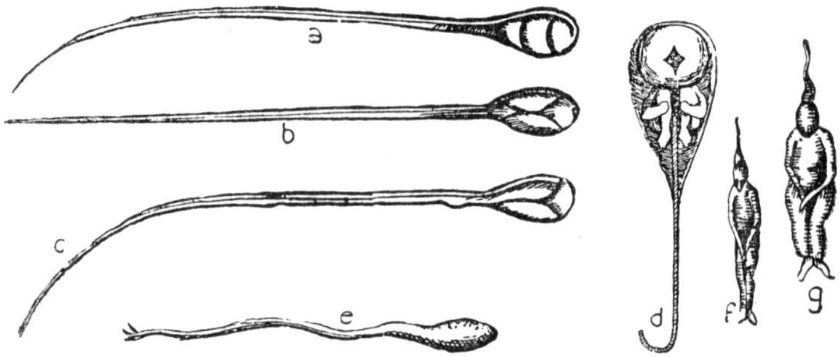

Fig. 3. Spermatozoans as seen in the 17th century by Leeuwenhoek (a—c: of dog, 1679); by Hartsoeker (d: homunculus, 1694); by Dalenpetius (e—g: e: intact, f and g broken, homunculus, 1693). From Singer, *History of Biology*, p. 499).

Thus, one seminal animalcule impregnates the egg and gives to the egg's nutritive substance the embryo, which is in the head of the sperm. This male determination is supported by a case of anticipated Mendelian dominance in rabbits. In the progeny of a crossing, only the male colour appeared in the progeny, which were in fact so like the male parent that they were sold as wild rabbits. Yet Leeuwenhoek refutes energetically the theories of Plantade (1699) and of Hartsoeker (1703), according to which the future embryo was already contained as miniature homunculus within the sperm's head. He thinks that the seminal animalcules perhaps reproduce and multiply by a kind of parthenogenesis.

This discovery of the spermatozoa was, however, only the very first step towards the understanding of their real origin and function. Vallisnieri (1721), who held them to be independent organisms or parasites, believes that they have the task of preventing the clotting of the semen and of keeping it fluid. Linnaeus (1746) denied their animalcule character altogether and called them inert corpuscles. Jones (1745) and others declared them to be the mere products of putrefaction. Buffon (1749, 1777) and Needham (1749) declared them to be aggregates of ' living organic molecules ' derived from the mucilaginous part of the semen and from its filaments, which is another way of assuming them to be products of putrefaction. Hill (1752) classifies the spermatozoa near *Vorticella* in a special genus, *Macrocercus*. Blumenbach (1776) describes them as Infusoria under the name *Chaos spermaticum* (Cercaria spermatica). Spallanzani (1770), that great observer, sides with Leeuwenhoek against Buffon: he thinks they may originate within the blood. Yet in 1780 he fails to note their importance in fecundation and shows, in a series of experiments with faulty technique, that seminal liquid without sperms induces fertilisation in toads, and this in spite of another series of experiments which showed that seminal liquid when heavily filtered was unable to fertilize the eggs. As the unfertilised egg already contains the foetus, the semen adds nothing in its production.

Only in 1824 did Prévost and Dumas demonstrate that the filtered male seminal liquor is deprived of its fertilising power, while the residue retains it. Thus, the animalcules are real and perhaps the exclusive factors in generation. They are not Infusoria and they are not parasites, but the true product of the male testes, within which their histogenesis is for the first time tentatively described. Their observations are confirmed by Bory (1824), who ascribes to them the role of giving the proper mixture to the various substances necessary for generation, though they are not fecundating. In 1827, however, he describes them as cercarias. Also C. v. Baer (1826), the discoverer of the mammalian egg, holds them to be Infusoria-like parasites or Entozoa of the semen, which play no leading part in the process of generation. It was he who gave these simple ' cercarias ' the name *Spermatozoa*.

Peltier and Dujardin demonstrated definitely, about 1827, that these spermatozoa are an organic product of the lining of the seminiferous tubules of the testes. Wagner, in the same year, gave good descriptions of sperms of all classes of metazoans. Yet Carus

(1839) still ' discovered ' internal organs in these ' parasites ', and others followed him, especially Pouchet (1847), who must have possessed a fingertip-feeling which led him to support every wrong theory. Even Cuvier (1817) still classifies the spermatozoans as Cercarias, in his *Règne Animal.* Only the monographs of Kölliker (1841) and of Wagner and Leuckart (1849) brought the testal origin of the sperm to victory, as well as the theory that the spermatozoa characterise the semen just as the erythrocytes do the blood. In some way they were known to fecundate the ovum by contact, and still their exact role in generation remained uncertain.

The development of modern knowledge of the animal egg. Since the beginnings, the role of the egg in generation was more conspicuous than that of the male semen. In some Polynesian tribes the role of the father in the production of the child is still entirely unknown, while that of the mother cannot possibly be overlooked. Thus most authors, from Plato through the Fathers of the Church to the advance of modern observation, have accepted, in contrast to Aristotle, preformation (or evolution). This view negates any true generation: the foetus as a complete miniature shape existed already, encased for all past and future generations since the act of creation, and embryonal development is a mere growth and unfolding. For many centuries, this ovulist school, according to which the egg is the bearer of the embryo, the sperm being only of additional or no value triumphed; but since Harvey and Leeuwenhoek, we find also an animalculist school of thought.

The first return to Aristotelian epigenesis was made by William Harvey (1651) the author of the *Ex ovo omnia*, or *Ovum esse primordium commune omnibus animalibus* (see Cole, 1930, p. 137). Harvey's definition of the ' egg ' remained vague, however. He regards the mammalian ovary as a venous plexus concocting a fluid for the mixture of the parts. He denied that any changes occur within the ovaries during the period of sexual activity, and finds no place for them in generation. In contrast to Leeuwenhoek, Harvey did not find traces of male semen in the genital ducts of the female. The male semen, hence, can have no part either in generation, except that its effluvia may make the female ' egg ' fertile. Yet he expressedly admits his ignorance of the mechanism of fecundation. In the egg, " there is no part of the future foetus actually in it, but yet all the parts of it are in it potentially." Or elsewhere: in the foetus the form is organised "*ex potentia materiae praeexistentis*". This pre-existence—in contrast to the pre-delineation of the typical preformism of that day—agrees well with modern concepts. Generation is caused by a ' Generative Principle ' which has the power of initiating growth. " This principle is inherent in certain diverse corporeal substances which exhibit intrinsic vitality, and are potential living organisms." Such substances are the eggs of animals, the seeds of plants, the ' conceptions ' (read: blastodermic vesicles) of mammals, and the larvae and even the pupae of insects (' imperfect eggs '). Thus, Harvey's ' *ova* ' differ structurally from each other according to the type of animal life to be produced by them, and they only agree in possessing the generative principle. The eggs are a widely varying secondary product of the primordial generative principle (Cole, 1930, p. 139). Such various beginnings cannot be a common beginning, which actually is not the *ovum*, but the Generative Principle of Harvey.

Notwithstanding his oft-repeated statement that " all animals whatever, even viviparous also, nay Man himself to be made of an egge: and that the first conceptions of all living creatures which bring forth young are certain egges," Harvey did not mean—as a critic has said—'*omnia*' and did not know what an '*ovum*' was. First, he does by no means rigidly exclude the possibility of the spontaneous generation of certain animals from putrefaction. He creates the term '*epigenesis*' for an undifferentiated and homogenous egg which develops through gradual differentiation and growth. Of this epigenesis, Harvey distinguishes two different types: (1) Epigenesis proper, where one part is made before the other. In it, development starts from a central point, like the *cicatricula* in the fowl's egg: " It is plain that the chick is built by epigenesis, or the additament of parts budding one out of the other." (2) Metamorphosis is that type of development in which all parts are formed *simultaneously* out of previously ' concocted ' material. The whole pre-existent material is given form in one brief operation, such as occurs in the sudden birth of an insect from its pupa.

This ' precipitation ' theory of insect metamorphosis was justly criticised by Swammerdam (1669) and by Réaumur (1734), who demonstrated the long preparation of each following stage in the preceding ones. From the encasement in insects, Swammerdam proceeds to conclude (from *Hebrews* VII: 9/10) that encasement exists also in man (1672). Croone tried in the same year to demonstrate the corporeal existence of a preformed foetus in the unfertilized egg of a fowl.

De Graaf (1672) describes, in his famous monograph on the female organs of generation, his dissections of rabbit females at various intervals beginning half-an-hour after copulation. He found a definitely formed embryo only ten days later, appearing like a *vermiculus* attached to the placenta by a thin cord. Two days later, head, neck, rump and limbs can already be recognised. The essential merit of Graaf's analysis is the tracing of the 'eggs' from the ovary, down the Fallopian tube, to the uterus. The number of ovarial cicatrices (read: empty follicles) generally agreed with the number of 'eggs' in the uterus. Thus, the ovary (the female testis of the ancients) is essential to generation, being its *fons et origo*. The 'ovarian vesicles or hydatids' (i.e. the Graafian follicles of our days) resemble the ovarian eggs of birds. The ovaries and eggs of birds and mammals are homologous, in the modern sense. De Graaf thought that fecundation occurs in the ovary, and that the eggs become detached from the ovary only after fertilisation. We realize well that de Graaf never saw the mammalian egg, either within the ovarian follicle or within the uterus. He mistook the ovarian vesicle for the egg, and what he saw in the uterus were early embryos, which he realised to be much smaller than the ovarian vesicle. He did find fertilised and cleaving eggs in the tuba and the uterus, on the second day after copulation.

Malphigi (1697, posthumous works) describes the big Graafian follicles of the cow and suspects that the much smaller egg is enclosed within them, their protective envelopes. His great merit is the detailed and accurate description of the epigenetic development of the chick (1672), in spite of which experience he remained a preformist. He describes, e.g., how, at the 36th hour of incubation, the region of the *umbilicus* is covered with small vessels forming a complete network, interrupted by longer and shorter gaps. These, however, are not real gaps, but are only invisible because of their want of red blood. The entire plexus is already present in the *cicatrix*, but becomes manifest only after the penetration of red blood into it. All parts are formed at once in fecundation by a kind of precipitation from materials already assembled in the egg.

This preformistic philosophy of the great epigenetic observer makes Wolff (1759) the initiator of modern epigenesis. In plant-seeds and in the chick, he demonstrates step by step the building up of organ after organ of the seedling and embryo, without the slightest evidence of any preformation. In view of Aristotle's observation that the heart, the *punctum saliens*, is the beginning of development, the discovery of Prévost and Dumas (1824)—that the spinal cord (as *sulcus neuralis*) appears always well before the other organs (translucent, and hence not easily visible), and that around this centre the various other systems of organs are successively built up—was important. Another important statement of these authors, soon to be confirmed and enlarged by von Baer, was that even a most experienced observer is unable to distinguish at an early stage of development the foetus of a dog or rabbit from that of a chick or duck of the corresponding stage. Carl von Baer (1828) discovered the true mammalian egg within the Graafian follicle, and refuted encasement. The younger the embryo, the less complex is its constitution. Every adult tissue is finer and more differentiated than the corresponding one in the chick. The early embryo is almost built up of paving-stones and granite blocks. On this account, the investigation of the embryos of the higher animals hardly ever requires a very great magnification; and if this is so, it is manifest that there can be no preformation.

Landmarks of further research in generation. After von Baer and Kölliker had established the origin and general character of the sexual products, we now may trace the great outlines of further development. Important was the foundation of the cell theory of Schwann and Schleiden, which established the egg as well as the spermatozoon as single cells: the former, big, immobile and full of stores for the future embryonal development; the latter, small, but extremely mobile. Newport (1853) closed a decade of research which showed the penetration of the zoosperms into the egg. He confirmed again in the frog, by filtration experiments, that the active agent in fertilisation is the sperm and not the *liquor seminis*. He thought that a number of sperms penetrate the egg, and not through any special orifice but at any part of the egg-surface with which they come into contact, part of which imbeds itself into the vitellary membrane. In 1854, he declares them also to penetrate by their own movement into the yolk. The first cleavage plane of the egg, observed already by Leeuwenhoek and Spallanzani, was declared by Newport (1854) to establish the median plane of the future frog.

The final stages of research are too well known to be described in detail. The establishment of haploidy in the matured sexual cells through meiotic division; the establishment of the fusion of the haploid chromosomes in both parent-cells at the moment of fertilization by Strassburger (1875) and Fleming (1882), the penetration of a single sperm only (O. Hertwig 1875); and the separation of the phenomena of fertilization from those of the initiation of development (e.g. by natural and by artificial parthenogenesis, see J. Loeb, Driesch, and others), are points to be mentioned.

The modern aspects of preformation versus epigenesis. The antagonism of preformism versus epigenesis dates back to Greek antiquity. The former dominated in the Middle Ages; and the early epigenetic theories from Harvey (1651) through Réaumur (1744) to Wolff (1759) found little echo in their time. Later contributions to epigenesis, by Blumenbach (1780) and others, followed by Kant (1790), were based upon the regeneration of *Hydra;* while the experiments of Geoffroy St. Hilaire (1822/26) regarding the origin of monsters led to the conclusion that no germ can be a predestined monstrosity. We have also mentioned the somewhat independent precipitation theory of Buffon (1749), according to which the embryo does not exist before fertilisation, and is formed suddenly and mechanically immediately after it by the male and female organic molecules of the two seminal fluids. A quantitative calculation of the implications of the encasement theory leads, within only a few generations, to absurd dimensions.

In preformism, we have differentiated between the pre-existence of a Leeuwenhoek (1683) and the pre-delineation (encasement) of a Plantade (1699) and a Hartsoeker (1694), among the animalculists who ascribed the embryogenetic power to the sperm. André (1714) declares, that the encased animalcules are, of course, smaller than the encasing ones. Belated adherents to animalculism are Gautier (1750), Ledermüller (1758), Astruc (1765) and Pouchet (1847). But ovism dominated in the 18th century, through Haller, Bonnet and Spallanzani. A. v. Haller professed himself in 1758 as an ovulistic preformist, basing himself upon the encased generations observed in parthenogenetic aphids and in *Volvox.* Regeneration he regards as a special case of parthenogenetic reproduction. And Bonnet (1745) describes the confirmation of the beautiful theory of encasement (' préexistence en miniature ') as one of the most striking victories of the understanding over the senses.

With the discoveries of von Baer and Kölliker, the preformation theories had lost their actuality, but not their *raison d'être.* In modern biology the old antagonism seemed to reappear in new clothes: preformism, in the chromosome- and gene-theories of genetics; epigenesis, in experimental embryology. Yet deeper analysis revealed that ontogenetic development is actually epigenetic in shape, but determined by an invisible preformism. Never will toads hatch from frog eggs; never ducks, from those of ravens. This invisible predetermination is nothing less than the *causa finalis* of Aristotle. Both theories are thus evidently essential in generation. If a biologist of our day chooses one of these attitudes, he is evidently unjust to one essential aspect of generation. This does not mean that with further advance the old antagonism may not be renewed.

Such a development is typical of many antagonistic biological theories. The struggle of vitalism versus mechanism of the ancient Greeks was strongly renewed in the modern era. Yet new aspects of life, making of the cell a very complicated structure composed of organic molecules (which are not a new conception!) has taken the old meanings out of this conflict. The creation of life, if conceived as the artificial production of these big organic molecules composing the living cell, is, actually within the reach of the possibilities of biochemical research. The problem of the soul and of its creation is transposed to philosophy, the proper place for all theories beyond the reach and cognition of present-day science.

(*b*). *Some facts about the Revolution of the Renaissance.*

Revolutions arise often from petty circumstances, but always from deep causes. One of the most important revolutions of the human mind, the consequences of which we still feel, is the Renaissance. An appeal to reason unsupported by authority was entirely alien to the mediaeval mind. This liberation from authority is the great event of that period. It manifests itself in all compartments of the human spirit (Randall, p. 262) :

" The scholar stresses the revival of ancient letters, the refounding of humanism, and the new orientation of thought. Those of a religious mentality centre their ideas upon its by-product, the reformation, and the momentous results that it effected. To the artist and the architect it opens the golden age of painting, the submergence of Gothic, and the rebirth of Romanesque architecture. The publicist emphasises the political ferment and the foundation of the sovereign state, the rapid decline of feudalism, and the extension of capitalist enterprise. No one would deny or even minimise the importance of the changes that marked the transition from the Middle Ages to modern times, but they all depended for their full effect upon the revolution in the basis of thought. That was produced by the circumnavigation of the globe (Magellan,

1522) and the promulgation of the Copernican theory *De Revolutionibus Orbium Celestium* in 1543, the same year as Vesalius' fundamental *De Corporis Humani Fabrica* founded modern anatomy. The discovery of the earth and of the opening of the modern mind to observation and experimentation as the main bases of thought—in opposition to written authority—produced rapidly the most characteristic of revolutionary changes, namely that of modern scientific method." The unique phenomenon of Leonardo da Vinci (1452-1518) is one of many illustrations in which a happy ignorance of bookish knowledge was mixed with a directness of observation, a questioning of nature in all its width and extent, and a co-ordinating of observations into lines of thought which led to new observations or experiments (cf. Paré and others). No other approach than the direct questioning of nature was admitted, the literature of science being reduced to auxiliary interpretation. Leonardo, who had so little influence upon the development of human thought and of science, remains for us the prototype of the new man, bred by the renaissance.

Science is only one of the many facets of the human mind. It will be informative to have a look at some of the other facets. From the moral point of view, the Italian Renaissance was an entirely immoral, or perhaps better amoral, development. Randall (p. 287 ff.) points out that it is an error to assume that the dispersal of Greek learned men after the conquest of Constantinople led to the rapid revival of Greek writing and learning. He points out that, after the visit of Manuel Chrysoleras to Florence in 1396, the sages of Italy learned Greek with the greatest enthusiasm, reviving philosophical thinking and scientific speculation, which indirectly led, through Ptolemaeus, Erathostenes, etc., to the discovery of America. This enthusiasm spread far and wide the knowledge of, and the interest in, Greek. It laid the foundation for the rapid penetration of humanism half-a-century later. It is important to point out this thorough preparation for the releasing event, which we find in all historical development, the study of which is not less interesting, informative and important than the sudden spread and victory of a certain thought or attitude. Similarly, Darwin would never have been able to conquer the world within five years after this appearance of his *Origin of Species*, if Lamarck, Malthus, Lyell and many others had not prepared the way for him. The foundation of the modern type of school, in the steps of Vittorino da Feltre (1425), took place during this period, another instrument ready to spread rapidly in the new humanistic learning released by the conquest of Constantinople. Historical criticism begins with Valla's revelation, in 1440, that the *Donation of Constantin* was a forgery. This proof was so thoroughly founded, that no later writer has ever again claimed its authenticity. The bringing to fruition of many earlier abortive beginnings, like those of the Hussites and Albingenses, led to the reformation, with its widespread cultural and political consequences. In politics, Machiavelli created modern theory; Grotius, the first notions of international law; and the law of nature was strengthened against the principles and the authority of Roman law.

These few hints regarding some of its facets show us the Renaissance as a fundamental change of mind and thought. And it would be futile to regard certain events, like the discovery of America, the theories of Copernicus, or the spread of Greek learning following the conquest of Constantinople, as the motive or causes of the spiritual revolution. Obviously the new spirit appeared before these events and prepared men's minds, and only because of this preparedness could the releasing conditions and events lead to the sudden spread and victory of new theories and attitudes. And it is just about these preparative decades that our knowledge is usually most restricted or blank. The burning of a heretic sometimes reveals some of the early sparks of new thought and new mental attitudes, but these are accidental revelations, the documents of which are often seen very one-sidedly, through the eyes of the authorities of the time.

(*c*). *The long History of the Interpretation of Insect Parasitism, an illustration of the Accumulation of Knowledge.*

Science is almost the only branch of human activity in which humanity benefits from the past by the accumulation of knowledge. We do not often realise the great difficulties of the proper interpretation of any series of facts. We once studied the development of our understanding of the facts of insect parasitism (Bodenheimer, 1931):

Every pupil has a great knowledge about natural phenomena, not from his own experience, but drawing upon the store of the cumulative knowledge of many centuries. Every child knows now, that the yellow cocoons surrounding a dead caterpillar of the Cabbage White are not the eggs of the caterpillar; that they are the pupal cocoons of parasites (*Microgaster glomeratus*), which developed in the caterpillar's body and which develop into tiny wasps, and which again lay their eggs into other young caterpillars. Thus insect parasitism ranges now through banalities familiar to everybody. Yet centuries of observations and interpretations were needed before this banality gained its proper interpretation. As long as *generatio spontanea* was accepted for lice and maggots, there was no reason to wonder why, from the same caterpillar, now a butterfly, now a number of tiny wasps or of flies were hatched. Aldrovandi (1502), Mouffet (1634), Merian (1679), Redi (1686), and Malpighi (1688, in his unpublished diaries) noted, observed and illustrated the facts without making any commentary. The Belgian painter Jan Goedart (1662) was one of the first to express his astonishment, stating that it is against nature that from the same animal two or three different species should develop. Only in Swammerdam's *Biblia Naturae* (published in 1752, seventy years after his death) do we find the first statement, in the name of Marsilius, that flies deposit their eggs into caterpillars; that their maggots develop as parasites in the caterpillars' interiors; and that the same kind of flies hatch from these maggots, or their pupae, as those which were observed to lay their eggs into the caterpillars. He added observations on a number of other insect parasites. The English physician Lister published (in 1671, in the *Transactions of the Royal Society*, followed in the same year by F. Willoughby) a good and clear paper on the development of the parasitic Ichneumonid wasps. Insect parasitism is well described by A. Vallisnieri (1700). Thus, only around 1700 was the true nature of insect parasitism fully and widely recognised. Réaumur (1736) and De Geer (1760) have thoroughly studied the entire life cycles of a great number of insect parasites, which have been followed by an ever-increasing number of studies, especially since Howard founded upon them the famous theory of a natural equilibrium produced in insect populations by their parasites.

This development illustrates well the great difficulties which even a series of good and true observations oppose to their proper interpretation. Once the truth is known, it spreads rapidly. Thus science gropes from ignorance to error, and the truth of today becomes either a banality or an error in the eyes of tomorrow.

8

ERRORS OF TRADITIONS IN THE HISTORY OF BIOLOGY

IN a general history of science, we should have to point out that some of the classical experiments at the threshold of modern science never took place. They were rather constructed experiments of thought. To this class belongs, for example, the fall of two unequal weights from the leaning tower of Pisa and reaching the ground simultaneously, ascribed to Galileo.

Here we will discuss briefly an experiment which is generally regarded as one of the milestones in the history of biology, the so-called synthesis of urea by Wöhler in 1828. The actual paper *On the artificial formation of Urea* is, as McKie (1944) was the first to point out, a very modest and unpretentious paper, the main part of which is reproduced in our *Reader* (Part III). Wöhler produced urea by the reaction between lead cyanate and ' liquid ammonia '. In the process of purification, a white crystalline substance appeared, identical with the already known urea. The discussion of the chemical properties of that substance is without interest for us here. McKie points out that the process is no synthesis (from elements), but a mere transformation, and that one of the bases of the transformation, the cyanate, was derived from organic matter (dry blood, hoofs, horn, etc.). Wöhler himself gives no hint of the basic or general importance of his ' synthesis '. In a letter to Berzelius on February 22, 1828, he reports that he could make urea without a kidney or a living creature, but continues himself with the observation that for the production of cyanates, as well as ammonia, organic substances are still needed. Neither Gerhardt (1842) nor Berzelius (1849), in their great treatises, were aware that vitalism had been eliminated from organic chemistry.

It was Wöhler's pupil Kolbe (1845) who made the first real organic synthesis (apart from CO_2), namely that of acetic acid. It was M. Berthelot (*Chimie organique fondée sur la Synthèse*, 1860) who, in a textbook, subsequently refuted the *raison d'être* of vitalism in organic chemistry. The heroization of Wöhler's discovery was begun in Wöhler's obituary by Hofmann (1882), who described the " synthesis of urea from its elements (*sic*!) as a joyful message for his generation " and—erroneously of course—as having achieved the unification of inorganic and organic chemistry. Nothing of that kind is recognisable: neither did Wöhler himself make any claim of this kind, nor does there exist any other document showing that the chemical world of 1828 to 1850 regarded his paper as an historical event in chemistry or natural philosophy. McKie concludes: " It is difficult to make any comment on Hofmann's extraordinary assertion; it appears to have been written without any sense of history or of restraint." The lesson which we draw from this story is that *one* wrong statement is sufficient, sometimes, to be accepted by generations of textbooks as a fundamental, proven fact, without, or almost without, any dissenting opinion.

We wish to point out another wrong tradition from recent times. J. Piveteau (1950) has recently analysed the famous discussion which took place in 1830 before the Académie des Sciences at Paris. Largely under the influence of Goethe's interpretation, it was for long regarded as a battle between the constancy of species versus evolution. Actually this conception is a misunderstanding of a debate on purely morphological problems.

Cuvier, the founder of modern comparative anatomy, had reduced the morphological structures of all metazoa to four basic types: Vertebrata, Articulata, Mollusca, and Radiata. Within every one of these four types, the shape of every limb and organ is determined by the habits of the animal, which determine its function: an animal with predatory teeth *must* have claws for holding the prey and by general structure be able to get at it. These obligatory correlations permit us to reconstruct an entire skeleton from the fraction of a bone.

Geoffroy St. Hilaire agreed with the reduction of all vertebrates to a uniform type. In his *Philosophie anatomique* (1818) he stresses that the only criterion for homologisation is the morphological position of an organ in the body—" les connexions ou analogues "—and not its functions. Geoffroy's important contribution to the study of the type of the vertebrates was the study of the bones, especially of the skull, in the embryos, where conditions are clearer than in the adult skull, and their use for homologisation. Later on, Geoffroy pushed his unity of plan still further, in joining the Arthropoda with the Vertebrates: He claimed the segmentary exoskeleton to be the ' analogue ' of the vertebrae, the legs that of the ribs. Cuvier, though obviously disapproving of these ideas, remained silent for over ten years, while Geoffroy promoted his theories at the meetings of the Academy.

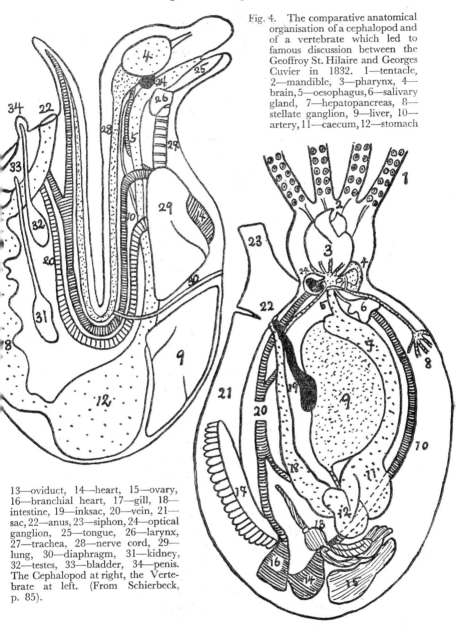

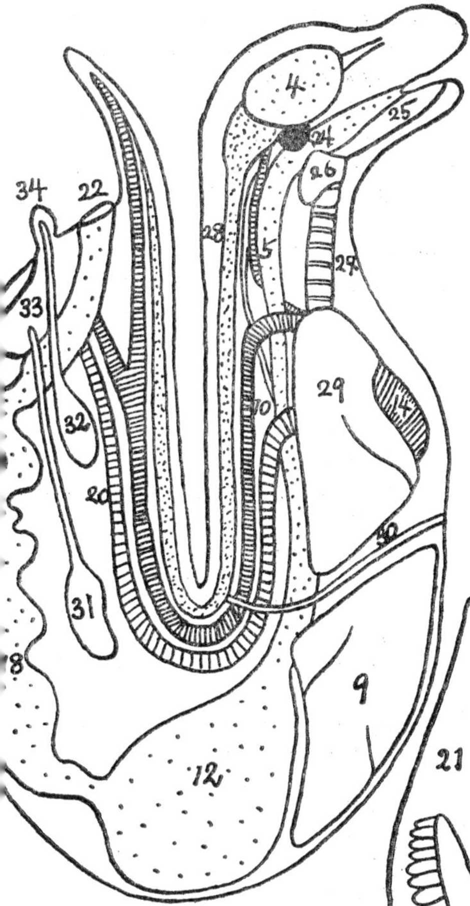

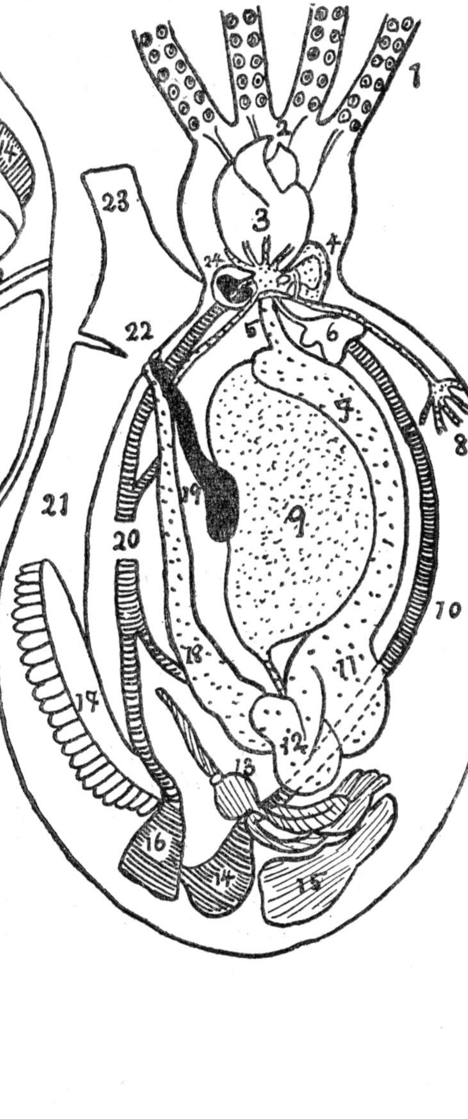

Fig. 4. The comparative anatomical organisation of a cephalopod and of a vertebrate which led to famous discussion between the Geoffroy St. Hilaire and Georges Cuvier in 1832. 1—tentacle, 2—mandible, 3—pharynx, 4—brain, 5—oesophagus, 6—salivary gland, 7—hepatopancreas, 8—stellate ganglion, 9—liver, 10—artery, 11—caecum, 12—stomach 13—oviduct, 14—heart, 15—ovary, 16—branchial heart, 17—gill, 18—intestine, 19—inksac, 20—vein, 21—sac, 22—anus, 23—siphon, 24—optical ganglion, 25—tongue, 26—larynx, 27—trachea, 28—nerve cord, 29—lung, 30—diaphragm, 31—kidney, 32—testes, 33—bladder, 34—penis. The Cephalopod at right, the Vertebrate at left. (From Schierbeck, p. 85).

Yet, on February 15, 1830, Geoffroy introduced a paper written by two young zoologists, Meyranx and Laurencet, who tried to demonstrate that the cephalopod body is also of the vertebrate type, if the rump is regarded as folded back, so that the pelvis touches the neck. Geoffroy added, that this case shows the absurdity of rigid separation between the various animal types. Actually there exists only *one* type of morphological structure throughout the animal world.

This general attack was answered by Cuvier at the meeting of February 22, 1830, where he read a paper: " Considérations sur les mollusques et en particulier les Céphalopodes ". He easily demonstrates that the torsion of the rump assumed by the two writers does by no means modify the general relations between the organs, which remain different, and that there is no identity of plan, apart from the general analogy which exists among all animals, insofar as they need organs for locomotion, feeding, digestion and excretion, reproduction etc. Yet these mere analogies are no proofs for the identity of the structural plan of a polype and a man.

Geoffroy answered on the 1st March, 1830, with a general paper: " De la Théorie des Analogues pour établir sa nouveauté comme doctrine et son utilité pratique comme instrument ", in which he proposes the principle of the unity of composition. Yet in his discussion there is no question any longer of a unique structural plan for all animals, nor even of one for Arthropoda and Vertebrata, but only of the *os hyoidale* of the upper vertebrates, which in all its manifoldness is yet claimed to be of a single structure. To this lecture Cuvier answered on the 22nd March in a detailed discussion of these bones, concluding: " By which art will one convince us of the identity of the connexions between the hyoid bones, part of which are suspended from an area of the os temporale, whilst others are twisted around the skull? " Geoffroy read at the same meeting a paper on the " Application de la Théorie des Analogues à l'organisation des Poissons ", in which he criticises certain of Cuvier's interpretations of the skull of the fish.

Yet Geoffroy answers Cuvier's attack regarding the *os hyoidale* only on the 29th March, in a paper " Sur les os hyoides ", in which he defends his earlier views and attacks Cuvier, who considers only facts which can be observed. On the 5th April Cuvier again considers the *os hyoide* of the vertebrates living in water. He points out, among others, that Geoffroy does not always refrain from taking function into consideration, as he does actually in his treatment of the sternum. He gives, at the end, a summary of his general views.

This was the formal end of this famous discussion, which was not taken up further in the debates of the Academy. Its background, the principles of comparative anatomy, could not have been possibly of interest to the public at large. Possibly it was the splendid style of Geoffroy which attracted the press to take sides in the issue: *Temps* and *National*, for Geoffroy; *Journal des Débats*, for Cuvier; the German *Naturphilosophen*, with Geoffroy.

With regard to the point: one versus (at least) four basic types of animal structure, we all share the view of Cuvier, even if Geoffroyan theories continue to crop up (Semper, Steinmann). With regard to the rigidity of the unique plan within the vertebrates, opinions are still divided, without arousing a wide interest. Nothing in the discussion designates Geoffroy more as an evolutionist than Cuvier or substantiates the claim that Cuvier by this debate impeded for decades the march of evolution. As Piveteau puts it: Those who claim an active role for the organism, a domination of functional influences, will favour Cuvier; while those who believe in chance variation, independent of environment and function, will prefer Geoffroy. But these are only afterthoughts, which were not promoted in the discussion itself.

We ask ourselves: If the contents and objects of a relatively recent discussion, all the documents regarding which are published, could have been so thoroughly misunderstood, what certainty do we possess in the history of science with regard to earlier controversies, the general scientific and biological background of which is still farther away from our present mentality, and the documentation of which is sometimes almost entirely wanting?

9

ANALOGY AND HOMOLOGY IN THE HISTORY OF BIOLOGY

CERTAIN ideas repeat themselves again and again in the history of science. A closer analysis, however, reveals that they reappear in ever-changing contexts, meanings, and connections. They are usually mere analogies, but not homologies.

The history of the evolutionary theories offers a series of good illustrations for this phenomenon. A number of such analogies to modern evolutionism are found in the early Ionian philosophers or in the Roman epicuraeans. Alexander of Milet (611-546 B.C.) describes the origin of life under the influence of the sun rays upon the primordial mud. The first organisms were vesicular beings, and they slowly developed into fish-like animals. Some of these were thrown upon the beach by chance. Many of them died, but a few survived, inspiring the air and developing lungs, abandoning respiration through gills. Those first terrestrial animals continued their slow development into higher and higher forms and into man. The slowness of this development was wise: the lower animals go in search of their food immediately after birth, but this is denied to the higher vertebrates. If their sucklings had appeared suddenly, there would have been nobody to take care of them and they would have died.

The sentence of Heraklitus of Ephesos (500 B.C.): " Struggle is the father of everything " is often regarded as an anticipation of Darwin's struggle for existence.

For Empedocles of Agrigent (490-430 B.C.), the hates and the loves of the molecules led to the creation of the first organic molecules which steadily improved by further fusions of molecules. Rumps of higher animals and plants were ejected from the inner fire within the bowels of the earth, together with various limbs and organs. These combined into all possible combinations. Most of these perished, being unable to live. The wise and useful structure of the body is thus the result of a slow and permanent perfection. The recent forms remain, as we should say today, through the survival of the fittest, while the overwhelming majority of the combinations perished because they were unfit for life. Here we find the striking combination which Darwin also used 2200 years later, namely, an explanation of the utilitarian structure of the organisms which does not call for its establishment by the activity of directing teleological or theological forces. Yet the chance meeting of compatible and incompatible parts and their integration into an organism, without the slow development of related forms, is something so utterly foreign to modern thought, that we see in the theory of Empedocles only a formal analogy for the non-directed origin of utilitarian body structures.

The notion of Diogenes of Apollonia (450 B.C.) that no changing being can differ from another one, without having been like it before, looks at first glance like a progressive step beyond Empedocles. Yet here we have not even a philosophical or cosmological speculation, but only a digression into the methods of logic, and no contact whatever with evolutionary problems.

Lucretius Carus (100-50 B.C.), the Roman epicuraean, sings that the creative earth once gave birth to plants and animals, whereas now its creative power is ended, as with the womb of an old woman; apart, of course, from the normal *generatio spontanea*, such as we still find in maggots coming out of the corruption of the dung. The struggle for existence is described in sentences which approach our modern treatment and style. Lucretius is anti-teleological in the extreme. Not the eyes are created for vision, but vision followed the earlier development of its optical instrument, the eye.

Similar analogies crop up in the 18th century. We mention here only the so-called ' ladder of nature ', or ladder of morphological progress in the organised world, so popular with Bonnet, Lesser, Durham, Oken and so many others, which referred only formally to a logical arrangement of the various kinds of organisms in the Aristotelian sense, without taking into consideration even any transformation or evolutionary affinity.

Speculation of this sort develops further in the Telliamed or Benoît de Maillet (1743). The first germs of life dropped from beyond the earth into the ocean, where slowly they developed into fishes, continuing slowly through the amphibians to the higher vertebrates. This view is supported for the first time by arguments from comparative osteology. In describing the development of the birds from the flying fishes, he makes use of the *homology* of the bird's wing, the fish's pectoral fin, and the human arm. Here is the first true rudiment of modern argumentation, but the form is still very crude.

Yet the basic difference from modern evolutionary thought and these older ideas is that none of them is based upon an empirical knowledge of the abundance of animal and plant species, upon their variation, and upon the constancy of their differences. In the Greek and Roman philosophers, their ' evolutionary ' notions are mere speculative corollaries to wise, yet crude cosmological theories. In the 18th century, they still remain speculative thought, which foreshadows, however, by its use of arguments from comparative osteology, the approaching period of modern evolutionism. We agree entirely with Ostoya (1951) that no history of evolutionary thought exists before Lamarck (1809).

The later history of the theory of evolution offers yet other instances which, from more than one point of view, are of the greatest and principal interest to the historian of science (see Ostoya, 1951).

10

PROBLEMS OF PRIORITIES

GREAT trouble is often encountered in distinguishing priorities, nor is such trouble by any means always in vain. Nothing is more inspiring and fascinating than the study of any phase of human activity or thought from its very first sources to the early, isolated precursors, until—after many emergences and submergences—the full-fledged discovery makes its successful appearance. But how rarely do we have sufficient and adequate documentation to write such a full story! Wherever the quest for priorities is mainly aimed at the establishment of national or personal priority claims, it deviates obviously from the true aims of science and its history. Priority by itself is nothing, truth and sincerity everything. Uncritical priority studies easily base themselves upon void or formal analogies, which never intended to express what a later interpretation may read into them. Thus, we have found in the partial study of the history of blood-circulation all the transitions from mere analogies to experimental proof, to the 'calcul', and, finally, to the last step, the discovery of the capillaries. It would be wrong, in our opinion, to accept anybody's claim of priority, anterior to the laying of a solid factual basis, in our case by Harvey and Malpighi.

One unpleasant aspect of modern priority claims is the restriction to the national horizon. Whether somebody accepts B. Grassi or R. Ross as the discoverer of the mosquito carrier of the *Plasmodium* will depend actually today upon the birthplace of the student. The Italian will accept Grassi, the Britisher Ross, as the discoverer. Careful analysis, however, reveals that both these great students came to their conclusions entirely independently, so that actually both claims are justified. It is not only a narrow chauvinism which is expressed in such situations, but a still more serious condition: the English reader knows only the English scientific literature; the Frenchman, the French; the German, only the German. The reading of the literature index of almost every Progress Report of any branch of science will confirm this statement. This leads to a restriction of horizon, to a falsification of perspective, to injustice, and to an entirely undesirable situation—in a world which should be one at least as far as science is concerned.

Recently Russian science has often been ridiculed because of its many priority claims for Russian scientists within the last 200 years.

Yet this is only the attempt—in principle justified—of a cultured nation whose language has been unknown and neglected in the world of science, to prove that, within the said period, Russian scientists participated in the solution of all the problems of European science. As soon as this justified nucleus is recognised, the detailed discussion as to which of the claims is correct and supported by

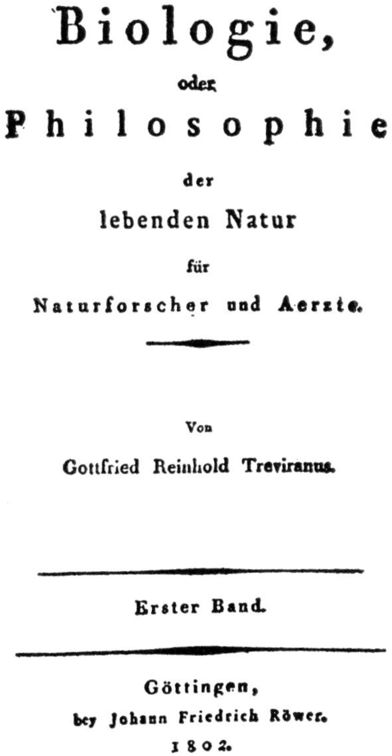
Biologie,

oder

Philosophie

der

lebenden Natur

für

Naturforscher und Aerzte.

Von

Gottfried Reinhold Treviranus.

Erster Band.

Göttingen,

bey Johann Friedrich Röwer.

1802.

Fig. 5. The first printed application of the term Biology by G. R. Treviranus (1802), describing the Philosophy of Living Nature.

experiment and observation, or whether both are independently justified, can then take place *sine ira*.

We may add another aspect. The term ' Biology ' was independently proposed in 1802 by G. R. Treviranus in his *Biologie oder Philosophie der lebenden Natur*, and by J. B. de Lamarck in his *Hydrogéologie*. Treviranus includes in biology the various phenomena and forms of life, the conditions and laws which rule its existence, and the causes which determine its activity. But, as Grassé has correctly pointed out, this good and pretentious programme is followed by six volumes of systematic descriptions of the animals and plants known at that time, without being followed or preceded by any attempt at ' biological ' synthesis. It is doubtful whether Lamarck in 1802 did go further in his interpretation of biology; he repeatedly varied his definition. But Grassé (1944) has published a manuscript from the Bibliothèque du Muséum of Paris, which—written in 1814—is called *Biology or Commentary on nature, the faculties, the development and the origin of organisms.* This manuscript clearly defines Biology as the study of the general phenomena which are common to living matter, in contrast to lifeless matter.

Lamarck had in 1803 already announced his—unfulfilled—intention of publishing a Biology, in order to prove that the nature of an organism has all the special qualifications for produce of itself everything which we admire in it.

If we let ourselves be guided by the name ' biology ' only, including a more or less vague definition, then Treviranus and Lamarck are the joint founders of the term Biology. But greater justice is done when we accredit Lamarck, who, in 1803, 1812, and in the unpublished manuscript of 1814, laid a solid basis for the modern conception of Biology, which was accepted in the following decades by French scientists and spread all over Europe.

11

AN INTRODUCTION TO THE BIBLIOGRAPHY OF THE HISTORY OF SCIENCE, FOR BEGINNERS

By far the best introduction to the study of the history of science until the end of the 14th century, into its general background as well as into its general bibliography, is the monumental work of

G. Sarton, The Introduction into the History of Science (3 vols., Baltimore, 1927-1948).

Good further introductions are:

G. Sarton, A Guide to the History of Science. 1952. Waltham.

F. Dannemann, Die Naturwissenschaften in ihrer Entwicklung. 4 vols. 2nd edition. Leipzig, 1907-1923.

T. S. Hall, A Source Book in Animal Biology. [From 1500 to our days]. New York, 1951.

The first of these gives ample information on the organisation of the History of Science all over the world and concerning all its bibliographical organs. The other two are the two best readers available for the history of zoology, so far. For information on the current bibliography of science the only available source is *Isis* (Cambridge, Mass., 1913-). In the *Archives Internationales de l'Histoire des Sciences* (Paris, 1947-), current and often stimulating reviews are given on important new books (but not on articles in periodicals).

These hints should be sufficient for the budding Historian of Science, to help him get his first orientation to the general background of a special personality, a special period, or a special historical problem. It must, of course, be assumed that he is already familiar with a number of the current textbooks on the history of biology, zoology, and botany, on which he can get information through the kindness of the Librarian of any good good public library.

If the exact title of an ancient writer is looked after, the following books are warmly recommended:

Botany.

G. A. Pritzel, Thesaurus Literaturae Botanicae. 2nd edition. Lipsiae, 1872.

Not quite so good is :

B. D. Jackson, Guide to the Literature of Botany. London, 1881.

Zoology. For older literature may be recommended:

J. L. R. Agassiz, Bibliographia Zoologiae et Geologiae. 4 vols. London, 1848/54.

For later literature the standard work is:

W. Engelmann, Bibliotheca Historico-Naturalis. Vol. I. Leipzig, 1861. From 1700 to 1846. Vols. II and III appeared until 1923 and continue the literature up to 1880.

Often useful are the Catalogues of great museums, like that of the British Museum, the annual volumes of the Zoological Records (London), and others. Also:

C. A. Wood, An introduction to the literature of vertebrate zoology. Oxford and London, 1931.

A guidance for the ancient literature of divers animal groups is summarized in:

F. C. Sawyer, Books of Reference in Zoology, chiefly bibliographic. J. Soc. Bibliogr. Nat. Hist. 3, 2. 1955. pp. 72-91.

12

REFERENCES

F. S. Bodenheimer, The Concept of Biotic Organization in Synecology. Bull. Res. Counc. of Israel. 3. 1953. pp. 114-122.
I. H. Woodger, Biological Principles. 2nd Impression. London, 1948.
B. Russell, Human Knowledge. London, 1948.
A. Arber, The Mind and the Eye. Cambridge, 1954.
A. Standon, Science is a Sacred Cow. London, 1952.
I. Needham, A History of Embryology. Cambridge, 1934.
F. S. Bodenheimer, Aristote Biologiste. Paris, 1952.
R. Taton, Causalités et Accidents de la Découverte Scientifique. Paris, 1955.
J. R. Baker, The Growth of Biological Ideas. From: Mees, The Path of Science. New York, 1946. pp. 146-172.
G. Senn, Die Entwicklung der Biologischen Forschungsmethode in der Antike. Aarau, 1933.
L. Darmstaedter, Handbuch zur Geschichte der Naturwissenschaften und der Technik. 2nd edition. Berlin, 1908.
G. Sarton, The Study of the History of Science. Harvard, 1936.
W. J. Lütjeharms, Zur Geschichte der Mykologie. Das XVIII. Jahrhundert. Amsterdam, 1936.
J. Sachs, Geschichte der Botanik. München, 1875.
G. Sarton, The Life of Science. New York, 1948.
J. Needham, Relations between China and the West in the History of Science and Technology. Actes VIIe Congr. Intern. Hist. Sci., Jerusalem, 1953. pp. 132-185.
C. Singer, History of Biology. 2nd edition. New York, 1950.
M. Steinschneider, Die Hebraeische Übersetzungen der Mittelalters. Berlin, 1893.
P. Sergescu, Coup d'oeil sur les origines de la Science moderne. Paris, 1951.
H. Metzger, La Philosophie de la Matière chez Lavoisier. Paris, 1935.
C. Elgod, A Medical History of Persia. Cambridge, 1951.
E. Schroedinger, What is Life? London, 1941.
C. Semple, Influence of Geographic Environment. London, 1911.
C. Semple, The Mediterranean Region. London, 1933.

L. Rosenfeld, Le genèse des principes de la thermodynamique. Bull. Soc. Roy. Sci. Liège, 1941. pp. 199-212.
S. Lilley, Social Aspects of History of Science. Arch. Int. Hist. Sci. Paris, 1949. Repr. 70 pp.
R. K. Merton, Science, Technology and Society in seventeenth-century England. Osiris 4. 1938. pp. 360-632.
I. Berlin, Historical Inevitability. London, 1954.
N. Weiner, The Human Use of Human Beings. Boston, 1950.
W. Pagel, Med. Bookman and Historian 1948, Oct.-Nov. 4 pp., Brit. Med. Journ., 1950. II. p. 621; Isis 42, 1951. pp. 22-38; Journ. Hist. Med. 6. 1951. pp. 116-124; Sudhoffs Archiv. 37. 1953. pp. 319-326.
C. C. Gillispie, The formation of Lamarck's evolutionary theory. Arch. Int. Hi. Sci. 1956. p. 323-338.
B. Dibner, Galvani-Volta. A Controversy that led to the Discovery of Useful Electricity. Norwalk, 1952.
E. J. Cole, Early Theories of Sexual Generation. Oxford, 1931.
H. J. Randall, The Creative Centuries. London, 1945.
F. S. Bodenheimer, Zur Frühgeschichte der Erforschung des Insektenparasitismus. Arch. Gesch. Math. Nat. Wiss. u Technik 13. 1931. pp. 402-416.
D. McKie, Wöhler's ' Synthetic ' Urea and the Rejection of Vitalism: A Chemical Legend. Nature 153. 1944. pp. 608-609.
J. Piveteau, Le débat entre Cuvier et Geoffroy Saint-Hilaire sur l'unité de plan et de composition. Rev. Hist. Sci. 3. 1950. pp. 343-363.
P. Ostoya, Les Théories de l'Evolution. Paris, 1951.
P. Grassé, ' La Biologie ' Texte inédit de Lamarck. La Revue Scient. 82. 1944. pp. 267-276.

PART II

A Short Factual History of Biology

1. Antiquity
2. The Middle Ages
3. Modern Times
    (*a*) Taxonomy
    (*b*) Early Microscopy
    (*c*) Morphology
    (*d*) Embryology
    (*e*) Cell Theory
    (*f*) Physiology
    (*g*) Microbiology
    (*h*) Biogeography and Ecology
    (*i*) Evolution and Genetics
    (*j*) Literature for further Reading

1

ANTIQUITY

WHILE still in the late animal stage himself, man came every day in intimate contact with plants and animals—as game, as gathered food (fruits, roots, insects, shellfish), in fishery, in the early stages of the domestication of plants and animals as far back as the neolithic and perhaps even in the mesolithic age), as enemies, as material for clothing (furs, fibres, bark), as firewood, as materials

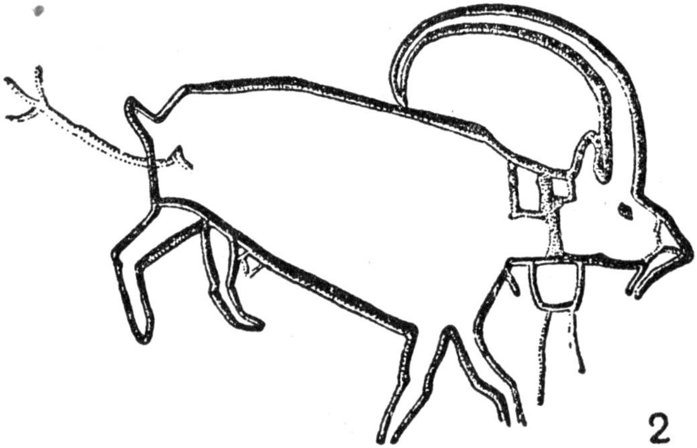

2

Fig. 6. Ibex with magic arrow from Kilwa. (Bodenheimer).

for tools (bones, wood, etc.), and so forth. The empirical use of medical herbs or of narcotics or stimulants must be extremely old. Yet it would be wrong to assume that man's knowledge of nature was built up only by empirical approach. Very complicated cosmological systems, much more complicated than the early theories of totem and taboo suggested, connected a certain herb with a certain bird, tree, mammal, musical and kitchen instrument, ornament, clothing, a certain star-constellation, and what not, combining these into groups and systems of objects which had nothing to do with empirics nor with a wild and primitive animism and mysticism. Magic connections were later derived from such 'natural' groups. Magical practice became widespread, such as the magic arrows of the wonderful animal drawings in palaeolithic caves (fig. 6) or the extremely complicated annual totem-rites of the Australian and other aborigines, which are essentially fertility rites, to the importance of which G. Frazer apparently first turned our attention.

In all the three ancient river civilizations considerable knowledge of animals and plants was accumulated (Mesopotamia: Thompson, Landsberger, Bodenheimer; Egypt: Keller, Lortet, Keimer). We find its manifestations on the seals and other ornaments of the Ganges civilization; in the wall-paintings, reliefs, statues and animal-mummies of ancient Egypt, and by an overwhelmingly rich evidence in Mesopotamia. There the oldest books on zoology and botany were written. The *Har-ra=Hubullu* and other bilingual sumero-accadian lexicons arrange the names of plants and animals not alphabetically, but according to related groups, all members of one group being characterized by a common prefix. Thus, the major groups of terrestrial animals on plate XIV of the *Har-ra=Hubullu* are according to the prefixes:

No.	*Prefix.*	*Zoological Group.*
1-46	mush	snakes
48-60	am (sun, silam)	aurochs
61-98	ur (sal)	dog, wolf, etc.
108-115	sa	small wild cats
121-136	ug	lion
145-153	lu (dara, mash)	ungulates
159-198	shakh (sal, lab)	pig
184-198	pesh	small rodents
200-204	nin	mungolike carnivora
207-215	erne (kun)	lizards
225-226	ilu	crab
228-244	buru	most Orthoptera
253-272	uh	vermin
273-279	za (ush)	caterpillars (or beetles?)
283-289	mar	earthworms
296-303	girish	Lepidoptera
304-332	num	Diptera, Hymenoptera
340-345	mul	Mollusca
352-359	kishi	ants
361-370	gir	scorpions
376-379	ne	Amphibia
387-409	u, nig, gar	animals in general

In addition, a great number of tablets and steles inform us about royal hunts of lions, elephants, aurochsen, etc. A special class of priests derived omina from the behaviour of animals, birds, serpents, insects, etc.; others from the variations in the structure of the liver of sacrificial sheep. Some of the Assyrian kings brought new plants and animals from their military expeditions and tried to

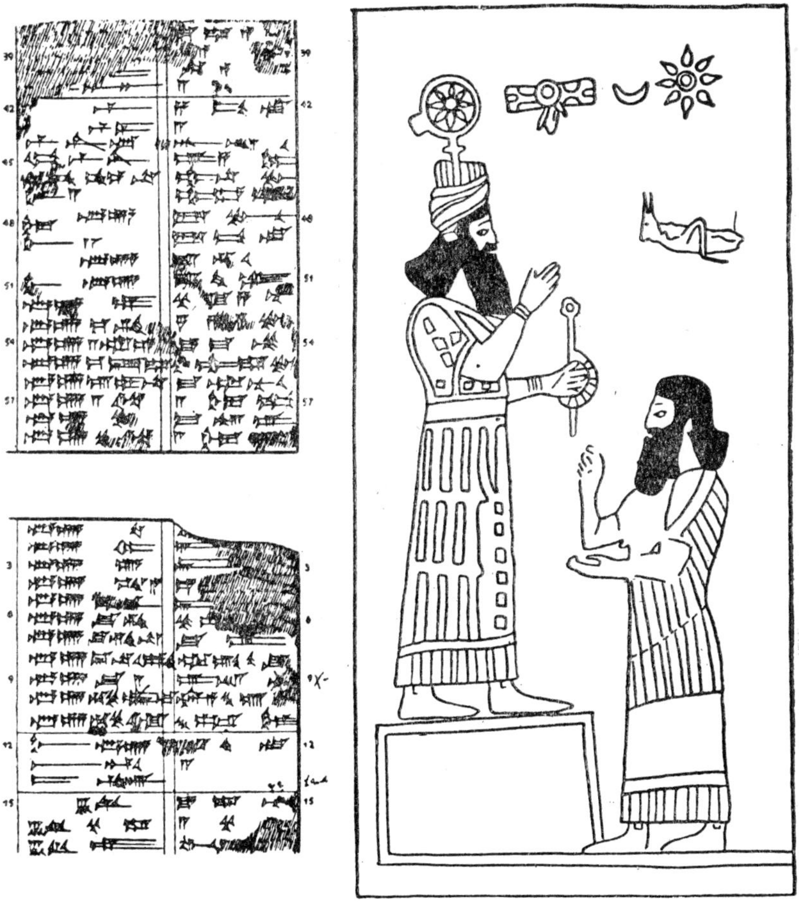

Fig. 7. Assyrian Zoology. At left : Part of a plate of the bilingual Har-ra-Habullu; at right: Locust prayer. (From Bodenheimer. 1956).

acclimatize them in Ashur (e.g. the bee); and others to show them to the people in their parks. An abundance of animal figurines and reliefs permit identification of the races of domestic animals of the various periods and of the big animals. Thus a gazelle may be identified which today is restricted to Tibet.

In the creto-minoan culture, drawings of sea-animals abound.

According to the generally accepted tradition the ' Great Miracle ' of the creation of science in the Western world began at the end of the first half of the first millennium B.C. We are growing increasingly sceptical in this respect; learning more and more about the early contacts between Mesopotamia, Egypt and Greece, and realizing that it was not only empirical knowledge, i.e. facts, but also generalised notions and theoretical speculation, not based entirely upon magics which spread from East to West helping to produce this ' miracle '. But it remains the uncontested glory of the people of the Aegean shores, not only to have catalogued scientific knowledge and speculation, but also to have brought into it method and system, especially with regard to ' natural causes '. We are best informed about their cosmology which combined still largely astronomy with theological mythology. Yet we cannot assume that the stupendous and unique work of Aristotle, his penetrating analysis of biological thought, his vast treasure-house of observations, was created *ex nihilo*. A long series of observers and of thinkers must have preceded him, most of whom are forgotten altogether, others mentioned in paragraphs of polemic in Aristotle's writings. We only sketch here the *Corpus Hippocraticum*, the writings of the ancient medical school of Kos (500-350 B.C.), where we find advice for the study of embryology of the chick; good chapters on human ecology, and definite ideas on heredity, describing minute particles migrating from all organs of the body into the sperm, and later on out of the sperm, each building up the organ from which it itself originated. In the discussion of diet, as the basis of all therapy, many animals and plants are mentioned. True, the pathology diagnostics, prognosis, and therapy of that school are built up from mere observation, and exhibit the most primitive ideas of anatomy and physiology; the more must we admire their achievements. Singer calls the treatise on 'the Divine Disease' (epilepsy) the *Magna Charta* of Science, since there ' for the first time ' observation is regarded as superior to superstition and magic, and the need for natural connections instead of speculative or theological ones is stressed. We should also mention Empedocles of Agrigent (484-

424 B.C.) and Democritos of Abdera (460-370 B.C.), whom the master mentions in his writings. Nor can we pass over with the usual silence Xenophon (445-355 B.C.), whose observations on the life history of the hare are so outstanding that even today they provoke our admiration (*Cynegeticus*, ch. V).

In Aristotle (384-322 B.C.) biology reached a peak to which it has never returned since. Whilst his philosophical, ethical, political and physical works have dominated their respective fields for many centuries, the biological ones were largely forgotten. This is the more incomprehensible as his biology is the backbone of his philosophy and method. On that basis Aristotle made observation and induction the foundation of scientific knowledge, always checking the theories based upon intuition and deduction. This was in strict contrast to Plato (429-347 B.C.) whose deductive philosophy with its sovereign despising of all induction and observation, barred the development of science in certain periods. The Aristotelian static cosmos and physics are now replaced by dynamic notions, but in biology he has remained *the* master, as has been recognized by men like Cuvier, Darwin, Singer, Russell, Needham, and many others. He was apparently the first who not only observed details of structure, behaviour and environment, but also connected them all together with morphology, anatomical dissection and development, into one comprehensive whole. He developed the relations between structure and function of the various animals into a comparative science. His biological books have come to us in the redaction of Andronikos of Rhodos (*ca.* 85 B.C.), to whom we also owe the survival of the works of Theophrastus. His main biological books are: *The Natural History of Animals* (*Historia Animalium*, H.A.); *On the Parts of Animals* (*De Partibus Animalium*, P.A.); *On the Reproduction of Animals* (*De Generatione Animalium*, G.A.); and *On the Soul* (*De Anima*, D.A.).

The epistemology of Aristotle was: the facts are real, but the ideas which we have of them may be wrong, as is shown by the colours as they appear through a coloured glass or to one suffering from jaundice. Every theory has to be in agreement with the facts, and if new observations contradict a theory, the theory has to be abandoned and not the facts.

Aristotle was no taxonomist, in the sense that he did not arrange the animals into a system. Yet his comparative anatomy and ethology proceed from group to group, so that we are able to reconstruct his system, which is the first one comprising the entire animal kingdom. Reconstructed, it runs approximately as follows:

THE ARISTOTELIAN SYSTEM OF THE ANIMAL KINGDOM (RECONSTRUCTED):

I. ANIMALS WITH RED BLOOD (VERTEBRATA).

[A]. *Quadrupeds with hair, producing living young:* *MAMMALIA*

1. Man and monkeys (as affiliated groups) . . . . *Primates*
2. Monodactyla, with incisors in the upper jaw . . . *Ungulata Perissodactyla*
3. Didactyla, ruminants, without incisors in the upper jaw *U. Artiodactyla*
4. Polydactyla . . . . *Carnivora, Rodentia, Insectivora*
5. Aquatica . . . . *Cetacea and others*
6. Varia: It is uncertain if elephant, hippopotamus, camel and others are included into the above-mentioned categories.
7. Volantia . . . . *Chiroptera*

[B]. *Quadrupeds without hair which lay eggs:* *REPTILIA, AMPHIBIA*

1. Without scales . . . . *Amphibia*
2. With scales and with legs . . . . (*most*) *Reptilia*
3. With scales, without legs . . . . *Ophidia, Lacertilia* (*ptm.*)

[C]. *Two-legged birds with two wings and with feathers, lay eggs:* *AVES*

1. Predatory birds with claws . . . . *Accipitres, Striges*
2. Insectivores . . . . *Passeres* (*ptm.*)
3. Granivores . . . . *Passeres* (*ptm.*)
4. Eating wood worms . . . . *Pici*
5. Pigeons . . . . *Columbae*
6. Long-legged waders . . . . *Grallae, Ciconiae*
7. Palmipeds . . . . *Lamellirostres*
8. Varia: Also the *Gallinacei* (remaining generally on the soil), swallows, ravens, etc., are recognised as separate 'genera.'

[D]. (*Primary*) . *legless animals with two pairs of fins* (*analogous to the extremities*), *with scales, laying 'imperfect' eggs:* *PISCES*

1. Cartilaginous fishes . . . . *Selachii*
2. Bony fishes . . . . *Teleostomi*

Various 'genera' are recognised within each group.

II. ANIMALS WITHOUT RED BLOOD (EVERTEBRATA).

[A]. *Soft-bodied with eventually an inner bone, with perfect eggs* *CEPHALOPODA*

[B]. *The soft body covered by a flexible outer skeleton, with perfect eggs* *CRUSTACEA*

[C]. *The soft body covered by a hard shell, with generative slime, spontaneous generation, etc.:* (*Alia*) *MOLLUSCA, ECHINODERMATA*

1. Shells of one valve . . . . *Placophora*
2. Shells with two valves . . . . *Lamellibranchiata*
3. Spiralled shells . . . . *Gastropoda*
4. Sea-urchins . . . . *Echinoidea*
5. Sea-stars . . . . *Asteroidea Ophiuridea*
6. Without shell, free-moving . . . . *Holothuroidea, Medusae*
7. Without shell, sessile, with spontaneous generation *Spongia, Actinia, Ascidia*

[D]. *Terrestrail animals with articulated body and incomplete eggs (scolex):* *INSECTA*

1.	With elytra	*Coleoptera, Orthoptera*
2.	With four wings and a sting behind	*Hymenoptera*
3.	With two wings and a sting anterad (forward)	*Diptera*
4.	Butterflies, moths and caterpillars	*Lepidoptera*
5.	Cicadas	*Homoptera*
6.	Wingless insects, parasites . .	*Asiphanoptera, Aphaniptera, Argulidae, etc.*
7.	Wingless insects with many legs	*Myriapoda*
8.	Wingless insects with eight legs	*Arachnoidea*
9.	Worms; which group, as in Linnaeus, includes all other hitherto unnamed groups	*Vermes, etc.*

The species of Aristotle are approximately those which we recognize today; of higher taxonomic categories he knows only the genus (eidos), and the same word refers, on the one hand to smaller groups (the genus, or family, of today) and on the other hand to higher ones (orders).

One of his great discoveries is that of the principle of homology, i.e., the architectural identity of organs, in opposition to the purely functional analogy or accidental structural resemblance. Other important principles of Aristotelian biology are the rule of the correlation of organs, also called that of economy in nature (' Nature does nothing to no purpose '), explaining e.g. the absence of upper incisors in horn-bearing ungulates (P.A. 665b); the relation between body-architecture, behaviour, and environment in birds (P.A. 693a); the organisation of the basic elements into tissues (*homoeomera*) which are combined into composite organs (*anhomoemera*); and the principles of zoogeography and of ethology (H.A. VIII, IX). Many of his splendid observations have been rediscovered only in the last 100 years, such as the brood-care of the sheat-fish *glanis* (*Parasilurus aristotelis*), the male migrating and fertilizing arm in certain cuttle-fish; the electric ray (*Torpedo*); the angler (*Lophius*) and its way of hunting; the types of locomotion of the *Nautilus*; the pseudo-placenta of the smooth shark (*Mustela*); a wealth of observations on marine mammals and bats, and on fish migration; the description of the four stomachs of the ruminants, etc., etc. His intimate knowledge of marine life resulted from a prolonged stay on the island of Lesbos. Many tales of fishermen, sailors, travellers, hunters, peasants, shepherds, etc., are discreetly used.

Aristotle is the founder of the vitalistic school in biology: there is no life without ' soul '. The plants, the lowest degree of life, have only a vegetative soul, taking care of nutrition, growth and reproduction. The animals have, in addition to this vegetative, an animal soul taking care of sensation and locomotion. Man alone has, in addition to both these souls, a rational and intellectual soul, which is the ' organ ' of thinking and reasoning.

Apart from his fundamental studies in comparative anatomy and ethology, Aristotle paid much attention to the influence of the environment, becoming thus the father of animal ecology, which,

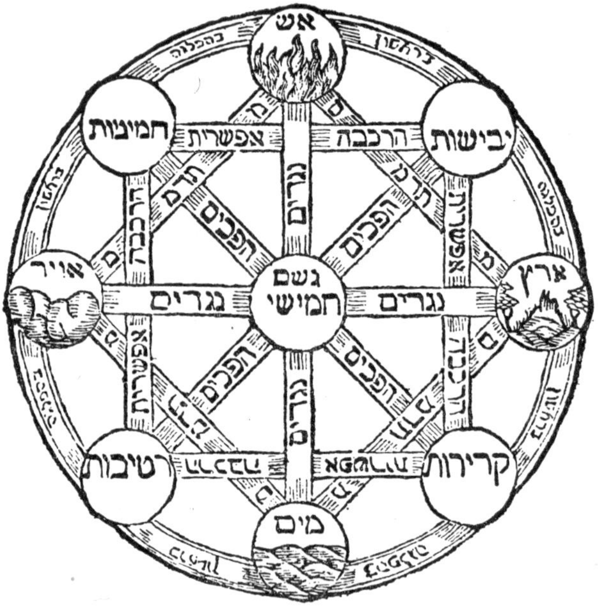

Fig. 8. The four elements, humours and temperaments of the sublunar world of Aristotle. (From a Hebrew text of Tobiah ha-Ropheh).

with respect to man was already well developed by the Hippocratic school. Ecological zoogeography; migrations of animals; hibernation; observations on bird territories; the comparative study of the comportment of solitary, gregarious and social animals, etc., are relevant here. His description of a vole outbreak (*Microtus guentheri*) remains a classic to our days (H.A. 580b).

Physiology is perhaps the weakest part of the Aristotelian biology. Its general background is the theory of the four elements (fire, air, water, earth), the four qualities (hot, dry, humid, cold) and the four humours (green bile of the liver, black bile of the spleen, blood of the heart, white humour of the brain) which compose the organism; the theory of innate natural heat or cold (in degrees) of every species; the heart as centre of sensation; the notion of respiration as a refrigeration. All these notions were already accepted by the Hippocratic school, and no important changes were made until the school of Alexandria and Galenus.

In his theory of reproduction, Aristotle claims a *generatio spontanea* from dead matter for many of the lower animals. He is the promoter of the epigenetic theory, namely that all organs develop, some sequentially others simultaneously, but all from an undifferentiated general nutritive matter in the egg. With regard to heredity, he refutes the Hippocratic idea of pangenesis, described above. His arguments include, that in this way the heredity of the voice, of body attitudes, etc., cannot be explained—neither why the hair of parents and offspring both being not white, assume that colour in senescence; that the composed (*anhomoeomerous*) structure of each organ and body part is not easily understood by this theory, the offspring of mutilated parents usually being born without these mutilations; etc. The egg contributes the nutritive substance, and the semen shapes this nutritive substance, just as the rennet coagulates the milk. The sex is determined according to which is stronger, the male or the female substance (or their respective innate temperatures). The sex is thus determined at the moment of fertilisation.

The biological philosophy of Aristotle is based upon his theory of the four 'causes': the formal, the material, the efficient and the final one. These causes are partly conditions and not causes in our sense—such as the formal and the material ones, neither of which produces an effect; but all four of them together produce the specific effect, and all four are needed for the formation of each specific organism. The material and the formal causes are the constituent elements of the process. The efficient cause gives the impetus; the final one concerns a plan which has not yet taken shape but which directs the embryo to develop into an individual of a given species. This immanent final cause produces the typical shape. In modern biology this process is called 'directive development'. With regard to all characters those which are essential, i.e., characteristic for the species, such as the size and structure of the eyes, are established by the final cause, whilst the accidental characters, such as the eye colour, may show variation. In this respect Aristotle polemicizes strongly against the theory of chance, such as was presented by Empedocles and Democritus. He declares development to be a fully determined process regulated by these four causes. This *Entelecheia* of Aristotle is regarded by many modern biologists as the exact equivalent of the views of modern experimental embryology.

Fig. 9. Drawings of Plants from the Herbal of Crataeus (100 B.C., the present copies from about 500 A.D.). *Aristolochia pallida*, at left; *Adonis aestivalis*, at right. (From Singer, *History of Biology*, p. 56).

In every respect Aristotle remains the outstanding personality in biology unto our days.

The most important disciple of Aristotle was Theophrastus of Eresus (372-288 B.C.) who is called the 'father of botany' with the same right by which the master is called the 'father of zoology, and/or of biology'.

Theophrastus mentions over 500 species of plants in his two important botanical works on the *Causes of Plants* (*De causis plantarum*) dealing with reproduction, diseases, etc.; and *The Natural History of Plants* (*Historia plantarum*), an analogy to the *H.A.* of Aristotle. He also did not propose a system, but unites many species into ' genera '. In this list of plants we find many exotic ones, usually those which are useful for man, such as cotton, datepalm, banana, rice, frankincense, etc. He describes the shaking of the male ' dust ' (pollen) upon the female inflorescenses in the date-palm; and some plant associations, such as the mangrove forest. He makes important contributions towards a causal plant geography, describing the various parts of plants and their function; some of his names, like pericarp, have survived to this day.

Theophrastus states that the plant has the 'power of germination' in all its parts. They may reproduce and grow from a seed, a root, a twig, a leaf, a trunk, etc., or by spontaneous generation. In his later writings he adds, however, that many details of reproduction evade observation up to a certain point and that therefore there may possibly exist still other ways of reproduction. With great clarity Theophrastus describes the dicotyledonous type of germination (from two opposite poles of the seed as root and stem) of the leguminous plants, and the monocotyledonous type of the grasses (root and stem sprouting from a common growth out of one point of the seed only).

The afore-mentioned remark in his later writings concerning observation and reproduction leads us to the basic difference between the epistemologies of the teacher and his pupil. The mature Theophrastus avoided all teleological assumptions, basing his conclusions only upon observed facts. While this was good in a period where observation was rare and speculation luxuriant, it is bad as a general principle. Observation and induction alone can never lead to general principles in science. Deduction and intuition may do so, but they are of value only when they are checked by observation and induction. This limitation was well known to Aristotle. The negation of the epistemological trinity in science by Theophrastus, if it was really a principle, is thus not—as Senn believes—a progressive, but a backward step.

Aristotle and Theophrastus did not initiate a new flowering of biology, but remained the established peak. Scientific progress centred now in Alexandria, where important progress was made, especially with regard to human anatomy and physiology.

Herophilus (*ca.* 280 B.C.) dissected human bodies and compared their structures with those of big mammals. He recognised the brain, which he described in detail, as the centre of the nervous system and as the seat of intelligence. He also distinguished between arteries and veins: the former as pulsating, the latter as non-pulsating vessels, the pulsation being due to a proper movement of the arterial walls. Also Erisastrus (*ca.* 280 B.C.) studied the brain, distinguishing between the main brain or *cerebrum* and the lesser brain or *cerebellum*. The convolutions of the former, showing greater complexity in man than in animals, are correlated to man's higher intelligence. In model experiments on animals, he showed that the anterior nerve roots of the spinal cord convey the impulse of motion to the muscles, whilst its posterior nerve roots convey the impressions from the body's surface, amongst them the sensations. He taught that the important parts of each organ are its vessels: arteries, veins and nerves (then still regarded as hollowed tubes conveying a ' nervous fluid ').

About the same time lived Nikander of Colophon (*ca.* 275 B.C.) who is remembered for his two didactic poems: the *Alexipharmaca* and the *Theriaca*, both about poisons from animals and plants and their antidotes, the latter being devoted to the preparation and use of the Theriac, the most famous antidote against snake poison for many centuries.

The further development of botany in antiquity leads us to Crataeus of Anatolia (1st century), who designed an illuminated book of medical herbs. What amounts to almost exact copies of this herbal are preserved in some very old codices of Dioscorides of Anazarba (*about* 60). The latter's book *On Medical Materials* (*De Materia Medica*) was *the* book of medical plants and still in the 16th century voluminous illustrated commentaries, such as that of Matthioli, were widespread. It contains some 600 plants, each one briefly described, arranged not in an alphabetical order, but according to ' genera ' or to the medical qualities of the plants, followed by the enumeration of the characters and medical applications of the plant. As already mentioned the illustrations in some of the oldest illuminated codices of Dioscorides go back to the drawings of Crataeus. Until the revival of botany towards the end of the 15th century, the *De Materia Medica* remained the only serious available book on botany.

From Roman antiquity we have, apart from Pliny, only a few books to mention. We begin with the fascinating didactic epos *On the Nature of Things* (*De rerum natura*) by the Epicurean philosopher Lucretius Carus (98-55 B.C.), imbued with a thoroughly materialistic spirit, and leaving the determination of development to chance. Much more typical of the Roman mentality is a series of books on agriculture, all of which contain descriptions of crops and domestic stock and their treatment, including all branches of farming and husbandry. The more important are:

Marcus Porcius Cato (234-149 B.C.): *De re rustica.*
Marcus Terrentius Varro (116-27 B.C.): *Libri tres rerum rusticarum.*
Maro Vergilius (70-19 B.C.): *Georgicae* (a didactic poem).
Lucius Columella (1st cent.): *De re rustica.*

A good medical textbook *De re medica*, usually ascribed to the senator Aulus Cornelius Celsus (1st cent.), is by some believed to be merely the Latin translation of a book written by an anonymous Greek physician living in Rome.

Few books have undergone a wider range of judgement than the compendious *Natural History* (*Naturalis Historia*) of Cajus Plinius Secundus (23-79 A.D.) who died while exploring the famous eruption of Vesuvius. Actually, the *Natural History* is neither as bad nor as good as some have judged it. It has no pretention to present original observations or to deal with the analysis of problems. As Pliny explains himself in his foreword, he wishes to offer a compilation of 20,000 facts from 2,000 books of 200 writers, and as such it is a very good book indeed, for his time. If not a few errors have spread and been preserved in public opinion till our day from his sixteen books of botany and four books of zoology, these were the errors and superstitions of his time. What Pliny regarded as facts, were, at least, given in a clear way and style, and this Plinian compendium thus preserved throughout the Middle Ages the bulk of the natural history of the ancients. If he did not preserve similarly the biology of the ancients, this is because philosophy and analysis were far beyond his scope. Pliny does not deserve the harsh judgement so often passed upon him. He was honest in his aims and in his method; he never intended to compete with the work of Aristotle or Theophrastus.

We now return to the Greek writers of the late antiquity. Galenus of Pergamon (129-201 A.D.) is the last great biologist of antiquity. An eclectic borrower from all schools of philosophy and medicine, he is *the* physician of late antiquity. While still following the Aristotelian teleology and the fundamental teachings of the four elements, qualities and humours, and of the *homoeomerous* and *anhomoeomerous* parts of the body, Galenus was much advanced in anatomy and physiology beyond Aristotle. He studied both these subjects most intensively in monkeys, transposing these results indiscriminately to man, and thus introduced a great number of minor errors into his human anatomy, which only began to be eliminated and corrected with Vesalius. His physiological achievements were stupendous. By his experiments he demonstrated, for example, the stopping of the pulse of an artery below a ligation, the stopping of the motility of a number of breast-muscles after the cutting of the fifth cervical nerve; the dependency of the heart beat upon the cerebral nerves; the continuation of arterial pulsation in an embryo whose maternal *arteria umbilicalis* is ligated, but the stoppage when both *arteria* and *vena umbilicalis* of the mother are ligated; the stoppage of the urine secretion into the urinary bladder when both urethers are ligated; and so forth.

In his basic theories, Galenus was much influenced by the Alexandrinian school. His three basic pneumas correspond to the basic three vital forces: the *pneuma psychikon* in the brain, spreading through the (hollow) nerves; the *pneuma zootikon* in the heart, producing the pulse and spreading in the arteries; and the *pneuma physikon* in the liver spreading through the veins. The two latter were held to meet and mix through pores in the septum of the heart, a view held until Harvey. Other subordinated or partial forces characterize each organ. The blood is formed (' boiled ') within the liver from the food, the innate heat in the heart, thought, sensation, and reasoning in the brain. Fig. 10 illustrates this physiological scheme, which, of course, did not indicate a circulation of the blood, but only a tidelike ebbing and flooding through the vessels, with no clear notion of when and whence the blood returns.

Splendid descriptions of nature are found in the poems of Oppian (perhaps two different writers, both anyhow in the second century A.D.), in his Hunting poem *Cynegetika;* and in that on Sea-fishery and marine life, the *Halieutika*, with its observations on marine ecology. Of a third poem, the Hunting with and of Birds, the *Ichthieutika*, only an extract in prose has been preserved.

A late fragment of an *Animal Book* of Thimothy of Gaza (5th cent.) may be mentioned. It leads us into the anecdotal style of animal books, a mixture of superstition and curious ' facts ', devoid of observation and judgement, which was initiated by the *De Animalibus* of Claudius Aelianus (d. *ca.* 200 A.D.) and surviving into the

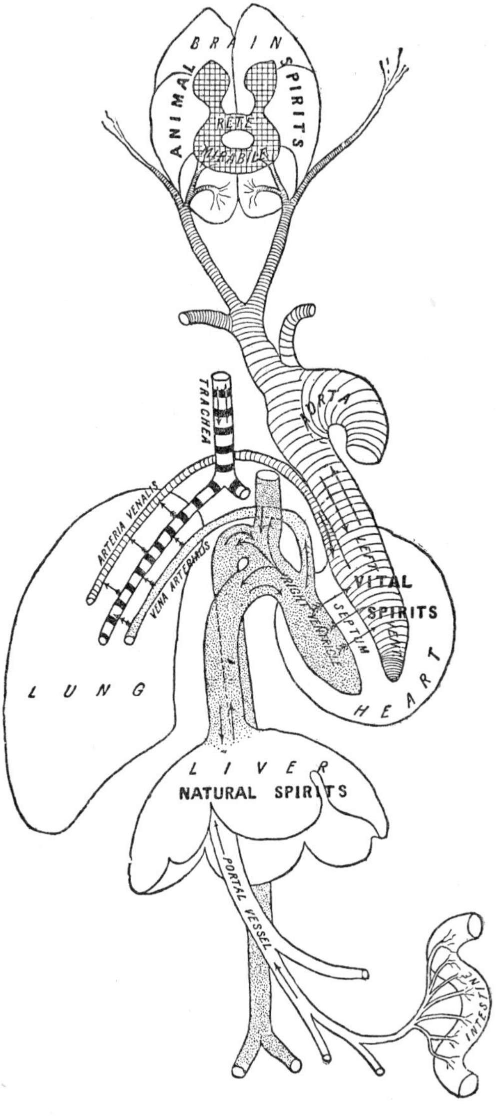

Fig. 10. The physiology of Galen: The seat of the vital spirits (striated), of the blood (dotted), of the natural spirits (liver) and of the animal spirits (brain). (From Singer, *History of Biology*).

early Arabic encyclopaedias, such as the *Uyun al-Akhbar* of Ibn Qutayba (d. 889).

In the same period also appeared a new type of animal books which enjoyed the highest popularity throughout the Middle Ages in all European countries. We refer to the *Physiologus*, or later the *Bestiarius*, the earliest prototypes of which were composed about 175 A.D. at Alexandria. The *Physiologus*, i.e. the ' Naturalist ', introduces an animal, plant or stone with a few words which are followed by a Christian moral. Some of the early eastern physiologi, such as the Syrian *Book of Natural Objects*, give more natural history information and may drop the moralising part, on which the main accent of the text was later laid. Descriptions of altogether up to sixty different animals, and a few plants and stones in addition, are contained in the physiologi-bestiarii, but each individual book rarely has more than twenty to thirty animals. In spite of their popularity we must regard the bestiarii as the lowest ebb of natural history knowledge, a mixture of superstitious tradition with absence of any observation, even with regard to common and well known animals.

The Byzantine period has no original records, but is of primary importance for preserving the writings of Aristotle. Of its few biological books the anonymous *Treatise on Agriculture* (*Geoponica; ca.* 940 A.D.); a treatise on falcons by D. Papagomenos (13th century); and a zoological poem by Manuel Philes (1295-1345) *On the Properties of Animals* are worth recording.

We should also mention that monks brought within hollowed pilgrim staffs to Constantinople (*ca.* 553 A.D. under the rule of Justinian) the eggs of the Chinese silkworm (*Bombyx mori*) and within a few years a flowering silkworm industry was established along all the coasts of the Levant.

2

THE MIDDLE AGES

THE greatest merit of Arabic biology is doubtless the revival of Aristotelian biology, so that it became actual and living knowledge. This knowledge had already penetrated into Arabic literature before Ishak ibn Hunein and his school prepared, in the 9th century, their important translations from Syriac into Arabic, which reached

PLATE I

ILLUSTRATED GREEK MANUSCRIPT OF MANUEL PERI ZOON (ON ANIMALS) IN THE BIBLIOTHÈQUE NATIONALE (PARIS).

(An illustrated page).

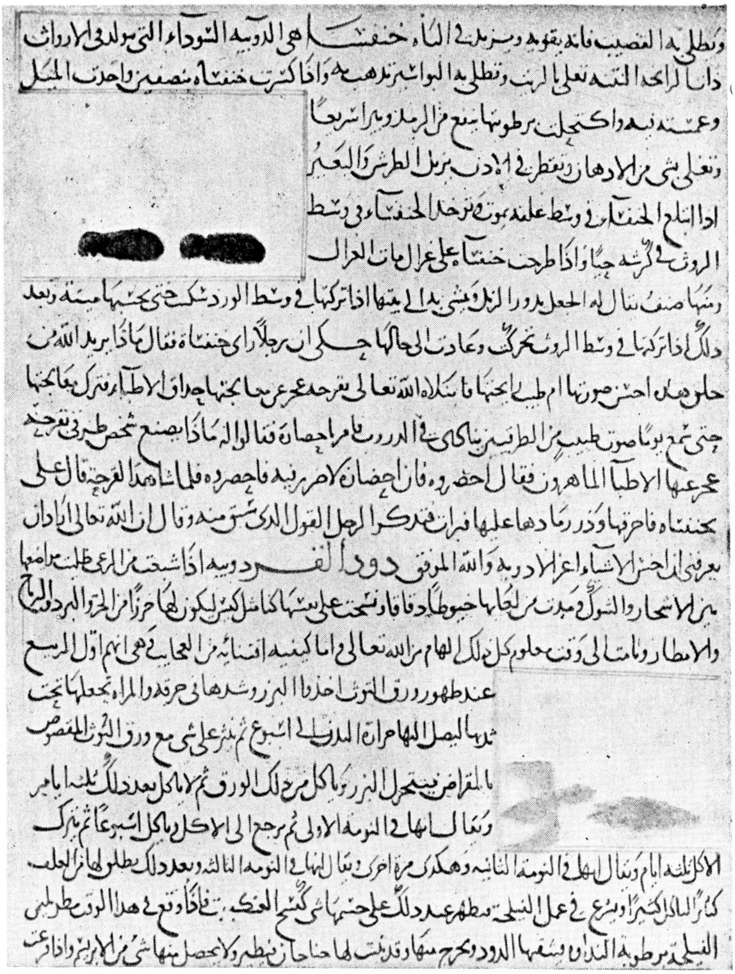

PAGE OF ONE OF THE VERY RARE ILLUSTRATED ARABIC MANUSCRIPTS OF AL-QAZVINI, PERHAPS WRITTEN IN HIS OWN LIFE-TIME.

This page deals with the dung-beetle (above) and the moth of the Chinese Silkworm (below). (From Bodenheimer, *Materialien zur Geschichte der Entomologie*. Vol. I. Berlin, 1928).

peaks of great scientific penetration and value in the al-Shifa of Ibn Sina (Avicenna, 980/1037) and the paraphrases and commentaries of Ibn Rushd (Averroes, 1126/1198). These were the main sources by which Aristotle reached the Latin and the Hebrew world in the 13th and 14th centuries, tinged by the Arabic philosophy and theology of that period. A detailed analysis of Arabic natural history and biology is an urgent desideratum.

The books on the hare, the camel, the honeybee, the datepalm, etc., of Al-Asma'i (d. 832 A.D.) are merely short philological treatises on Arabic animal names. A number of early encyclopaedias (Ibn Qutayba, d. 889 A.D.; Al Nuwairi, d. 1332; and others) contain animal lore of the type of Aelianus, but mixed with good observations of bedouins. The best book of Arabic zoology is without any doubt the *Animal Book* (*Kitab al-Hayawan*) of Al-Jahiz of Basra (767/868 A.D.), a mixture of animal lore and of splendid observation, from Jahiz's own travels as well as from the tales of bedouins, sailors, travellers and fishermen. His account of the language of ants, of fish migrations in the Tigris, and others could appear almost without change in a modern book.

Also in the neoplatonic Encyclopaedia of the Brethren of Sincerity of Basra (9th cent.) we find interesting chapters on zoology, botany, anthropology, etc. We may mention the stress laid upon the parallel between the microcosm and the macrocosm, man and the universe, not only as analogy but as a functional homology, which played such an important role in mediaeval philosophy and has perhaps nowhere been better expressed than in the *Guide for the Perplexed* (*Moreh Nebuchim I* : 72) of Moshe ben Maimon (Rambam, 1135/1204).

The great geographer Mustapha al Qaswini (d. 1283) described many plants and animals in his *Curiosities of the World.* We also find interesting notes and speculations about *generatio spontanea*, phototropism and other biological problems. The best book on agriculture is the Agriculture of Ibn Al-Awam (ab. 1175) of Spain, which is one of the few books very largely based upon observation and experience.

The zoological tradition finds its peak in the comprehensive compilation, the *Animal Book* (*Kitab al-Hayawan*) of Kamal al-Din Al-Damiri of Cairo (1349/1405). The animals are arranged according to the alphabet, so that the same animal often reappears under different names. For each animal quotations from the Hadith, the oral tradition of the companions of the Prophet, and from other Arabic sources, poetry and bedouin-lore are followed by paragraphs on dream-interpretation, the animal in religious law, and eventually a few tales on the life-history of the animal, of which those based upon tales of bedouins are often not bad. The book is a compilation, not a work of research based on experience.

In botany the *Plant Book* (*al Nabatat*) of Al-Dinawari (820/895 A.D.) is preserved only in fragments which, however, arouse our interest. The great compilation of Arabic botany is the voluminous *Kitab al-gami* of Ibn Al-Baytar (1197/1248). His arrangement is also alphabetical and one third of his 2324 names are synonyms. A still almost untapped wealth of plantlore and pharmacological and industrial experience is found in his book, which again is not based on original research.

Of the many excellent Arabic physicians we shall only mention Hasan Ibn Al-Nafis (1200/1288) who discovered the small (lung) circulation of the blood, and Ibn Al-Haitham (905/1038), the founder of physiological optics.

In the Latin sector the early Middle Ages were a definite low of knowledge and research. The small compendia or encyclopaedias of Isidorus of Sevilla (*Origines sive etymologica;* 7th cent.) and Breda's *De natura rerum* (8th cent.) represent almost a return to the name lists of the Har-ra = Hubullu. The medico-mystical writings of St. Hildegard of Bingen (1100/1178) are on even a still lower level. Yet, at about the same time, the gardens of many monasteries contained orderly collections of medical herbs. A scientific revival began only with the penetration of the Aristotelian tradition. The most important centre was, for many years, at the court of Frederick II of Palermo (1194/1250), who eagerly promoted Latin translations of Arabic books and paraphrases, of which those of Michael Scotus deserves special mention, and he was also the author of the first European book written in a thoroughly modern spirit, *The Art of Hunting with Birds* (*De Arte venandi cum avibus*). In this work he criticises Aristotle for not having always checked up on the tales of his sources, and gives an excellent general bird biology in the general introduction to this treatise on falconry. His illuminated manuscripts remain the joy of all connoisseurs.

The most important other centre was the Dominican Order. Three famous Dominican monks compiled compendious encyclopaedias: Thomas of Cantimpré (1186/1263) a *De natura rerum;* Vincentus of Beauvais (d. ab. 1264) a *Speculum Naturale.* But most outstanding is the third, by Albertus Magnus of Cologne (1193/

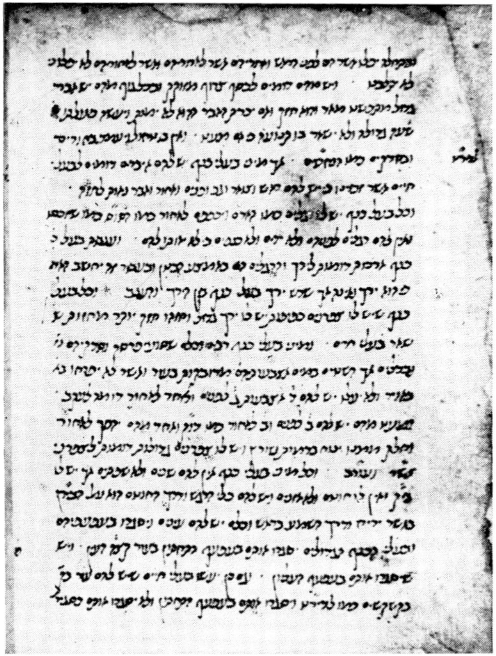

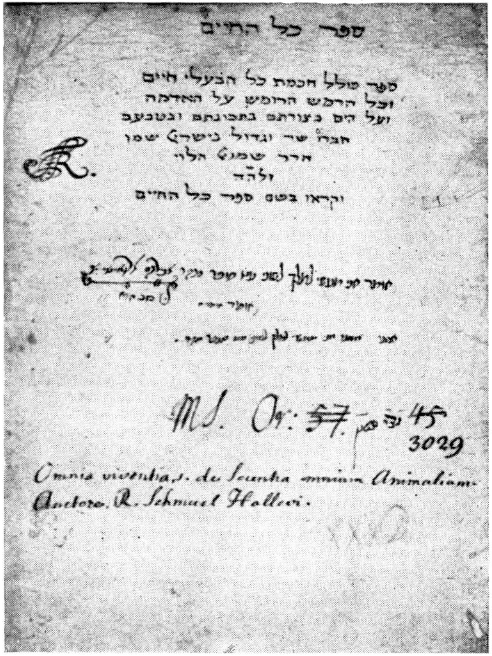

THE TITLE-PAGE AND A SAMPLE PAGE OF THE HEBREW MANUSCRIPT *Sepher kol Hahayim* (BOOK OF ALL LIVING BY SHMUEL HALEVI ABULAFIA OF TOLEDO (15TH CENTURY).

This is the first Hebrew zoology which is based not upon Arabic, but upon a Latin translation of Aristotle. (British Museum, London).

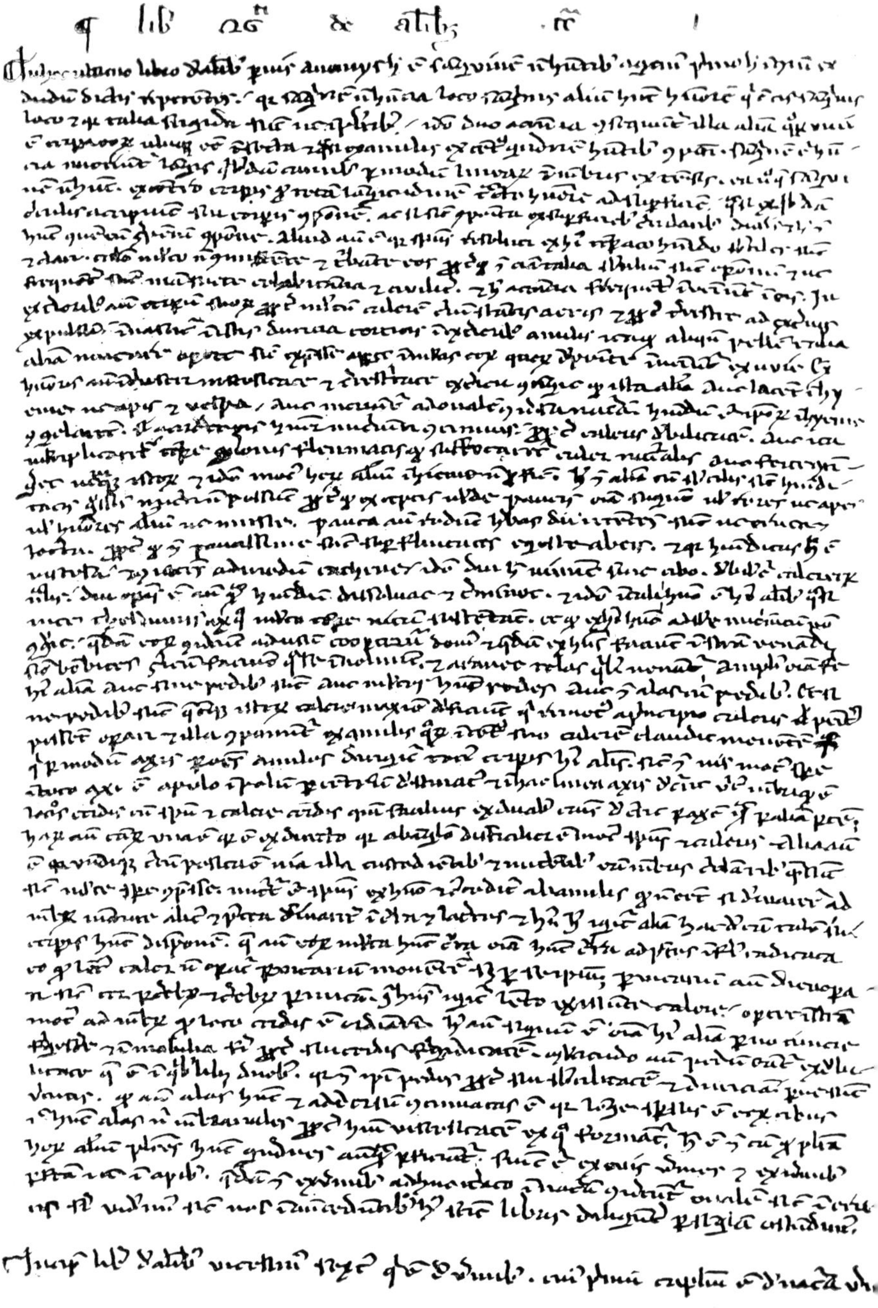

Seite aus dem Original-Manuskript des „De Animalibus" von Albertus Magnus (Culex etc.; aus dem Kodex im Kölner Stadt-Archiv).

FROM THE ORIGINAL MANUSCRIPT OF *De Animalibus* OF ALBERTUS MAGNUS IN THE STADT-BIBLIOTHEK OF COLOGNE.

(From F. S. Bodenheimer, *Materialien zur Geschichte der Entomologie*, Vol. I. Berlin, 1928).

1283). Albertus aimed in his *Opus naturae* to give a Latin version of Aristotelian natural history. Actually he gave much more. In his animal book we find an addition of four books devoted to the animals observed by him and actually known to exist at that time. This is the first attempt since Pliny to write a natural history. Also, in many other books of the *Opus*, we find not a few original observations, including a description of the leaves of *Citrus* or the first discovery of the ventral situation of the central nervous system in the arthropods, and so forth.

On a lower level apart from the *Bestiarii*, there are a few writers who began to write in their national languages, among whom we may mention:

Brunetto Latini, *Le Livre du Trésor* (1230/1294).

Gershon ben Shlomoh, *Sha'ar ha-Shamayim* (*The Gate of Heaven;* ab. 1300).

Conrad of Megenberg, *Buch der Natur* (1309/1374).

Jakob van Maerlandt, *Der Naturen Bloeme* (d. *ca.* 1300).

3

THE MODERN AGE

(*a*) *Taxonomy.*

THE spread of the knowledge of the zoology and botany of Aristotle and Theophrastus stimulated slowly the desire to know and to identify the animals and plants of which they wrote.

In botany, the early books were various herbals, of which the *Hortus Sanitatis* was the best known and the most widespread. Valerius Cordus (1515/1544) of Wittenberg described in his posthumous *Historia plantarum* (1561) over 400 species of plants from his country with a discussion of the structure of their flowers. The compendious *Commentarii ad Dioscoridem* of P. A. Matthioli of Siena (1500/1577) form illustrated descriptions of all plants (and a few animals and materials) mentioned by Dioscorides, and of a few more. Wrong identifications of the plants of Dioscorides, grew in proportion to the geographical distance from their habitats in the writings of the wrongly called early ' Fathers of Botany ', such as those of Otto Brunfels (1489/1534). Mainly philological was the voluminous, alphabetically arranged great herbal of the learned Leonard Fuchs (1501/1566), intended to be a guide to the medicinal plants, and maintaining in description and illustration, a standard much below that of Cordus.

This new desire to identify correctly the animals and plants of the Ancients lead now to travels, like those of Pierre Belon, the main aim of which is to arrive at a correct identification of the many Greek and Latin names of plants and animals in the scientific literature of antiquity. New is the desire to base these final identifications upon a thorough study and knowledge of the fauna and flora of the countries of antiquity.

Another new development is the progress of illustration from very crude woodcuts to better, finer, and more exact prints, especially of plants and flowers. This progress will be readily realised from the selected figures of this book. Yet we should not overlook the fact that this progress is not only a progress of better quality of work, it is largely based upon the transition from wood blocks to lithographic and copperplate printing. This will become very conspicuous when we compare, for example, the aquarel paintings of Ulysse Aldrovandi, which have served as models to the very crude woodcuts illustrating his foliants, now exposed in the University Library of Bologna.

In the 16th century there appeared also the first great books on the flora and fauna of both Americas and of the Far East. Among their authors we may mention Garcia de Orta, Christovam de Acosta, Francesco Hernandez, Charles de l'Ecluse, Prosper Alpinus (on Egypt), William Piso, and many others.

Amongst the zoologists we may regard Rabelais (1490/1553) as a precursor of the observational spirit.

Pierre Belon (1517/1564) was the first modern naturalist to visit the Levant including Palestine, mainly driven by the desire to identify the plants and animals of the Ancients. His books *Histoire Naturelle des Estranges Poissons Marins* (1551) and *Histoire de la Nature des Oyseaux* (1555) are the first modern zoologies based largely upon observation, especially the second one. There we find the oldest drawing of comparative vertebrate anatomy showing the homology of the bones of man compared with those of a bird, a really revolutionary step in his days. He studied plants as well and eagerly promoted the acclimatisation of foreign trees in France.

Belon's compatriot and contemporary Guillaume Rondelet (1507/1566), is known for his extended studies on the fishes, molluscs

Symphonia Plato

rls cum Ariftotele: & Galeni cū Hippocrate D. Sympho-
rianī Chāpeiij. Hippocratica philofophia eiufdem.
Platonica medicina de duplici mundo:cum cūfdē fcholijs.
Speculum medicinale platonicum:& apologia literarū hu-
manorum.

Quæ omnia venundantur ab Iodoco Badio.

Fig. 11. The syncretism of the Italian Humanism is excellently
represented on the title-page of the *Symphonia Platonis cum
Aristotele; et Galeni cum Hippocrate, Hippocratica Philosophia*,
of S. Champier (Paris, 1516). The woodcut shows the four
scientists playing the instruments of an orchestra in full
symphony.

and other vertebrates of the Mediterranean Sea, many of which he described and illustrated for the first time.

This brings us to the two great encyclopaedic naturalists, Conrad Gesner (1516/1565) of Zürich and Ulysse Aldrovandi (1522/1605) of Bologna. They both eagerly collected every literary note made upon natural objects in a kind of card-index and gummed all pertinent cards together before they began to write any special chapter. Observations of their own and even occasional dissections (see the intestines of a silkworm, the larynx of a swan, etc. in Aldrovandi) were added. The observations were supported by a great many beautiful and fairly accurate drawings in water-colours upon which later, often crude woodcuts were based. Both the water-colours and wood-cuts of Aldrovandi are still preserved in Bologna. The five folio volumes in 4500 pages of Gesner's *Historia Animalium* appeared from 1551 to 1621, the botanical and palaeontological ones much later (1751, 1771). It is not their voluminosity, but the new spirit they display based upon observation, which makes these books a starting point of modern science. Gesner, who died as physician during a plague epidemic, gave expression to a modern theme in a letter to a friend in which he described vividly the wonderful landscape of the Alps, and the stimulating effect of their ascent—an entirely new note and one not restricted to natural history!

Aldrovandi founded, in 1567 in Bologna, one of the first botanical gardens connected with a university. Beginning in 1599, he published eight more compendious folios. His volume on insects (1602) is rightly regarded as the foundation of modern entomology. Gesner's notes on insects were bought by E. Wotton, then by T. Penn, and again edited and augmented by T. Mouffet (d. 1604), but printed only in 1634 as *Theatrum Insectorum* by T. Mayerne. Their combined effort does not reach the standard of Aldrovandi, who had forty years more than Gesner to complete his work but who did not achieve the latter vivacity of style.

Andrea Caesalpino (1519/1560) expressed the belief in his *De Plantis*, that the root collar is the seat of the plant's soul, and that the marrow which produces its fruit is its noble part. For this reason, he chose the fruit as the character part upon which he built his taxonomic units.

Caspar Bauhin of Basel described already 6000 species of plants in his famous *Pinax Theatri Botanici* (1623), including short diagnoses of the genera, which he often arranged in classifications approaching a natural system.

A MEDIAEVAL LESSON ON ARISTOTLE. (DEVENTER, 1489)

Plate VI

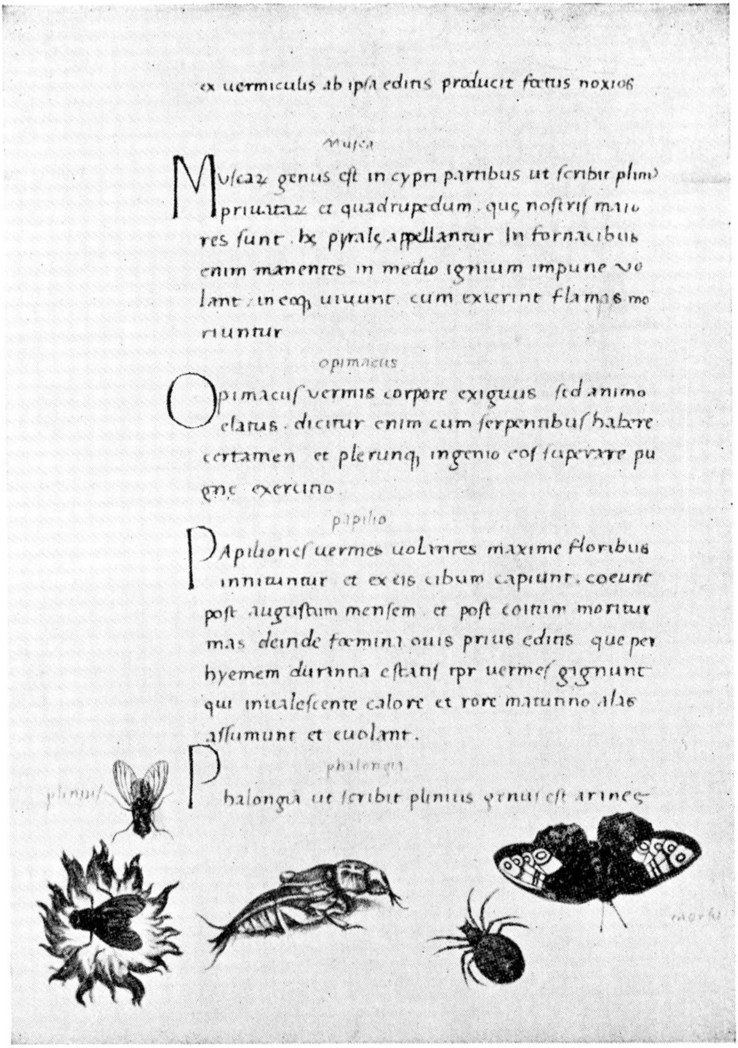

ex uermiculis ab ipsa editis producit foetus noxios

Musca

Muscaz genus est in cypri partibus ut scribit plinius priuataz et quadrupedum. quę nostris maiores sunt. hę pyralę appellantur In fornacibus enim manentes in medio ignium impune uolant, in eoque uiuunt. cum exierint flamas moriuntur

opimacus

Opimacus uermis corpore exiguus sed animo elatus. dicitur enim cum serpentibus habere certamen et plerunque ingenio eos superare pugnę exercitio.

papilio

Papiliones uermes uolitantes maxime floribus innituntur et ex eis cibum capiunt. coeunt post augustum mensem et post coitum moritur mas deinde foemina ouis prius editis que per hyemem durant a estatis tpr uermes gignunt qui inualescente calore et rore matutino alas assumunt et euolant.

phalongia

Phalongia ut scribit plinius genus est araneę

plinius

A page from an illustrated Latin Manuscript of the 15th Century, the *Codex Animalium* of Petrus Decembrus in the Library of the Vatican.

(From F. S. Bodenheimer, *Materialien zur Geschichte der Entomologie.* Vol. I. Berlin, 1928).

John Ray's (1628/1705) *Historia Plantarum Generalis* also compiles all data on known plants, and he defines the species as a constant unity. He also wrote about the system of mammals and reptiles, and of insects.

We should at least mention the names of the most important authors on agriculture in general, all empiricists with a wealth of important physiological observations and of the greatest benefit to their countries. Through the centuries—before the establishment of a scientific agriculture, which came into being only with this growth of modern chemistry—these are: Pietro Creszenzi (b. 1230 in Bologna), in his *Ruralium Commodorum Libri XII*, which appeared in 1478 in Italian, Olivier de Serres (1539/1619), whose *Théatre d'Agriculture et Ménage des Champs* appeared in 1600 upon the request of Henry IV, and which saw 20 editions within the 17th century ; H. L. Du Hamel du Monceau (1700/1782) who contributed much to the physiology of agricultural crops; and A. Thaer (1752/1828) of Germany whose contributions were mainly in the field of soil cultivation and crop rotation.

Réné Antoine Ferchault de Réaumur (1683/1757) is widely known because of a thermometer invented by him. He studied wide fields in biology with great success: from the artificial egg-incubation of domestic fowl, sterilization by boiling, the formation of pearls, the silk of spiders, the regeneration of crustaceans, the electric apparatus of the torpedo-fish, the nature of the corals, etc., to the introduction of a piece of sponge into the crop or stomach of birds, pressing the liquid from the redrawn sponge and experimenting with its digestive force upon various foods. Yet his fundamental contribution remain the six folios of his *Mémoires pour servir à l'Histoire des Insectes*. These, though unknown to most entomologists, formed until our day the base of our knowledge of the life-histories of most insects, apart from the many ecological observations contained in them.

Other famous illustrators and observers of the life-cycles of insects were Maria Sibylla Merian (1647/1717), August Roesel von Rosenhof (1705/1759) and Carl de Geer (1720/1778).

Unrivalled to our day is the remarkable, most minute anatomy of the caterpillar of the willow moth (*Cossus cossus*) with descriptions and illustrations, for example, of over 6000 individual muscles of that caterpillar, and with other organ systems described in similar detail (1740) by the Dutch notary, Pieter Lyonet (1707/1789).

Charles Bonnet (1720/1795) of Genève, a strict Calvinist in the period of the Encyclopaedia, was almost blind in the later part of his life. We need not discuss here his many speculations on natural philosophy which made a profound impression upon his time (preformation, the ladder of nature, the 'germ'-theory, etc.). But he had performed a number of important experiments in his early days, and he maintained a very extensive correspondence with scientists all over Europe. Among these experiments, we may mention the first experimental establishment by carefully isolated breedings of a sequence of parthenogenetic generations in an aphid, followed towards winter by an egg-producing generation of sexuals. He discovered the reproduction of the pupiparous Diptera. His experiments on regeneration are built on the work of his compatriot Abraham Trembley (1700/1784), who first established the stupendous power of regeneration in the freshwater polypes, the animal nature of which he also was the first to recognise.

Bonnet was also among the first to experiment with plant tropisms. The experimental technique of such experiments was perfected later by T. A. Knight (1759/1838) who proved the fact of geotropism by putting seedlings along the periphery of wheels rotating speedily in a vertical or a horizontal plane, on which the plants grew towards the centre, the roots growing outwards from the wheel.

Carolus Linnaeus (Linné; 1707/1778) of Uppsala devoted his life to the classification of all the three kingdoms of nature: animals, plants and minerals. The first edition of his *Systema Naturae* (1753) contained 11 pages (in folio), the 12th edition (1766) 2300 pages. His great gift was the development of short and precise diagnoses of species, genera and classes. In botany he based his system upon the structure of the flowers. Thereby he created for practical purposes as he well realised, a complete but artificial arrangement into which every flowering plant could easily be inserted. He also introduced the binary nomenclature, in which every form is unambiguously defined by the name of genus and species, such as *Musca domestica* for the common fly. This binary nomenclature was rigorously followed only with the 10th edition of the *Systema* (1758), from which date modern nomenclature begins. Linné regarded genus and species as natural, constant categories. He realized that variations occur, yet these do not deserve nomenclatorial treatment. '*Tot sunt species quot ab initio creavit infinitum Ens*'. (There are as many species as the Creator created at the beginning).

Plate VII

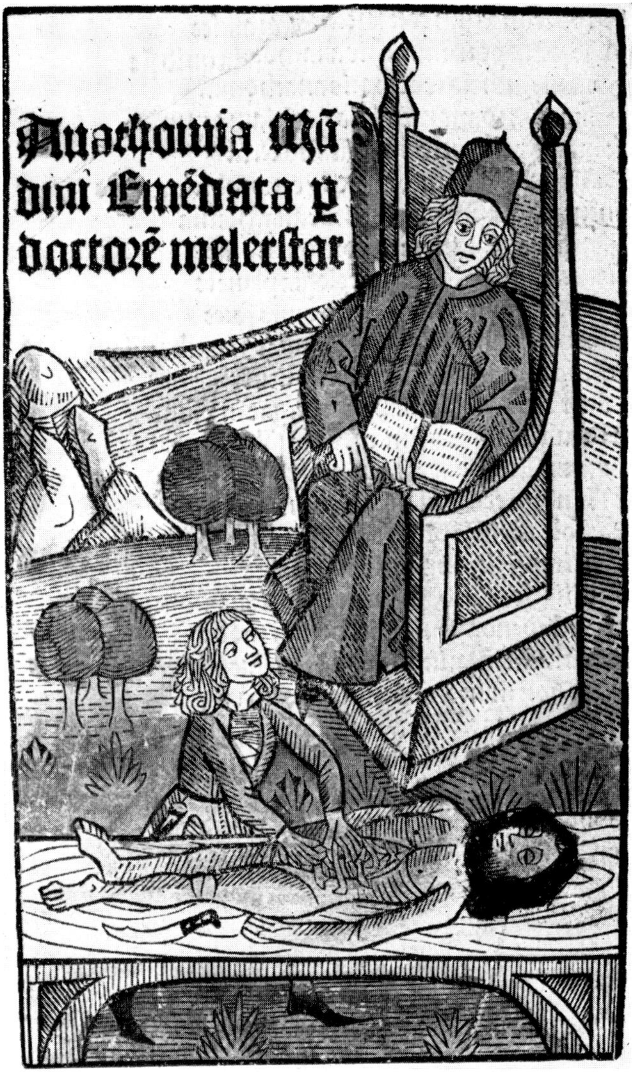

An early anatomical dissection. From the Anathomia of Mundinus. (Leipzig. circa 1453)

PLATE VIII

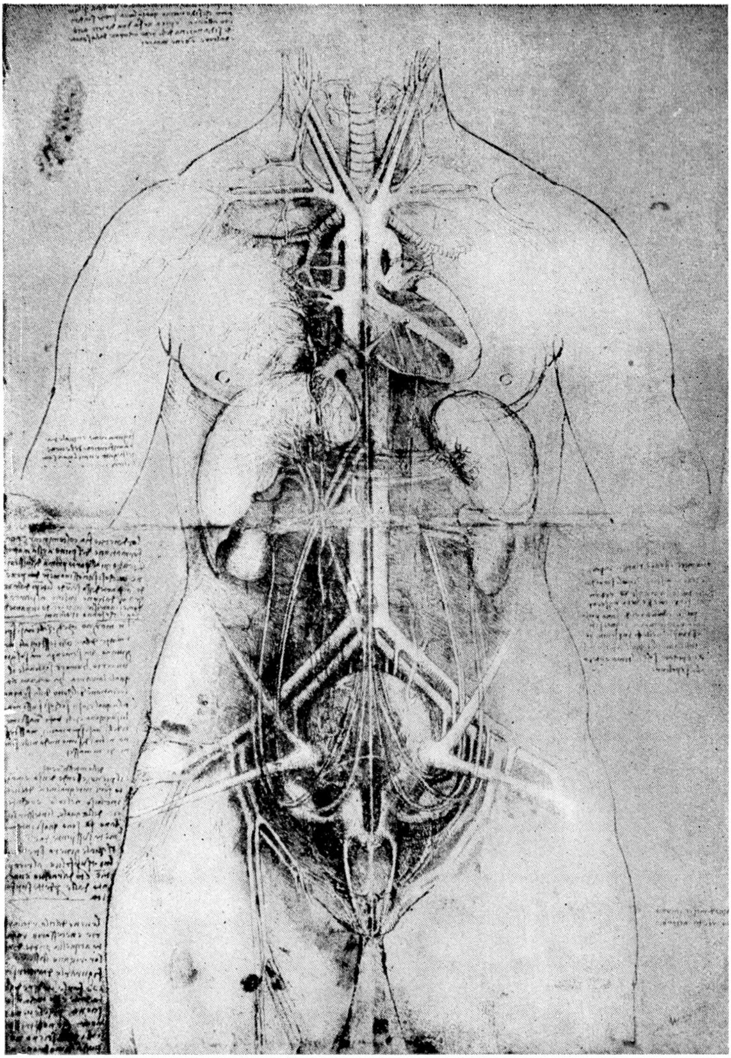

A DRAWING OF TOPOGRAPHICAL ANATOMY OF THE HUMAN INTESTINES IN SITU. FROM THE MANUSCRIPT Q12 OF LEONARDO DA VINCI.

Leonardo was the first to draw anatomical objects in natural perspective, in artistic design, and with great accuracy, as nobody has done before him and few after him.

This Linnaean system permitted the arrangement of the increasing anarchy of descriptions with vague names, constantly increased by the rapidly growing additions of new forms from all parts of the world. Linné himself sent many well prepared students out to explore and collect the natural history, for example, of West India (Ternström), of North America (Kalm), of the Levant (Hasselquist), of Dutch India (Osbeck), of Surinam (Rolander), etc. *The Systema Naturae*, with all improvements made necessary by the introduction of new characters, by a more intimate study, and by the inclusion of higher categories (families, orders), remains the foundation of modern taxonomy.

Linné's *Philosophia Botanica* is a textbook of general botany. His travel books reveal a deep ecological understanding. His *Nuptiae Plantarum* are written in a fascinating style. Linné was anything but a dry systematizer, as is amply proven by the fact that he is regarded as the initiator of the romantic school of literature in the Scandinavian countries.

(*b*) *Early Microscopy.*

Two opticians in Middelburg, the Janssens, in the last decade of the 16th century, combined lenses into ' microscopes '. These first and very primitive microscopes contained two fixed lenses, one at each side of the perforation of a metal plate, the object being stuck at the proper distance upon a fixed needle. These instruments were scarcely more powerful than a good modern pocket lens. Later they were improved by joining the two lenses in a tube, as is still done in modern microscopes.

One of the first to make use of these instruments was Galileo Galilei (1564/1642) who inverted them and turned them into telescopes. Yet he also made use of the inverted microscopes, studying insects, for example, and describing their facetted eyes.

Prince F. Cesi and his colleagues in the early Accademia dei Lincei in Rome made intensive use of these microscopes, discovering, among others, the ' seeds ' (=*prothalli*) of ferns; and Stelluti gave an enthusiastic description of the morphology of the honeybee (1630). M. Rooseboom has rightly remarked that it is not the invention of the microscope in itself that provided the great stimulus to biology, which followed in the 17th century. It is more proper to state that the time was ready to make use of the microscope for the solution of its problems, side by side with the immense pleasure which play with this instrument gave to the hundreds of dilettanti and amateurs.

The first great microscopist who gained fame by the intensive use of these ' looking glasses ' was Marcello Malphigi (1628/1694). He demonstrated in 1660 the existence of the capillaries in the lungs of the living frog; and a few years later, the vesicular structure of the lungs, explaining respiration as the exchange of air penetrating into the blood (not as a mere ' refrigeration '), the lungs being the place of the entrance of air into the blood. In his anatomy of all stages of *Bombyx mori*, written at the invitation of the Royal Society in London, he described the true function of the tracheal respiration and of all parts and appendices of the intestinal tract (1668). In his monograph on the development of the chicken embryo within the egg, he observed the formation and disappearance of ' gill-vessels ', the five cerebral vescicles as beginnings of the brain. He described the anatomy of plants and of many plant galls, showing the insect origin of the latter. He discovered the embryo sac and the endosperm, described germination of bean, laurel, date palm, and grasses, distinguishing well the differences of germination of the mono- and di-cotyledons. It is almost unbelievable that no adequate biography of this important pioneer in the history of biology has been written, and that his amply illustrated diaries, preserved at Bologna, have not yet been published.

Antoon van Leeuwenhoek (1632/1723), town clerk of Delft, gained a well earned world fame by the great extent and variety of his microscopical observations, for which he built well over 100 microscopes with his own hands. His observations, in consequence of their excellence, were all published as letters in the Philosophical Transactions of the Royal Society of London (later in four volumes as: *Arcana Naturae Detectae*). He describes a good many animalcules (protozoans, bacterias) from lakes, from rivers, and from infusions of hay and pepper, many of the protozoans being identifiable even today. Of his many other discoveries we shall mention only the striation of the voluntary muscles, the red blood cells, the spermatozoans of man and many animals, the lens of the vertebrate eye, the vivipary of aphids, and the insect character of the Mexican cochenille. He described a great quantity of small insects and mites, and their parts. All his observations were illustrated by primitive drawings explained in a reasoned text. There are not many available objects which evaded the curiosity of this remarkable personality, with his ever-open eyes.

Jan Swammerdam (1637/1680) of Amsterdam turned, after an initial study of the respiration in higher vertebrates, to the study of the life histories and of the anatomy of insects chiefly. With a fanatic enthusiasm so often found in people affected by phthisis, he tried to discover the most minute details of their structure as witnesses of the greatness of God's creation. His guiding principle was an attempt to show how, already in every larva (including those of frogs and flowers), the adult insect (frog, flower) is already pre-formed, needing only growth and liberation from its now unnecessary and unwanted (larval) wrappings. His *Biblia Naturae* (1737) was published, almost a hundred years after his early death, by Boerhaave. His treatises on may-flies, dragonflies, mosquitoes, the common snail and tadpoles are masterpieces.

Robert Hooke (1635/1703) gave excellent illustrations in his *Micrographia* (1665) which opens with one of his microscopes. He is known for drawing the cellular structure of cork—being the first to use the word 'cell'; the stinging mechanism of nettle cells of stinging plants; the development of various fungi and moulds; the first description of a polyzoan; and good illustrations of many insects.

Nehemiah Grew (1641/1712) was, with Malpighi, the founder of plant anatomy. In his *Anatomy of Plants* he gives mainly cross-sections of stems and roots of many species with great detail and accuracy. These anatomical studies led him to recognise the 'sponge-like' (= cellular) structure of the plant tissues, and to trace certain details, such as vessels and fibres. The flowers he regards as the sexual organs of plants (his conception erring only in detail), and he draws the pollen of many species. His term 'parenchym' has persisted in plant anatomy.

(*c*) *Morphology.*

The most widespread anatomy of the later Middle Ages in the Latin sector was the *Anathomia* of Mondino de Luzzi (1270/1328). Anatomical knowledge at this time was very inexact. It was based upon the three-day public dissection of a body, practised once a year, as a rule, the professor directing the whole procedure from his perch on a high chair. A lecturer or reader stood below, to read aloud a text (usually that of Mondino), whilst the dissector (of non-professional standing) opened the body, and the demonstrator pointed out with a stick each organ as it was 'read'. The chain of this speedy dissection was enforced by the stench of the decomposing corpse, in the absence of all methods of preservation. The illustrations of the books were extremely primitive and inexact, as were the descriptions in the text.

Leonardo da Vinci (1452/1519), one of the great passions of whose life was dissecting and anatomy, raised the level of illustration, especially in topographical and comparative anatomy, not only to great artistic heights, but above all to a high level of exactitude. He was the first to draw a human embryo in its natural intra-uterine attitude. Unfortunately, Leonardo did not produce a school, and the treasure of his drawings and manuscripts remained unknown and unpublished until 100 years ago. Thus, his anatomy, as well as his studies on physiological optics (in the sense of Al-Haitham), and especially his still unreached wealth of observations on the flight of birds, remained without consequences for the history of science. We may draw one consolation from this experience, however, namely that almost every mechanical or scientific discovery of Leonardo was already independently discovered long before the rediscovery of his manuscripts. This suggests that the suppression of a genius, for whatever reasons, may retard but not prevent the progress of science.

A few years after Leonardo's death the Belgian physician Andreas Vesalius (1514/1564) revolted against the inadequacy of anatomical instruction in the medical curriculum and began with more intensive dissections with great success. His *De Corporis Humani Fabrica* (1547) is a milestone in the history of biology. It offered an atlas of drawings after nature, and corrected many of the errors of the Galenian tradition, for example, the musculature of a monkey transposed into the human body, the liver of the dog, the uterus of the cow. This partial break with traditionalism was perhaps the greatest influence of Vesalius, who (despite respectful correction of details) conscientiously never left behind the teaching of Galen. Because of this exaggerated loyalty, he never denied the existence of pores in the septum of the heart, though he declared in his later edition that he could never see them with his own eyes.

In the wake of Galileo Galilei mechanical thought also entered biology (as Leonardo da Vinci's important pertinent enquiries were still unpublished). Nicolaus Steno (ab. 1650), who described the muscles not only as connective tissue or as auxiliary organs of touch, but as the active organs of locomotion acting according to mechanical and mathematical principles. This was the main topic of the iatrophysicians, among which Alfonso Borelli of Bologna (1608/1679) excelled. They raised the mechanical physiology of animals from observation and speculation to the level of experimental science. Borelli begins his *De Motu Animalium* (1681) with an explanation of the elasticity of the muscles which are activated by the nerves. It became a matter of mechanical analysis to determine how an animal walks, flies and swims; where the centre of gravity of its body at rest and in movement may be; and so forth.

Human and comparative anatomy also made great progress. Michael Servetus (1509/1553) discovered the pulmonary blood circulation and Fabricius of Aquapendente (1537/1619), the valves of the veins, also showing by experiment their function, namely to prevent reflux of the blood away from the heart. Gabriele Fallopio (1523/1562) described the anatomy of the nervous system and of the sexual organs; and the term ' placenta ' dates from him. Carlo Ruini published a standard anatomy of the horse (1598), and many more anatomists of Bologna and Padova did outstanding work. Of the foreign students of these universities we may mention the Dutch Volcher Coiter (1543/1576), the most important comparative anatomist of this period.

One of the most famous students of Padova is William Harvey (1578/1657) of London. Under his guidance the circulation of the blood was discovered and established. The pulsation of the arteries induced by the heart's systole fully explained the importance of Fabricius' vein valves, since a flux and reflux of blood within the vessels could no longer be assumed, and was replaced by the idea of a continuous and one-directional circulation.

Harvey is regarded as the founder of modern physiology. He was, however, not the master of experimental methods which he is often taken to be. His great achievement in the discovery of the circulation of the blood was not the result of experiments, but of thinking his calculus which revealed the conspicuous discrepancy between the known quantity of blood, in a horse's body for example, and the calculable amount of blood propelled by the heart during the 24 hours of the day. " Harvey modernized physiology by solving a complicated problem, proving that it can have no solution other than the proposed one, regardless of whether or not full experimental evidence can be provided " (Peller). The philosophical background of Harvey's thought has already been discussed.

The most important early quantitative experimental studies of blood pressure, the capacity of the heart, the diameter of the blood vessels, etc., are contained in the *Haemostaticks* of Stephen Hales (1677/1776). This is the same Hales, famous also as a temperance leader, who discovered for plant physiology the root-pressure of plants, general phenomena of the rise of water, the importance of air (read : CO_2) as a substantial nutrient of the plant, etc., in his *Vegetable Staticks.*

It is difficult to describe the impact of the first scientific Academies as a stimulation of scientific research in its full extent. Among the oldest of these Academies were: the Accademia dei Lyncei in Rome (1609, interrupted for some time in 1628 after the death of its founder, Prince F. Cesi). In the meetings the members occupied themselves mainly with microscopy, with whose early development it is connected. The Académie des Sciences at Paris was founded in 1635 (recognized by the King in 1668). In its early years a great number of dissections was performed, mainly of mammals, under the direction of Claude Perrault. Other similar organisations were: the Royal Society of London (1645, officially recognized as 1662). The Academia Naturae Curiosorum in Schweinsfurth (1652, now the Deutsche Akademie der Naturforscher at Halle), etc. These all began with weekly meetings of the members, who actively microscopized, dissected animals and plants, demonstrated chemical or physical experiments, as did Robert Boyle, and so forth. The first scientific journals appeared in connection with these societies. Intensive correspondence was maintained, in the Royal Society with and by the Secretaries; while in the Académie des Sciences a number of corresponding members were attached to every effective member to whom they had regularly to report upon the progress of their scientific work. Most academies charged certain members with important research, a famous example being the demand upon Marcello Malpighi by the Royal Society to describe the anatomy of an insect, followed one year later by the publication of his *De Bombyce.* The main effect of these academies was an enormous multiplication of scientifically active people from among the circle of those who before had been merely *Naturae curiosi.*

At the same time the first scientific museums also began to be formed from the many collections of curiosities collected with pride by many amateurs and Maecenases. Among the first such scientific collections, we should mention those of the Jardin des Plantes, forming later the Musée d'Histoire Naturelle at Paris; the collections of Nehemiah Grew and of Hans Sloane, which formed the first foundations of the British Museum of Natural History in London; and the famous collections of Carolus Linnaeus in Uppsala, largely now preserved by the Linnéan Society in London.

Leclerc, Comte de Buffon (1707/1788) initiated—by his compendious encyclopaedia, the *Histoire Naturelle*, famous also for its splendid style—many original theories, especially concerning synthesis; it also contains his descriptions of the cosmogony, his history of life on earth, and his observations on the structure and behaviour of animals. His ideas on the origin of life and on heredity are based on the theory of ' organic molecules ' as having formed by aggregation the smallest animals. These build the embryo within eggs and sperms, guided by a form-giving force by regional structural arrangement. He thus became an adversary of preformation. By degeneration (read: variation) of the few early organisms, the actual wealth of life-forms took their origin. Buffon regarded the organic world as a fluid continuum, in contrast with the strict taxonomic categories of Linné. He is one of the first to formulate the problems of zoogeography. To him the local quantity of organic molecules is just as decisive as climate and food.

George Cuvier (1769/1832) was in his time the recognized leader of zoology. In his big *Règne Animale* he states that all animals belong to four basically different types (embranchements), according to their vegetative (heart and circulation) and their animal organs (central nervous system), namely: I. Vertebrata, II. Mollusca, III. Articulata (including the Annelida), and IV. Radiata (all the rest!). The most important of his anatomical enquiries were devoted to the molluscs and to the fishes.

Cuvier founded modern palaeontology by rigorous application of the (Aristotelian) principle of the correlation of parts. Every organ of every species is fundamentally connected and correlated in structure with its mode of life. Hence, he thought that one bone of a living or fossil animal should be sufficient to construct its entire body and to describe its mode of life. This guiding principle grew in his hands and in his sober mind into a masterful tool.

His palaeontological studies revealed that in the history of the earth we observe at certain periods a radical change of the entire flora and fauna, which he ascribed to extensive catastrophes, after which the devastated areas were repopulated by plants and animals from undestroyed areas. His epigones interpreted the appearance of these new floras and faunas as new successive creations.

Together with Cuvier, there worked at the Musée d'Histoire Naturelle at Paris Jean-Baptiste de Lamarck (1744/1809) and Étienne Geoffroy St. Hilaire (1772/1844). The former, who was also an accomplished florist, to whom we will return when discussing the history of evolution, introduced the separation of vertebrates and invertebrates into systematics. He put much order in the arrangement of the latter, for example, by separating the spiders and crustaceans from the insects as independent classes; and by separating the protozoans, Coelenterata and Echinodermata from the worms proper, with which Linnaeus had still combined them.

Étienne Geoffroy St. Hilaire made ample use of embryology for the interpretation of the structure of adult animals. He attacked the four basic types of animal structures of Cuvier, trying to combine them all into one type. Famous is the prolonged and heated discussion in the Académie des Sciences in 1830, after Geoffroy had presented a paper of two of his students who attempted to show that the cephalopod structure is almost identical with that of a vertebrate rump when the latter is folded back so that the pelvis touches the neck. It was easy for Cuvier to refute this false theory, and it is difficult to understand today the wide and general interest in this discussion (see: Eckermann's *Gespraeche mit Goethe*). It is important to note that such a recent event, with all the documentation fully available, should be described in many modern histories of science as a battle for and against evolution, which *prima facie* it certainly was not. It was, however, near to the conceptions of Goethe, Oken, and others, as they were developed in the *Metamorphose der Pflanzen* and the *Metamorphose der Tiere* of the former, claiming and developing the unity of plan of all plants and animals and of their parts.

Among the early publications in pathological anatomy is the *Compendium Sepulchretum* (1679) of Théophile Bonet of Genève (1620/1689), in which 3000 abnormalities from autopsies are described. Yet that branch of science was really founded by Giovanni Battista Morgagni (1682/1771) of Padova, who clearly stated that every disease has its seat in an organ where it produces visible changes under the influence of pathogenous factors which are specific for every disease. This is the main thesis of his fundamental *De Sedibus et Causis Morborum per Anatomen indigatis Libri quinque* (1751).

About the same time, modern phytopathology was founded in France by the *Dissertation sur la Cause qui corromp et noircit les Grains de Blé des les Épis et sur les moyens de prévenir ces Accidents* (1755) by Mathieu Tillet (1714/1791), and by his successor, the abbé Tissier (1741/1837) who wrote the *Traité des maladies des grains* (1783).

PLATE IX

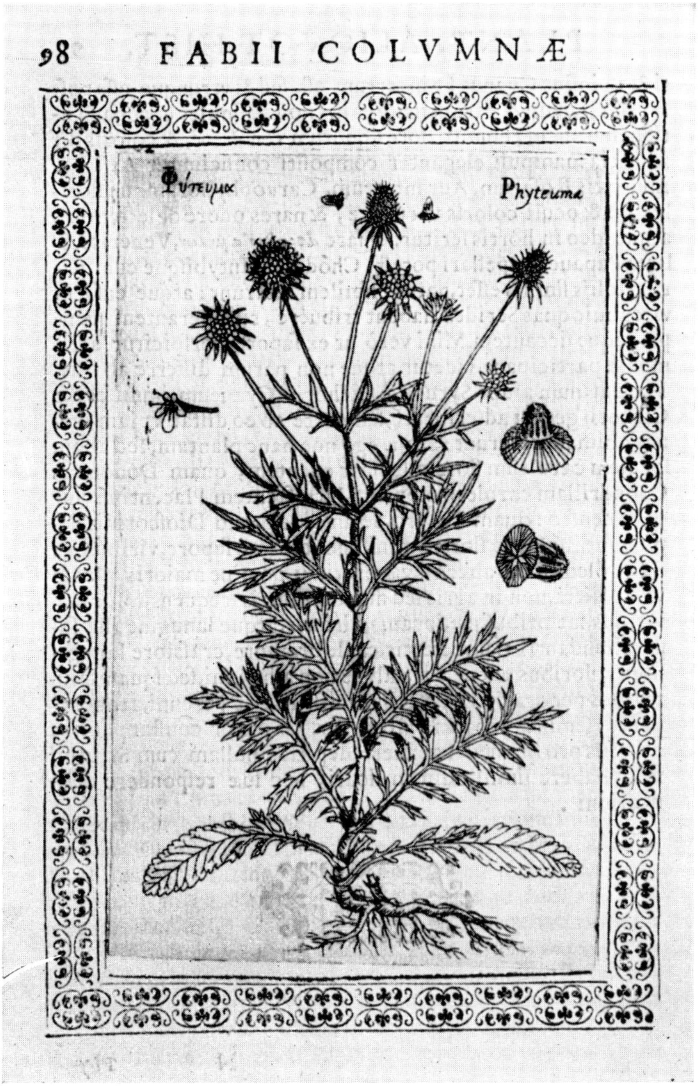

VERY FINE DRAWINGS AND COPPERPLATE ENGRAVINGS OF A THISTLE. FROM FABIUS COLUMNA, PHYTOBASANOS. (NAPOLI, 1592)

Portraict des Conches, qui produisent des oyseaux.

IN THE BEAUTIFUL HERBARY OF CLAUDE DURET (*Histoire admirable des Plantes*, PARIS, 1605) WE ALSO FIND AN ILLUSTRATION OF THE LEGEND OF THE BERNICLE GEESE, BORN FROM THE LEAVES OF A NORTHERN TREE.

Felici Fontana (1730/1805) of Firenze discovered the uredineo- and the teliospores in the diseases of the cereals.

The first attempts at agricultural entomology were also made at that period in France, where H. L. Du Hamel (1700/1782) devoted attention to the pests of the stored grain, especially in a monograph on the grain moth *Sitotroga cerealella* and the control of these pests.

(*d*) *Embryology*

The theory of preformation ruled the embryology of the 17th century (e.g. Bonnet, Swammerdam, etc.). It stated that the process of individual development is merely the unfolding or unwrapping of the preformed individual, and also that all later generations of the human race had been already preformed and incapsulated, so to speak, one into the other in the first men. There was a great debate in the period, as to whether this incapsulation was in the sperm or in the egg. A number of contemporaneous drawings show how a miniature man is already fully developed in the human sperm. In that century, also Regnier de Graaf (1641/1673) described the ' ovarial eggs ' of mammals. (Today these Graafian follicles are recognized as vesicles in which the much smaller egg is contained).

The most pronounced antagonist of preformation (after Aristotle and Harvey) was Christian F. Wolff (1738/1794) of Halle. He demonstrated in his *Theoria Generationis* (1759) that the growth of new organs of plants proceeds from a uniform, undifferentiated tissue at the tip of the growing shoot or root, and thus cannot possibly be an ' unfolding ' of preformed parts. Regarding the development of the egg of the chicken he says that " nobody has seen with a microscope more details than can be seen by the naked eye ". This theory of epigenesis soon replaced that of preformation.

The great period of embryology started with Karl Ernst von Baer (1792/1876). In 1827 he described the true mammalian ovum in the oviducts, and later within the ovarian follicles of the dog, as tiny objects. His *Über die Entwicklungsgeschichte der Thiere* (1828-37) is a classic of the subject. In it he established the *chorda dorsalis* as the most characteristic differential feature of all early vertebrate embryos.

His other two great conceptions were based upon observations of predecessors. Thus, von Baer stressed the resemblance of the early embryos of all vertebrates, basing this conception on the discovery by M. H. Rathke (1793/1860) of the gill-slits and the gill-arches in the embryos of birds and mammals, corresponding to a fish-like stage of evolution. Von Baer's biogenetic law states that, in development, general characters appear before the special ones, which are derived by differentiation from the former. In the course of development the embryos of different species and groups diverge progressively from one another. And, finally, a higher animal passes during development through stages which resemble stages in the development of lower animals.

One of the great biological discoveries of the 19th century was the germ-layer theory. Already Wolff had described certain layers (Keimblaetter) appearing in the early development of the egg. H. C. Pander (1794/1865) had already begun to sort out the various systems of organs developing from these germ layers. This process was completed by von Baer and by Remak (1845). The latter reduced the four original germ-layers to three (the mesoderm being a folded single one) and gave them their modern names: ecto-, ento-, and mesoderm. This theory is one of the most generalized in biology, claiming that in all vertebrates (and other metazoan animals) the skin, the nervous system and the sense organs are formed from the ectoderm, the mid-intestinal epithelium from the entoderm, and all the remaining organs from the mesoderm. With certain qualifications this theory has passed the test of time for the entire animal world. It permitted the conception of homology (structural identity) and analogy (functional identity) in comparative anatomy to be revived.

E. Haeckel (1834/1919) reformed Baer's biogenetic law, as follows: The individual development (ontogenesis) is a shortened recapitulation of that of the ancestral stock (phylogenesis) (1866). Von Baer never accepted this evolutionistic interpretation of his law, and modern biology has also returned to Baer's more sober statements. An important addition, however, of Haeckel was the gastrula theory, that most organisms pass very early through a gastrula stage composed of two germ-layers, in the form of an invaginated ball, the mesoderm appearing later by folding or by delamination from one of the former layers.

The poet Adalbert of Chamisso published (1819) after sailing around the world in the Russian ship Rurik, his discovery of the alternation of generation in the Salpae, where free-swimming forms alternate with chains on which salpae of another shape are germinating. Japetus Steenstrup of Copenhagen published in 1842 a book devoted exclusively to this alternation of generations, which he extended to the polyp and the medusal generations of Coelenterata and to certain parasitic worms, where the individual always resembles its grandparents, but not its parents. Carl Theodor von Siebold (1804/1884) devoted much energy to the experimental establishment of this phenomenon in parasitic worms, where it is connected with a change of the host. He found that the famous echinococcus cysts are a generation alternating with a small cestode in the dog's intestines, and the ' Drehwurm '-disease in the brain of sheep is caused by a generation of another cestode, also living in its other phase in the dog's intestines.

Around the same time began the enquiry of Wilhelm Hofmeister (1824/1877)—a bookseller promoted to a professorship in Heidelberg against the protests of the faculty—into the most complicated alternation of generations in the development of the cryptogamous plants (1851), which at once made of this hitherto neglected field a fascinating topic of primary importance for systematics. After his enquiry the world of plants was no longer seen as composed of flowering and non-flowering plants, but of: I. Thallophyta, II. Mosses, III. Ferns, IV. Conifers, and V. Flowering plants. Of especial importance was the discovery of characteristically shaped receptacles, the archigonia, containing the eggs which are fertilized by motile spermatozoids, these sexual forms alternating with a sexless (agamic) one.

The next step forward was experimental embryology or ' the mechanics of development ' as its founder, Wilhelm Roux (1850/1924) of Halle, called it. This branch of science tried to explain the part played by the inner constitution of the egg and by external influences upon early development. By burning with a heated needle one cell of a frog's egg which had just undergone its first segmentation, the remaining cell was made to grow into a lateral half-embryo, thus demonstrating that in the two-cell stage the regional differentiation of the embryo was already determined. This determination is, however, not irreversible, as later experiments showed that after the careful elimination of all remainders of the second cell or after the turning upside down of the remaining cell complete nanic embryos may be obtained. Oscar Hertwig (1849/1922) discovered a most suitable object for experimental embryology in the translucent eggs of sea-urchins, which are poor in yolk. When these eggs are separated into their single cells at the 2, 4, 8, 16 or even 32 cell stages (by shaking, for example), each one still produces a complete larva (pluteus)). Roux and O. Hertwig determined that the polar axis of the embryo is fixed by the arbitrary place at which the sperm enters the egg.

By centrifugation of frog eggs O. Hertwig obtained frog eggs that segmented only at the protoplasmatic pole, thus proving that the yolk, the storage food of the egg, is a merely passive partner in segmentation.

As soon as the organs of the embryo differentiate, functional relations within the embryo become conspicuous. After the main trunks of the blood vessels become established, their further branchiation—and the same is true for the nerves—depends upon the development of the organs or regions that they supply.

Hans Driesch (1867/1941), one of the strongest promoters of modern vitalism, was instrumental in demonstrating that in some eggs the first segmentation cells (= blastomeres) could develop into total, nanic embryos, while in others each blastomere developed into a partial, regional fragment only. In the regulative eggs each blastomere contained the potential for full and total development (as in the sea-urchin), while in others the future development is already fixed very early by the regional position of the blastomere (molluscs).

Another aspect of the subject was opened up by Driesch, J. Loeb (1859/1924) and C. Herbst (b. 1866), when they showed that many eggs, from those of sea-urchins to those of frogs, are able to develop normally (with one set of chromosomes only) after a great variety of mechanical and chemical stimulations. This proved that fertilization and initiation of developmenτ are different processes in essence.

Striking results were obtained with the method of embryonal transplantation by Hans Speman (1868/1941). When, for example, a part of the dorso-median ectoderm of an amphibian's gastrula, which normally forms part of the future development of the neural tube, was transplanted somewhere into the ventral ectoderm of the same or another individual, it developed normally according to the new locality, and vice versa. Yet a similar implantation from the upper lip of the gastrular opening was able to induce in the ventral ectoderm entirely unlocal development of the neural tube and of a chorda. Such results of dependent and independent development opened up for investigation new aspects of the integration of early embryonic development, the time and the extent of regional determination and induction, the mutual interactions of mechanism and chemism, etc. The results became still more striking when transplantations to different species or genera were successfully obtained.

Tissue culture was first used by Ross Harrison (1907) and perfected by Alexander Carrell (1873/1944). In suitable nutritive liquids the cells of small particles of an embryonic organ—the standard objects used are from the embryo of a chick—keep dividing until, after a certain time they stop to grow. When transported into an empty flask with fresh nutritive liquid, however, they grow without limitation as long as these transportations are continued. The cells from the heart of a chick-embryo planted by Carrell in 1912, for example, are still continuing to grow. These animal tissue cultures however, retain, within strict limits the character of the tissue from which they took their origin, while in certain plants, such as *Begonia*, entire organs or even plants may grow from single cells.

Also in Protozoans Emile Maupas (1844/1916), G. N. Calkins (1869/1943), H. S. Jennings, and others have shown that under optimal conditions which require the continuously reiterated transportation of individuals into fresh nutritive media, conjugations and other sexual phenomena can be suppressed in favour of a permanent series of vegetative cell-divisions.

The mechanics of the development of plants were also extensively studied by Karl von Goebel (1855/1935).

(*e*) *Cell-theory*

After the work of Malpighi, Grew and Hooke the cell-like structure of the flowering plants was well established, the thick cell-wall being their most conspicuous feature. No special importance was ascribed to this formal statement, however.

M. F. X. Bichat (1771/1802) was the first to propose natural ' units ' within the animal's body which like a ' tissue ' compose the various organs. Twenty-one such tissues were separated, such as bone, muscle, nerve, connective tissue, etc. These were understood to combine like a texture to produce the organs, which for their part combine into organ systems, and these again into the whole organism.

M. J. Schleiden (1804/1881) in 1838 proclaimed the cell as the building unit of every plant and animal. In addition to the wall, he recognized the slime lump—now called protoplasma (H. von

Mohl)—and the nucleus, which had been described by Robert Brown (1773/1858) as the most vital part of the cell. Schleiden even believed that new cells are formed as buds of the nucleus. In 1839, Theodor Schwann (1810/1882) enlarged this cell-theory considerably by studying the development of various animal tissues. The cartilage, for example, an apparently homogenous and undivided mass begins its development as a purely cellular tissue, and the cartilage is nothing but the secretion of these cells; and this process of development was confirmed for every tissue which in its established form does not show any longer the cellular structure of its origin. Schwann also denied the importance of the wall as an important feature in animals. His cellular theory, which in this simple form is valid until today, may be summarized as follows: Every plant and animal is built up from cells or from cell-derivates. Every cell has an individual life apart from its integration into the life of the organism as a whole.

Much of the further discussion centred around the nature of the cell-content, that lump of slime which was soon to be called the protoplasm. Most of the theories of this discussion concerning the foam-like, reticular, granular, etc., structure of the protoplasm, have become meaningless today in view of their dynamic interpretation as stages of colloidal processes, the protoplasm being recognized today as a colloidal body. Still, in 1861 M. Schultze knew no better definition than the one now generally accepted: A cell is a lump of protoplasm containing a nucleus.

A special discussion has centred, since Ehrenberg (1795/1876), around the cellular character of the protozoans and many green algae. However, even von Siebold's masterly paper on this subject in 1841 did not convince his contemporaries, a result which was only achieved by Bütschli in 1875.

In anatomy Kölliker (1844), Remak (1852) and others applied this theory to embryology and established histology as a science; and R. Virchow (1821/1902) developed the cell-theory into one of cellular pathology, as every pathological phenomenon either occurs within cells or is derived from them.

Important progress concerning the nucleus was due to the application of dyes in histology. Until that time cell-division was regarded as a simple division of the mass of the cytoplasm and of the nucleus. Now in botany E. Strassburger (1875), and W. Fleming (1882) for animals, discovered a succession of complicated changes within the nucleus, accompanying every cell division

—such as the appearance of two polar bodies, the apparent dissolution of the nucleus, the formation of a spindle, and the appearance and equatorial arrangement and longitudinal division of darkly stained chromosome bands which finally separate. The final division of the cytoplasm leaves two cells with an exactly corresponding number and structure of chromosomes, which become visible again only at the next cell-division or mitosis.

A feature which gained high importance in modern genetics is the fact that almost all body cells have a double set (2 N) of chromosomes, one of each derived from the father and one from the mother. But the mature sex cells, egg and sperm, have only one set of chromosomes (N), since at one of the maturation divisions of the egg a whole set of N (without longitudinal division of the chromosomes) is expulsed as one of the polar bodies, while in the sperm maturation in one division the longitudinal cleavage of the chromosomes is merely suppressed. This mechanism, which maintains the normal number of body chromosomes (2 N), provides at the same time a mixture of maternal and paternal chromosomes, which are the bearers of heredity. This type of division is called meiosis.

(*f*) *Physiology*

Francis Bacon (1561/1639) claimed inductive collection of the full facts to be the base of the ' scientific method ' a new and important concept of that period. He frequently used the word ' experiment ', but when we look up these experiments they are often either mere observations (such as: Go out in a warm July night and you will find glowworms!) or speculations as in the ecological chapters of his *Sylva Sylvarum*. His warnings against prejudices, which he called the " idols of the market, idols of the theatre ", etc. (*Novum Organum*, 1620), remain important.

Deeper and more consequent was Réné Descartes (1596/1650) whose *Cogito ergo sum* (I think, therefore I am) initiates modern epistemology (theory of knowledge). This great thinker's *Discours de la Méthode* (1637) with its requirements for the student of science, still retains its actuality. In biology he is important because of his strict separation of life from thinking, the latter being reserved as *anima rationalis* only for man. Life in animals is purely mechanical. Their nervous impulses go out from the brain through the nerves as impulses to the organs of action. Sensory stimuli may be reflected in the brain to a nerve, causing action in the absence of any thought as in a machine. He was one of the first followers of Harvey's theory of circulation, in a posthumous book on man, written about the same time.

Really consequential attacks against the existence of a separate *anima rationalis* in physiology was made only much later. De la Mettrie (1709/1751) denied its existence emphatically, starting from observations that parts of even the highest animals retain motility after their separation from the body. Noting during a fever that his psychic functions were influenced by his bodily condition; localizing the sensory functions in the brain; treating ideas as bodily phenomena, etc.—no place remained for a rational, immortal soul.

The physician Francesco Redi (1621/1697 of Firenze was one of the most courageous figures in the history of science. He showed independence of judgement, for example, in depicting a louse of a donkey (*pediculus asini*), after Aristotle had definitely stated that all animals have lice except the donkey. Much more important is his independent inquiry into the Aristotelian theory of the *generatio aequivoca* in insects. He demonstrated by simple and conclusive experiments that in fresh meat covered by muslin no maggots of flies develop, while in uncovered vessels flies develop from the maggots developing in the meat, exactly like those which entered the vessel for oviposition—or in *Sarcophaga*, for laying small maggots —upon the meat. He was severely rebuffed on this point by the famous and learned Athanasius Kircher of Rome (1602/1680), not only as contesting the authority of Aristotle, but also his own (i.e., A. K.'s) after the latter had published a number of prescriptions describing how, by different mixtures of flowers, soil, excrements, etc., it was possible to produce at will butterflies, various beetles, and other insects. Yet Redi was unable to prove the generation of insects developing in plant galls or of intestinal worms from parent insects or worms; thus he remained silent about the problem of their generation, which was only cleared up by Marcello Malpighi and Antonio Vallisnieri (1661/1730) respectively. It was only Louis Pasteur who, a hundred years ago, destroyed the last traces of a belief in a still persisting *generatio aequivoca* (in cellular organisms).

In the meantime Leeuwenhoek had discovered his ' animalculi ' in natural water and in infusions. Buffon and the Reverend J. T. Needham (1713/1781) appeared to have shown by experiment that in water sterilized by boiling, there appear a few hours or days after the boiling, an abundance of such animalcules (protozoans, proto-

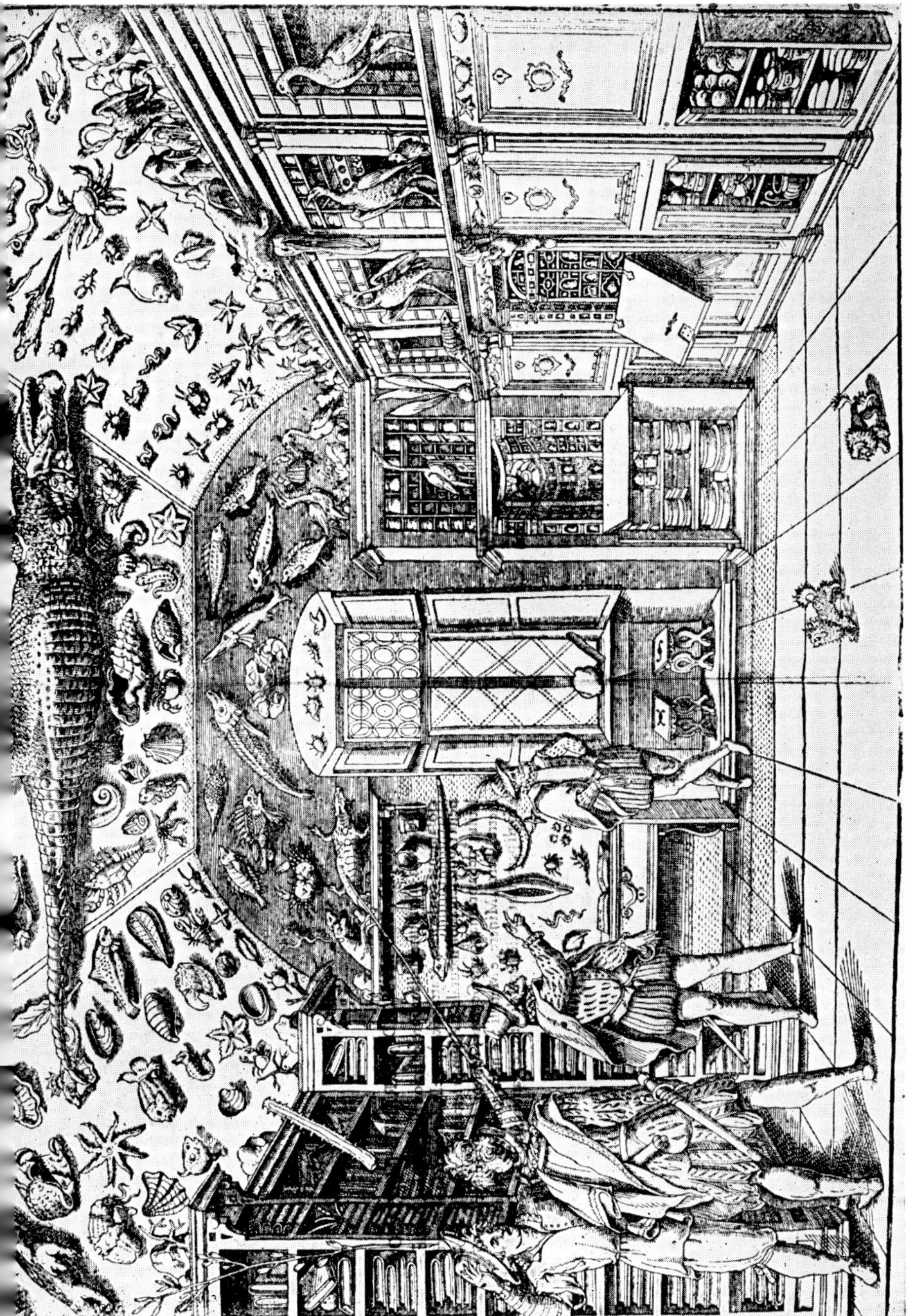

EW OF AN EARLY PRIVATE MUSEUM OF NATURAL HISTORY OF THE 16TH. CENTURY.
FROM FERRANTE IMPERATO, *Dell' Historia Naturale Libri* 28. (NAPOLI, 1599)

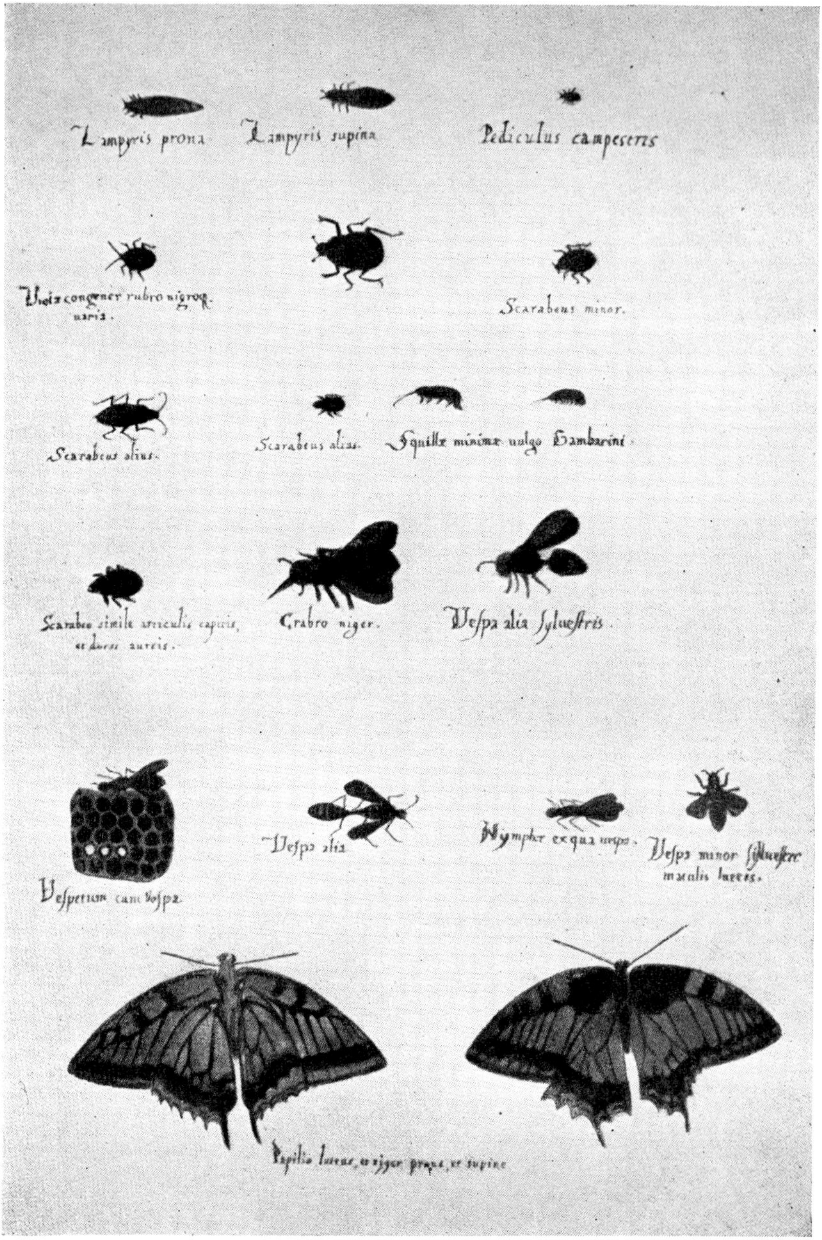

ALDROVANDI ILLUSTRATES SOME INSECTS IN WATER COLOURS, FROM WHICH THE WOOD BLOCKS WERE CUT.

phytes and bacteria of today) which must have been built up, in their opinion, from ' organic molecules ' within the boiled water. At this point we must insert some information on the founder of modern experimental biology, the abbé Lazzaro Spallanzani of Pavia (1729/1799).

Against this prevalent view of the origin of the animalcules was arraigned a whole phalanx of anti-spontaneists, such as Albrecht von Haller, Réaumur, and Charles Bonnet of Geneva. Yet an experimentum crucis was still needed as the latter school of thought objected to the former that ' sperms ' from the air had entered the vial after the boiling of the liquid and before its sealing with cotton-wool. Spallanzani sealed the vials with fire before boiling, and then never found any protozoans though a few bacteria still appeared (to be suppressed much later only by a process called ' pasteurisation '); and after much prolonged boiling even these smallest animalcules were prevented from appearing.

Of Spallanzani's varied experiments on the regeneration of earthworms, aquatic worms, amphibians, etc., perhaps the most astonishing were those concerning the regeneration of the heads of the snail. In the discussion following his publication, many thousands of snails were operated upon all over Europe, so that the fear was even expressed that these mass-experiments must lead to the extermination of the snails. Spallanzani was supported by Bonnet and Lavoisier, and most energetically by Voltaire who also had conducted pertinent experiments, and also had followed the complete regeneration of the snail's head in detail. Remarkable is Spallanzani's work on the resuscitation of rotatorians, tardigrades (which group he was the first to describe) and nematodes from dry moss and dust by wetting, after they had apparently lost all signs of life. By the introduction of perforated metal tubes containing food into the crop and the stomach of various birds he continued Réaumur's experimental study of digestion.

Spallanzani is responsible for the introduction of the frog as a common experimental animal in physiological laboratories. He definitely confirmed the fact in frogs and toads that, while copulation is not indispensable for the egg's development, the fertilization of the egg by the male sperm is indispensable. He even performed experiments on the length of time for which the sperm retains his power of fertilization; as well as the degree of thinning at which the sperms still retain this power. That the semen, and not a secondary chemical secretion, is responsible for fertilization was demonstrated by bathing eggs in water in which testicles just freshly opened were put, lead to fertilization. Spallanzani continued these, the first experimental fertilizations in the history of animal biology, even with higher vertebrates. He successfully inseminated an isolated bitch in heat. This artificial insemination has grown in the last decades to great importance in animal husbandry. By isolating the buds of flowers he discovered that there are species which need fertilization by pollen, whilst others are able to develop without it.

Recent experiments have fully confirmed Spallanzani's extended research on the orientation of bats in darkness, even with blinded eyes. He demonstrated that even blinded bats avoid touching not only the walls, but even any threads stretched across a room. He concluded that this perception must be a kind of hearing. Only in our days have the radar-like mechanisms involved been discovered: the bats emit high sounds, too high to be heard by man, and orientate themselves according to the returning echo of their sounds, the latter (in many bats, but not in all) being increased in number (from 10 to 60 per second) in a complicated environment.

Towards the end of his life, Spallanzani turned to the study of respiration. He showed that all parts of the body, muscles or liver, as well as lungs, are able to absorb free oxygen from the air and to give off CO_2. As this respiration continues after death, it is a chemical force and not a specific vital one; he had propounded the same theory before, concerning digestion. Other of his experiments and observations concerned the hibernation of mammals, the circulation of blood, bio-luminescence, etc.

Spallanzani fought all his life against vitalism, propagating theories of a physico-chemical determinism of the vital phenomena. It was for this reason that he was always trying to let parts of the body—separated from the latter—continue their vital activities for a more or less considerable time, to explore the dependence of biological processes upon temperature and other environmental factors, and so forth.

As a comparative physiologist and experimenter, as a scientist who knew by carefully planned experiments (of so great a number, that his adversaries reproached him, complaining that their abundance was intended only to veil and obscure the issue) how to explore the fundamental biological problems of his time, the pre-eminence of Spallanzani is uncontested in the history of modern biology.

Rudolph Jacob Camerarius (1665/1721) of Tübingen observed that isolated mulberry trees never produced normal fruits. To test this observation he isolated female plants of *Mercurialis annua*, and discovered that they produced fruits only when pollen of male plants was spread over them. Many similar experiments are described in his *Epistola de Sexu Plantarum* (1694).

Many cross-breedings of plants led J. G. Koelreuter (1733/1806) to the discovery of such phenomena as self-sterility, pre-Mendelian results of cleaving, etc. He definitely established the equal importance of both parents for heredity, and thereby established firmly the nature of sexuality (1761).

Christian Konrad Sprengel (1750/1816) of Spandau established the biology of flowers in his *Das entdeckte Geheimnis der Natur in Bau und in der Befruchtung der Blumen* (1793). He demonstrated that the structure of flowers is such as to allure to them insects which collect the pollen and transmit it to other flowers for pollinisation (i.e. fertilisation) for which service they are rewarded by the nectar.

Only in 1825 did Giovan Battista Amici of Firenze discover the pollen tubes and it was in 1846 that he described the entire process from the germination of the pollen to the development of the embryo.

Albrecht von Haller (1708/1777) of Bern—author of a well-known epos on the beauty of the Alps and of a book on their flora—was a great encyclopaedic personality. His main efforts were devoted to physiology, which he called *animata anatome* (enlivened anatomy). The capacity of the muscle fibre for contraction, its irritability (we say today: its contractability) is an inherent and characteristic force of the muscle, activated in the organism by a nerve force. In contrast to his predecessors, he denied that the nervous action is the flux of a nervous fluid within the hollowed nerves, assuming something nearer to our own conception of nervous conduction. In any case, the nerve continues to transmit stimuli even after the animal's death. Opposed to the 'irritable' organs are the sensitive ones. Thus Haller showed that only the epidermis in the skin is sensitive, but not the corium or the sub-cutaneous fat tissue. The muscles and ligaments are not sensitive, insofar as they are not furnished with nerves, and the sensation is felt in certain parts of the cortex of the brain, not in the place of the stimulus. Haller began his great *Physiologia* with a detailed discussion of the blood, and he ascribed to it a paramount importance. Thus, he says, that he has ignored the function of the thyroids, but it must be important because of their ample supply with blood vessels.

At Haller's time the naked-eye anatomy of the nervous system was fairly well known. Luigi Galvani (1737/1798) of Bologna showed in 1791 a nerve subjected to certain stimulations induces muscular contraction. Alessandro Volta (1745/1827) of Pavia proved the electric nature of Galvani's stimulation. Only about 1850 did Emil du Bois-Reymond—the same who confessed our intrinsic ignorance of the last problems of science (*ignoramus et ignorabimus*, a slogan which others answered by: *we wish to know and we will know*)—demonstrated that every nervous impulse is accompanied in its passage along the nerve by a progressive change of its electrical state which leads to the explosive flaring up of the chemical processes accompanying muscular contraction.

Sir Charles Bell showed (1791) that the upper (anterior) roots leaving the spinal cord convey sensations to the brain, while the lower (posterior) roots carry the impulses from the brain to the effector organs (e.g. muscles). This discovery opened the way for that of the animal reflexes or ' reflex actions ' (M. Hall, 1833). A peripheral sensory stimulation runs to the spinal cord and is there, either directly or by intercalation of one or more nerve cells, automatically connected with an effector nerve leading to a muscle or to a gland.

Kölliker's great discovery was that all nerve fibres are merely enormously elongated processes of the nerve cells with which they retain continuity (1889). These nerve cells are concentrated in the central nervous system. The most important further progress in the knowledge of the finer structure of the nerves is due to Camillo Golgi (1844/1926) of Pavia and to Ramon y Cajal (1852/1934) of Madrid.

F. J. Gall (1758/1828) of Vienna learned about the importance of the difference of the grey and white matter of the central nervous system. He tried to localize many functions in definite areas of the brain's cortex. This conception—by now established in great detail—was supported by the experiments of Fritsch and Hitzig in Berlin, who succeeded by the stimulation of certain areas of the cortex in producing regular contractions of certain muscles.

In our generation, Sir Charles Sherrington (1861/1950) followed Claude Bernard's studies on the vaso-motory regulation of the blood supply of the organs, and enlarged them into a most complicated general system of the nervous integration of all processes within the organism.

The scientists of antiquity regarded respiration as the refrigeration of the heated, ' boiled ' humours and of the innate body heat. Already in 1540, V. Biringuccio (1460/1539, posthumously in *De la pirotechnic*) had shown that the weight of certain metals increases under calcination by aggregated air. This was confirmed by J. Ray in 1630. But only John Mayow (1643/1679) demonstrated in his *Tractatus quinque medico-physici* that only a fixed part of the air—called by him spiritus nitro-aereus—participates in this congress and that combustion and respiration are identical processes. No animal can live in air in which this fraction is wanting. Robert Boyle, one of the most active founders of the Royal Society of London, gave a number of demonstrations there on the effect of the vacuum upon various animals, their fainting and dying and eventual revival when air was readmitted in good time.

The fixed air (our CO_2) of Joseph Black (1728/1729), and the dephlogisticated air (i.e. O_2) of Priestley (1774) and Scheele (1742/1786) prepared the way for Antoine-Laurent Lavoisier (1743/1794) in his *Expériences sur la respiration des animaux et sur les changements qui arrivent à l'air en passant par leur poumon* (1777), *Altération qu'éprouve l'air respiré* (1785), etc. Lavoisier assumes that in the blood the inspirated air undergoes, upon arriving at the lungs from the heart, a perfect but very slow combustion, which purifies it from scoria and restores its fresh red colour. The heat liberated during this combustion is responsible for the high body temperature. But if combustion were restricted to the lung only, it would—according to calorimetric measurements—burn the lung. Hence, oxydation must take place all through the body. This was well demonstrated by Spallanzani.

EARLY NAMES FOR THE BIOLOGICALLY FUNDAMENTAL GASES OF THE AIR.

OXYGEN (Lavoisier)	=	spiritus nitroaereus (Mayow 1674)
		fire air (Scheele 1777)
		dephlogisticated air (Priestley 1775)
CARBONIC ACID (Lavoisier and his collaborators 1787)		
	=	fixed air (Black 1756)
NITROGEN (Chaptal)	=	noxious air (Rutherford 1772)
		phlogisticated air (Priestley 1774)
		foul or vitiated air (Scheele 1777)
		mofette (Lavoisier 1777)
		azote (Lavoisier and his collaborators 1784)
		nitrogen (Chaptal 1790)

Heinrich Magnus made in 1837 the first quantitative analysis of the gases in the blood, confirming the progressive consumption of its oxygen with the progress of circulation. In 1845, J. Robert Mayer (1814/1878) evolved the notion that the oxygen of the blood is burnt, into his fundamental law of the preservation of energy, after his observation that the blood of the veins is lighter in the warm tropics (where less energy for homoiothermic maintenance needs to be expended than in Europe).

Leonardo de Vinci (MS. G.32v) had already stated that water is *the* food of the plants. He grew a side-root of a pumkin in water and produced a plant with sixty big pumpkins. By careful observation he noted an ample impregnation by the nocturnal dew. Van Helmont (1577/1644) made a similar experiment in soil and found no change in the weight of the soil at the beginning and at the end of the experiment.

Only Stephen Hales (1677/1761) recognized in his *Vegetable Statics* (1727) the importance of root pressure, by which food comes into the root from the soil and is transported by this ascending force upwards; and that, in addition, the plant absorbs part of the air. For the study of these gases, he collected them in vessels inverted over the plants for a sufficient time.

J. Priestley (1733/1804) discovered the absorption of CO_2 by the plants and their ' purifying ' action upon the air aroused great enthusiasm. Yet Scheele (1778) found that the plants absorb oxygen and emit CO_2. This contradiction was solved by Jan Ingenhousz (1730/1799) through his discovery of the different behaviour of plants in light and in darkness. In light they absorb CO_2 and emit O_2; in darkness the contrary is true. At night the leaves respire like the animals; but during the day they respire inversely (a process later called assimilation), and produce a considerable part of the vegetable substance. He observed the formation of the amylum granules in light. That the plants absorb O_2 and emit CO_2 (but in a lesser degree) even by day, was only recognized in 1857 by Julius Sachs.

Nicolaus Théodore de Saussure (1767/1845) of Genève concluded this line of research by proving that the carbon-requirements of the plant are satisfied from the CO_2 of the air and that water is used for the formation of dry substance; that leaves respire by darkness, i.e. they take in oxygen and emit CO_2; and that succulent leaves store organic acids. He was the first to establish respiration as a fundamental process in *all* organisms. This was at the dawn of organic chemistry and it needed the authority of people like Liebig, Boussingault and Sachs to establish this separation of the processes of carbon assimilation and of respiration, together with the already established need of the plants for salts within the soil.

The synthesis of urea (from ammonium cyanate) by Friederich Wöhler (1800/1882) was the first time that an organic substance was synthetically produced from inorganic components. But the experimental proof was not decisive, as some of the substances used were gained from organic substances. Neither Wöhler himself nor any of his contemporaries drew any far-reaching conclusions from this synthesis. Only in 1880 did Berthelot come to the conclusion that organic substances could take their origin without the presence of a *vis vitalis*, enlivening the organism. Mechanism became very popular during the period from K. Büchner, K. Moleschott, and Vogt to E. Haeckel.

Digestion was regarded in antiquity and in the middle ages as a 'boiling' of the ingested food in the liver to blood and into a still finer distillation within the heart. Opposing this theory in the 16th and 17th century, the iatromechanists regarded the mechanical triturition of the food as the major process of digestion, whereas the iatrochemists maintained that it is based upon putrefaction, i.e. its chemical decomposition—though the concept of chemical action was still unknown at that period.

A. F. de Réaumur (1689/1757) invented the method of introducing sponges into the crop and stomach of some animals, and studying the digestive effect of the retracted liquid. This procedure was intensively followed by Spallanzani (1777) by food within perforated tubes introduced similarly. The latter proved, for example, that the gastric liquid was not only not putrifying, but definitely antiputric and of an acid character, the occasional lack of acidity being a pathological symptom. It was only in 1823 that W. Prout recognized the acid of the stomach as chloric acid. Spallanzani was heavily attacked by the anatomist J. Hunter (1728/93), especially because the former stressed that digestion may continue also after the death of the body.

In 1822, W. Beaumont (1785/1853), an American army surgeon, had an Indian patient with a broad stomach fistula. We should mention that in 1664 De Graaff had already produced similar pancreas fistulas experimentally on animals. Beaumont used this opportunity for prolonged observations and experiments, mainly by introducing various foods for varying periods into the stomach of his patient. In 1833 he published fifty-one ' conclusions ' which gave a clear picture of the gastric digestion in man.

Further progress is connected with the name of the famous French physiologist Claude Bernard (1813/1878) in his famous *Du Suc Gastrique et de son Rôle dans la Nutrition* (1843), mainly by studying the intestinal sap from the mouth cavity to the rectum. This led, together with other experiments, to the discovery of the antagonism of the bile to the action of the gastric pepsin.

Justus von Liebig (1803/1873 in Giessen, and Jean Dumas (1800/1884) in Paris, established chemistry as the foundation of agriculture and plant- and animal-physiology. At the entrance to the former's laboratory was written: ' God has ordered all His Creation by Weight and Measure '. Most important for a rational agriculture was his principle of the importance of minimal factors for plant development, with respect to the various nutritional salts of the soil. Insufficient quantities of one necessary salt or element may prevent the utilisation of large quantities of all the other salts needed which may be present in the soil, and thus small additions of that salt in the minimum results in crop increases which are entirely out of proportion to the smallness of the addition. It was during this period that the nature of protoplasma was established as composed of a mixture of proteins, lipoids and carbohydrates. The animal food is divided, according to the presence or absence of nitrogen, as food for growth or for maintenance. The change of carbohydrates into fats within the body was also proven, as well as the origin of all animal heat from combustion.

Boussingault and Liebig had established that nitrogen can be absorbed by the roots—the latter believed this happened in the form of humus, the former demonstrated that nitrogen containing salts (ammonia) can fully replace humus. But it was an important progressive step for the conception of proper crop rotation in agriculture, when Marcellin Berthelot (1827/1907) showed in 1886 that certain soil bacteria are able to fix nitrogen from the air, and two years later Helriegel and Wilforth demonstrated this air fixation of nitrates to be in direct proportion to the amount of bacterial nodules on the roots of leguminous crops, and independent of the amount of nitrate or ammonia which eventually could and would be absorbed through the rootlets of the plants. Much progress is still being made in our knowledge of the association of other plants with other micro-organisms, or even of similar phenomena in

PLATE XIII

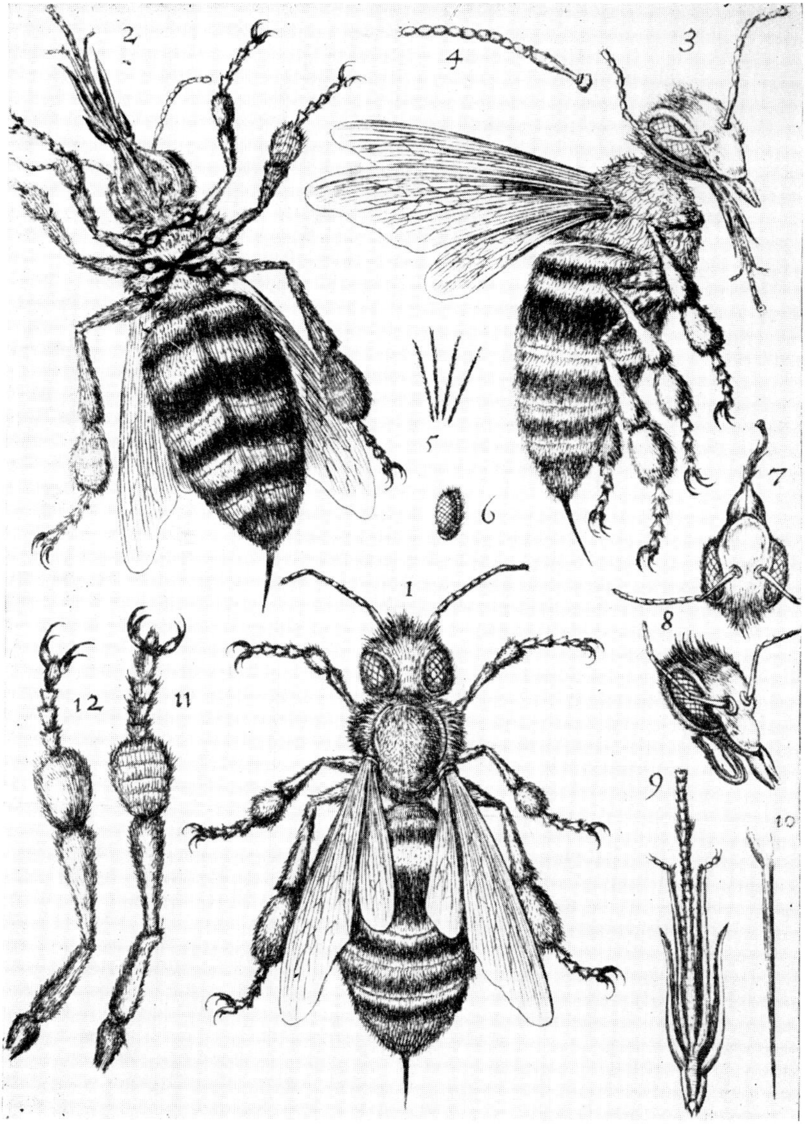

THIS FIGURE OF BEE-MORPHOLOGY IS ONE OF THE EARLIEST RESULTS OF MICROSCOPIC STUDY WITH THE MICROSCOPE BY THE ACCADEMIA DEI LINCEI OF ROME. ITS PRESIDENT, THE PRINCIPE CESI, PUBLISHED IT IN HIS *Apiarium* (ROMA, 1625).

Here it is reproduced from an appendix in F. Stelluti's edition of Persio. (Roma, 1630).

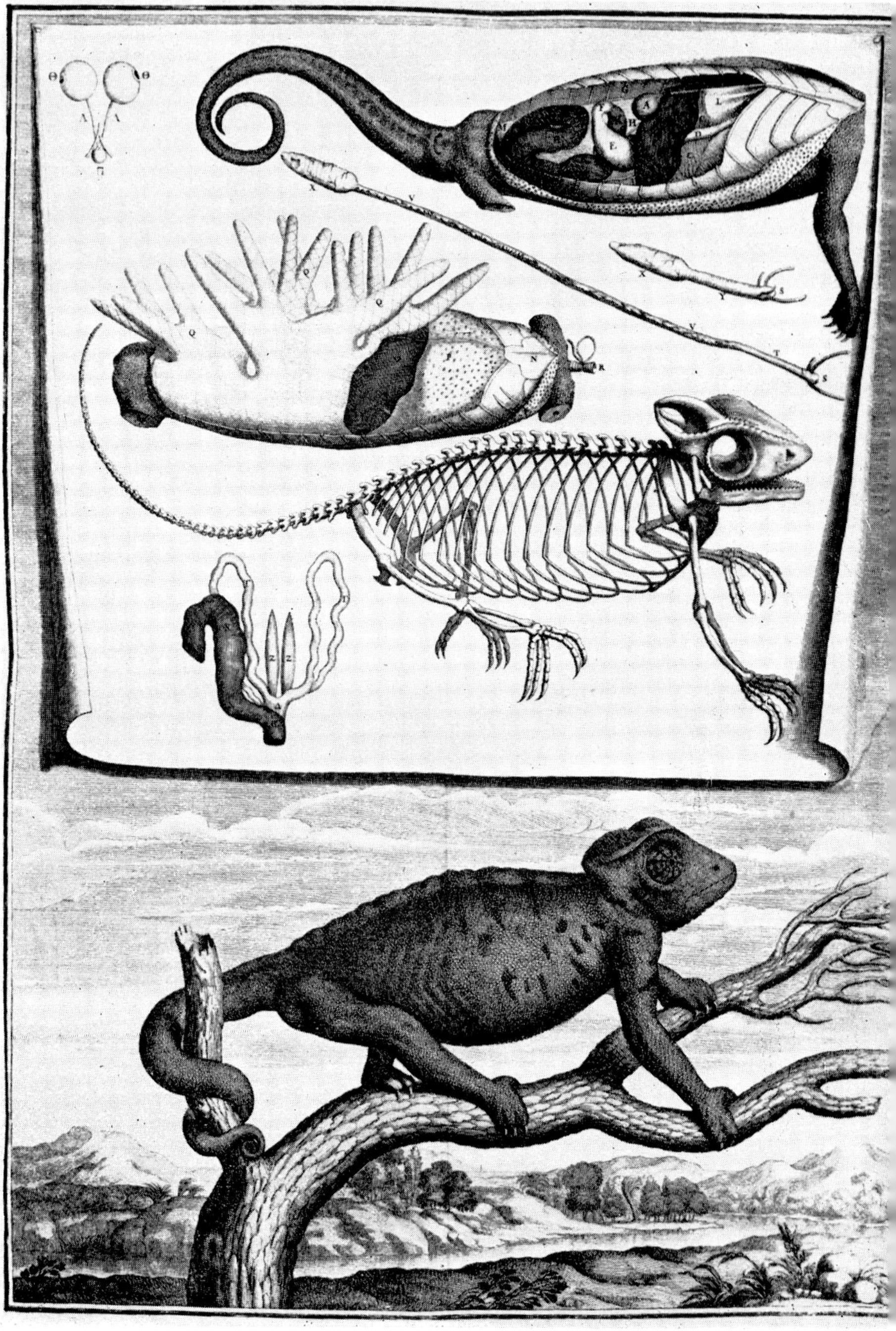

THE EARLY WORK IN COMPARATIVE ANATOMY OF THE ACADEMY OF SCIENCE PARIS, ILLUSTRATED BY THE DISSECTION OF THE CHAMAELEON.

From the *Description Anatomique d'un Cameleon, d'un Castor, d'un Dromadaire, d'un Ours, d'un Gazell* (Paris, 1669) of Claude Perrault.

animals (aphids, for example), which thus are able to fix nitrogen at least at certain stages of their development.

Among the early physiologists of the 19th century, we must mention Johannes Müller (1801/1858) whose great discovery was the principle of specific nerve energies. This explains that any sensory nerve may respond only with its specific sensory manifestation. The mechanical beating or an electrical stimulation of the eye, for example, provokes (*via* the retinal sensory nerves) only one reaction, the only one of which the eye is capable, namely seeing—normally produced by light. His many other enquiries included researches into the reflexes of the frog; into the ganglion cells of the brain as the bearers of the latter's functions; and a monograph on the anatomy of a primitive fish-family, the *Myxinoidea.* He also was one of the first to start plankton catches in the sea, as a result of which he described many new larval types.

Karl Ludwig (1816/1895) is known as the inventor of many important instruments of physiological experimentation, such as the mechanically rotating drum (kymograph), and the mercurial blood-gas pump. Hermann Helmholtz (1821/1894) should be named for his intensive study of the physiology of visual optics; of the ophthalmic mirror, etc.

Another great figure of that period was Claude Bernard (1813/1878), a genial and elegant experimenter, from among whose many achievements we can only mention the synthesis of glycogen in the liver and its breaking down when used, after storage; the pre-digestion in the stomach and the great importance of the gastric juices of the pancreas, including that for diabetes; the vaso-motory regulation of the blood-supply to the different parts of the body; etc. Life is characterized for him by the body's concentrated effort to maintain a stable '*milieu interne*', whose stability is life's pre-condition. Bernard thought mainly of the blood as this internal environment, and while we interpret it more largely, this principle as such still maintains its central importance in physiology and in ecology.

(*g*) *Microbiology*

Louis Pasteur (1822/1895), originally trained as a physical chemist, will remain one of the great figures of history of biology, mainly because of achievements of far reaching consequences in two fields.

In his studies on fermentation, from that of milk (1857) to that of wine (1866), he became a strong antagonist to the ruling opinion of Liebig, that putrefaction and fermentation are purely chemical processes. Pasteur proved, by step after step of experiments, that no fermentation is found in nature in the absence of typical micro-organisms which produce it. During these enquiries, he also discovered that certain of these micro-organisms grow and develop only in the absence of oxygen (anaerobiosis). The final test was his experimental campaign in 1859/1862—with a paper crowned by a prize of the Academy of Science in Paris in 1860—by which he demonstrated that in a liquid boiled for a long time within a flask ending in a long, horizontal, S-shaped neck no fermentation occurs and no micro-organisms are found after hours, weeks, and even months. Yet, very soon after breaking this neck, fermentation began, after micro-organisms from the air had dropped into the liquid. This was the final settlement of the age-old dispute whether primitive life originates in suitable liquids from 'organic molecules' contained in it, or not. Pasteur decided in favour of the negation of such spontaneous generation by saying: " Omne vivum ex vivo." This sentence has retained its validity on the cellular level of organisation, but modern students are attempting to build up at least the great organic molecules from which every protoplasm must be composed, well realising that even success in these attempts does not yet mean the generation of life.

A silkworm epidemic in southern France, endangering the existence of that important industry, lead to the first contact of Pasteur with an infectious disease. He discovered two micro-organisms responsible for two different diseases, the Pebrine and the Grassery. He studied their life cycles and found ways to prevent the outbreaks of further epidemics to a certain degree (1862). This was the first time that a specific micro-organism could be defined as the cause of a specific disease, and Pasteur did not fail to recognize the general importance of this discovery, which directed the trend of all his future work.

Just before Pasteur and Koch turned to the intensive study of bacterial diseases, Ferdinand Cohn (1828/1898) of Breslau stated that bacteria are not, as assumed hitherto, one or a few species of extremely variable shapes, but that quite a number of different groups of bacteria exist with definitely different forms, even if highly variable to some degree, for which he proposed the first system. Today, in bacteriology, physiological systems replace his morphological one, but the latter played an important part in the initial phase of bacteriological research.

This splendid period of bacteriology, immunology and preventive medicine, initiated by Pasteur, Robert Koch (1843/1910), Emil von Behring (1854/1917), Elias Metschnikoff (1845/1916), and many others, had, of course, its precursors. Among them we mention the great achievement of Ignaz Semmelweis (1818/1865) in discovering the role of septic infection in puerperal fever, which deadly disease was reduced enormously by the application of antiseptic treatment; and also Joseph Lister's (1827/1912) fight against the terror of small-pox by immunisation through inoculation with a weak virus which stimulated the development of immunising antibodies. The details of these researches, however, belong to Bacteriology and Immunology. Also, the discovery of the malaria cycle by Laveran, Grassi and Ross, as well as the history of the romance of virus research, belong to modern research. They are discussed in every text-book.

(*h*). *Biogeography. Ecology*

We have already mentioned a number of earlier travellers of the 16th and 17th century who paid special attention to the flora and fauna of the countries they visited. The list of those men, who enlarged considerably our knowledge of the natural history of South and Tropical Africa, Arabia, the Far East, etc., increased rapidly in subsequent centuries and cannot possibly be enumerated here. The first intensive explorers of Australia's nature were Captain James Cook (1722/1779) and J. Banks and R. Brown. A pioneer in plant geography was Alexander von Humboldt (1769/1859), who, on his extended travels in S. America (1799-1804), was the first to give descriptions of types of vegetation instead of mere lists of species. He distinguished sixteen such types, among them heather, grass, palm, banana, cactus, etc., landscapes. He also inquired into the relation between plant life and climate and the influence of altitude on floral changes; and applied the isotherms (lines of equal average, maximum or minimum temperature) upon the distribution of these types of vegetation.

We have also mentioned before that Buffon not only delimited barriers to faunal and floral distribution, but raised a number of biogeographical problems, such as the presence of the tapir in Malaya and in S. America.

Charles Darwin's cruise around the world in the *Beagle* (1831/1835) became a markstone, as his visit to the Galapagos Islands opened his eyes to the peculiar character of its insular flora and fauna as well as to the importance of isolation in time and space for speciation (i.e., the origin of species).

P. C. Sclater's pioneer work in zoogeography (1859) was soon followed by the monumental *Geographical Distribution of Animals* (1876) of R. Wallace, in which he established the Palaearctic, Nearctic, Ethiopian, Oriental, Australian and Neotropical kingdoms. His long stay in the Malayan Archipelago brought forth his splendid *Island Life*, where he also established a border line between the islands of Bali and Lombok—now called the Wallace line. This line opened one of the most fertile discussions, which still continues, concerning the character and definition of biogeographical borders in general. Among the many important conclusions, we may mention that of the geological structure, at present and in the past; the difference of distribution in groups of recent speciation, such as mammals, birds, butterflies, and those of most plant groups and of animals of an older standard, such as reptiles, many insects, scorpions. Not less important is the lesson that no analysis of mere faunal and floral lists will ever replace the immediate impression of nature.

A. H. R. Grisebach (1813/1879) of Göttingen proposed similar areas of distribution for plants, paying much attention to the influence of climatic factors (*Pflanzengeographie*, 1878). He also extended the irano-turanian region of the Palaearctic far into Asia Anterior. Most of the plant geographers who followed left regional geography behind for ecology, especially for the detailed study of climate and competition: E. Warming (*Ecology of Plants*, 1895) and A. F. W. Schimper (*Pflanzengeographie auf ökologischer Grundlage*, 1898), for example.

Historical plant geography centred largely around the great floral changes since the late Tertiary period in Central Europe. A. Engler (b. 1844) of Berlin described these changes, starting from warm-climate flora in the late Tertiary to the establishment of a cold-climate steppe in the ice ages, some of which are now occurring in discontinuous areas in high mountains (Alps) and in the Arctic region. The present flora of Central Europe is formed by immigrants derived from the eastern steppes of Siberia (Angara) and from relics of the earlier floras which survived the ice age in especially protected habitats. The history of the post-tertiary flora has been much refined by pollen analysis. The pollen of the dominant trees is preserved in successive layers in moors or in the bottom strata of lakes, thus permitting a rather accurate history of local floral changes (Lagerheim 1902).

Similarities of floras and faunas in now widely separated areas, such as Guinea and Brazil, were explained in two different ways: either by landbridges which connected these areas directly or indirectly in the geological past (e.g., J. D. Hooker, 1847) or by the theory of continental drift (1912) of A. Wegener. The latter assumes all continents once formed one continuous land-mass which split and drifted apart, now being separated by the oceans.

A most unusual experience was the volcanic eruption in 1883 of Karakatao, a small island between Sumatra and Java, which destroyed all its life. Dammermann (1948) described the history of the first fifty years of its resettlement by animals and plants. The analysis taught much about the spreading of organisms into insular areas.

J. C. Willis' theory of *Age and Area* (1922) tried to explain the differences in the size of areas of species as a consequence of their age. Young species have, according to him, a small—old ones, a large—area. This theory was abandoned in favour of an ecological determination, mainly by the ecological valence (vitality) of these areas. In 1922 appeared R. Hesse's *Tiergeographie auf ökologischer Grundlage.* Animal ecology, especially its theory of populations, developed mainly from applied zoology, such as applied entomology, hydrobiology and human demography.

Actually, the discussion in animal ecology is mainly about the role of climatic and edaphic (soil) factors on the one hand, and competition and co-operation with other organisms on the other hand. Many ecologists now assume a high supra-organismic integration of all species living together in one habitat (biocoenosis) with high power of self-regulation. Others doubt that the available facts permit such a conclusion. In botany this school of research is named plant sociology. The most competent manuals of ecological thought are by F. E. Clements and V. E. Shelford (*Bio Ecology*, 1939) and W. C. Allee, A. E. Emerson, and others (*Principles of Animal Ecology*, 1949).

Exploration of life in the oceans was accomplished by means of a number of special expeditions of which two were of outstanding importance. That of the Challenger (1872/76) apart from its great contribution to oceanography, brought to light a surprising number of deep-sea animals. The Deutschland expedition (1911) discovered the nannoplancton, an overwhelming abundance of flagellates which escaped through all the nets of earlier explorers but are found in masses in the intestines of Ascidians. They explained some great lacunas which hitherto existed in our knowledge of the circulation of organic matter in the sea (J. Lohmann).

J. Schmidt of Copenhagen succeeded in extended enquiries (1904/1921) in solving the old riddle of the life history of the eels which mature in European rivers and return for egg laying to a limited area east of the Bahama Islands, whence the larvae, which differ markedly in shape from the adult fish, slowly return to the European rivers.

Animal migrations, especially those of birds, retain their fascination for modern science, as none of the basic factors inducing the orientation and determination of the directions of their flights have been satisfactorily solved, so far.

(*k*) *The Theory of Evolution*

We cannot possibly regard the natural philosophy speculations of the Ionian philosophers like Anaximander and Empedocles, or of the Stoic Lucretius Carus, and others—concerning the origin of life in the sea, the progressive development of the higher forms of life from the lower ones, and some kind of struggle for life and the survival of the fittest as factors producing transformation or survival, as homologous to the modern theory of evolution; we miss in all of them a clear understanding of slow progressive transformation through long periods based upon a good knowledge of the forms of animal and plant life. Benoît de Maillet, Maupertuis, Didérot, Erasmus Darwin, Buffon, Lacepède, Treviranus and Oken, all promoted in the 18th century speculations with an evolutionary tinge.

But modern transformism begins with Jean Baptiste de Monet Lamarck (1744/1829) in his *Philosophie Zoologique* (1809). He was an excellent taxonomist, whose ideas were based upon a good knowledge both of animals and plants. He accepted the term 'biology', first published by Treviranus in 1802, and is responsible for its general acceptance. Lamarck declared the linear 'ladder of nature' to be the expression not only of formal classification, but also of actual progressive transformation, each step differing but little from its neighbours. Wherever in this system wide gaps are present, future palaeontological discoveries will bridge them. This continuity of life leads at its roots to a common origin of plant and animal life. The high variability of domestic animals is proof that no species is definitely fixed. The environment, and the organism's reaction towards its changes is the cause of progressive transformation. A species remains unchanged as long as the environment remains unchanged. Every change of the species is brought about by hereditary changes of qualities acquired by the use and disuse of various organs or body parts. Every organ which is much used grows stronger by this permanent exercise, whereas an organ which is constantly not put to work degenerates and dwindles. The disappearance of the eyes in animals living in permanent darkness illustrates the latter case; the former is given as the cause of the development of the giraffe's peculiar body shape. Thus, antelopes, in an environment where herbs and grasses were insufficient for food, attempted to reach the leaves of the tree-crowns of their environment, and stretching their necks persistently through generations, the present long neck and high legs of the giraffe were hereditarily established. " The new characters facilitate exercise of those habits which have produced them by heredity."

Lamarck gave a masterly exposition of the facts and the theory of evolution. We know, however, today that inheritance of acquired characters does not become manifest within a few generations. But it was not this or any other error that was the reason why Lamarck's book remained without the success it amply deserved; the minds of the contemporary biologists were not yet ready for its contents. Today a great number of leading biologists, such as Caullery, Jennings, and many others, declare that, while no factual proofs are still available, the existence of some kind of Lamarckian transformation is a ' postulate of practical reason '.

The inconstancy of species and the transformation of homologous organs was much in vogue with the German romantics and is also well expressed in Wolfgang von Goethe's *Die Metamorphose der Pflanzen* (1799) or in the comparative anatomy of Etienne Geoffroy St. Hilaire (1772/1844).

The Reverend T. R. Malthus (1766/1834) in his *Essay on Population* promoted the idea that among animals, plants and within human society, there exists a permanent struggle for existence with the survival of the fittest produced by an ever increasing disproportion between the food present or produced and the numbers of offspring born in such a multitude, so that they cannot possibly all develop. His main application was to the social problems of humanity.

Another important development was Charles Lyell's (1797/1875) discovery, in his *Principles of Geology* (1833/35), that all geological changes of the past can be explained by the same factors shaping geomorphology today, without any need to assume catastrophes of great extent separating the various geological ages—if only those factors act over sufficiently long periods, such as actually were at their disposal in the past history of the earth.

It was Charles Darwin (1809/1882) who succeeded where Lamarck had failed, fifty years before. His *Origin of Species* (1859) brought about a revolution not only in biology but in all spheres of cultural life. When Darwin started as a naturalist in 1831 to 1835 to accompany the world cruise of H.M.S. *Beagle*, he was ill prepared for this job. Yet Lyell's *Principles* were in his library. This voyage was the most important event in his life. When they arrived from the South American mainland to the not very far distant volcanic Galapagos Islands, Darwin was struck by the very high endemicity of the plants and animals of the archipelago: genera as well as species were in their overwhelming majority restricted to just the Galapagos, but in character and affinity every one of them was clearly related to the flora and fauna of the nearest American territories. Furthermore, in most islands of the archipelago not the same but closely related species, usually of the same genus of birds, turtles, etc., occurred. And, equally striking, was the fact that one family of finches (*Geospizidae*) had acquired the beaks and other characters of grain-, seed-, herb-, insect-eaters, of ground-, tree-, or of soil-birds, etc. Had the Creator created all these forms so closely related and neighbouring for bureaucratic reasons only? Malthus and Lyell helped him to find an answer to this question. Beginning in 1837, Darwin began to enter notes into a long series of note-books which grew with the years into a great heap, where he entered all these facts and ideas related to this problem. So far only his friends were aware of this steady work. When in 1858 the excellent naturalist Alfred Russell Wallace (1823/1913) sent in from the Malayan Archipelagos a paper to the Linnean Society of London which actually came, as the result of a rich body of observations, to the same conclusions as Darwin, his friends forced him to publish, together with Wallace's paper, a first resumé of his own work. When, in November 1859, the *Origin of Species by means of Natural Selection* appeared the edition of 3000

PLATE XV

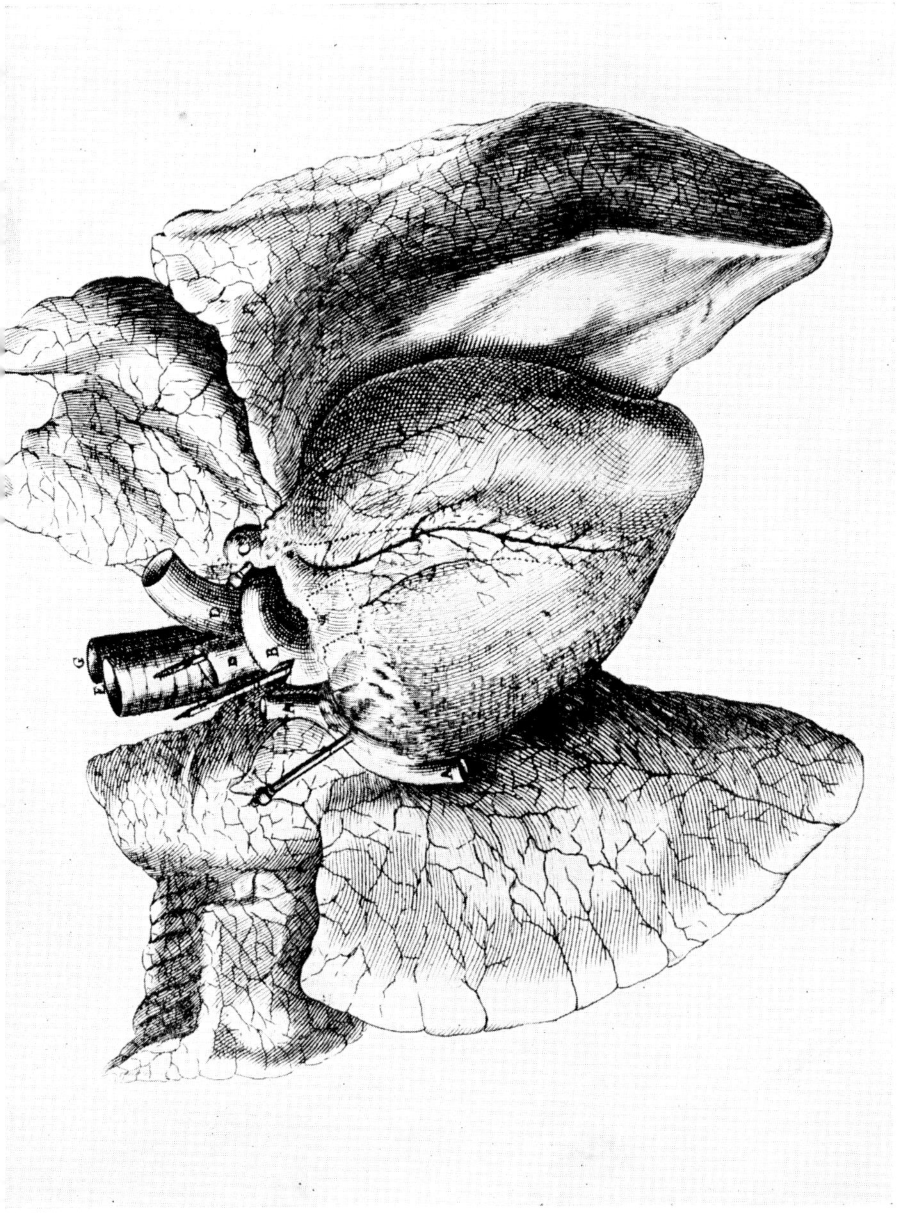

Rene Descartes gives in his posthumous work *De Homine* (Leiden, 1662) a sketch of human physiology. He retains the heart as the boiling and rejuvenating motor of the human machine.

He is one of the first to accept Harvey's discovery of the circulation, but being aware of the missing link, namely the capillaries, which were only discovered by Malpighi in 1661.

books was sold out within a week, probably the first scientific book in the history of the book trade with such a success.

Darwin submitted a very well organized material of proofs for the theory of transformation, from the fields of comparative anatomy, comparative embryology, biogeography, and, last but not least, from palaeontology. The latter arguments were rather suggestive, even if the *Archaeopteryx* was only discovered in 1861 in the slate of Solenhofen. Arguments from rudimentary organs and from parasitic degeneration, etc., were forcefully advanced. In addition, he offered a new interpretation for the mechanism of evolution by natural selection. His basic assumptions are that all organs or instincts are, however slightly, variable. The enormous over-production of seeds and offspring necessitates a struggle for life in which those variants survive which fit best into their environment. Of Darwin's two theses: that of the common descent of all animals and plants from more primitive forms and finally from a common root, and that of natural selection, the former was immediately widely accepted after some heavy public discussions, in which men like T. H. Huxley (1825/1895), Herbert Spencer (1820/1903), and Ernst Haeckel (1834/1919) took a leading part on Darwin's side. Without hesitation, Darwin extended his theory of evolution to the descent of man from monkey-like ancestors, a development almost as revolutionary in human thought as that of Kepler, who dethroned the earth from its position as the centre of the Universe.

The validity of Darwin's second thesis, that of natural selection, derives from the inheritance of the small fluctuating variants appearing in nature. Darwin regarded such heredity as self-evident in the same way as Lamarck had assumed it for acquired characters. In 1868, Darwin developed a theory of ' pangenesis ' which is extraordinarily similar to the Hippocratic theory of heredity. Tiny particles migrate from all organs into sperm and egg, and hence are able even to transfer to them the characters of the parents, including the acquired ones. ' Pangenesis ' is utterly untenable in the light of modern research, and it is an unusual compliment to Darwin's maturity of mind and to his judgement—unsurpassed in the last centuries—that his theories in general still hold their ground, of course with new genetical arguments, the development of which we now have to study.

The utter lack of any sound and factual knowledge on heredity which is evident in Darwin's great book, its restriction to empirical rules of mass phenomena and the study of variation in domestic plants and animals—these were corrected by future research as a response to the many problems evoked by the *Origin of Species*.

August Weismann (1834/1914) separated in principle the germ-cells of all the other body or somatic cells. The latter all are mortal, the former being potentially immortal and maintaining the species through the sequence of its generations. This potentially immortal substance of the germ cells is called by him 'germ-plasm'. Variability is thus produced by the different germ-plasm of the parents. Nothing new can appear, and the theory of pre-formation is revived in a modern form, still maintained by orthodox genetics today. Experiments in which the tails of white mice were cut off for over one hundred generations showed no influence on the offspring. Weismann regarded this experiment as a definite proof against the inheritance of acquired characters, but it is only one against the inheritance of mutilations. He regards the *ids* of heredity as small particles within the chromosomes, each *id* having the virtual potency for the development of a specific individual. Weismann's work had the greatest influence upon the early generation of geneticists.

Gregor Mendel (1822/1884), abbé at Brno, conducted from 1857 to 1868 a long series of experiments concerning the hybridization of various races of peas, differing in shape and colour of seed, pod, etc., by pollinizing the female flowers of one race with the appropriate character with the pollen of the alternating one with the artificial exclusion of any further chance of free pollination. The results were recorded quantitatively, with the following results :

1. The hybrids of the first generation were all uniform: in yellow (Y) and green (g) hybrids they all were yellow, in round (R) and wrinkled (w) seeds all round. Those characters which are manifest, are called the dominant variants, those which apparently disappear in the first daughter-generation (F_1) the recessive variants. The former are written in capitals, the recessive ones in normal letters.

2. When the hybrids of F_1 are self-crossed, a mode of fertilisation possible in peas, the following generation (F_2) has 25 % of individuals manifesting the recessive character. And this group, when crossed amongst themselves, will give the recessive character and none other in all future generations. The remaining 75% of F_1, when all individuals are self-fertilized, yielded another 25% of the total number, which when self-crossed, gave pure dominant characters in all future generations; while the remaining 50% of the total,

when self-crossed, segregated in every following generation in the following way: 25% pure recessives, 25% pure dominants, and 50% with dominant manifestation, with future segregation in each following generation in the same proportions.

Whenever two individuals with two different and alternating characters were crossbred, such as yellow/round and green/wrinkled seeds, the result in F_1 is a combination of both dominant characters, even if each parent has one of the dominant and one of the recessive characters, whilst in F_2 the combinations are more complicated:

		Germ-combination in the haploid egg-shell.			
		YR	Yr	YR	yr
Germ-combination in the haploid sperm cell	YR	YR YR	Yr YR	yR YR	yr YR
	Yr	YR Yr	Yr Yr	yR Yr	yr Yr
	yR	YR yR	Yr yR	yR yR	yr yR
	yr	YR yr	Yr yr	yR yr	yr yr

Thus, in F_2 appear 16 different combinations of the elementary units of heredity, but only four are manifest :

9 are yellow and round
3 are yellow and wrinkled
3 are green and round
1 is green and wrinkled.

These sixteen different combinations of the hereditary units represent the genotype, the hereditary character of the individual organism, while the four manifestations—and one such manifestation only is possible in one individual, as the seed can be either yellow (or round) *or* green (or wrinkled) and not both together—is the phenotype. Both these concepts are fundamental in modern genetics and they spring logically from Mendel's experiments.

The body of each organism is thus a mosaic of hereditary units which combine independently of one another. Of a pair of alternating characters or allies, such as yellow/green, only one manifests itself in F_1. (Intermediary heredity in F_1 as a much rarer case has been discovered only since Mendel.) Mendel was fortunate in his choice of an object for experimentation: first, because self-fertilization is possible and really exact results could thus be obtained of genetically known parents; second, by chance all the eight pairs of characters which he studied in peas were located in 16 different chromosomes of the pea, each one thus being transmitted actually independently of the others, whereas if two or more of the units would have been situated (as many characters are) together in the same chromosome, they would have been transmitted not independently, but always in combination.

In spite of their perfect presentation these results of Mendel, today the uncontested foundation of modern genetics, remained unknown and neglected, and had to be rediscovered independently in 1900 by three students: De Vries in Amsterdam, Correns in Berlin, and Tschermak in Prague.

Modern biometrics were developed by Lambert Quetelet (1796/1874) and awoke great enthusiasm, so that even Florence Nightingale described it as the most dignified form of veneration of God, as it permits us to recognize fully the wisdom of His creation. Francis Galton (1822/1911), a cousin of Charles Darwin, and his pupil Karl Pearson (1857/1936) followed this line of research and of enthusiasm, creating the notion of correlation and studying especially the mathematical expression of Darwin's fluctuating variability and its influence on heredity (*Natural Inheritance*, 1879). Their approach to the analysis of the data from mass-statistics were much cruder than the individual experimentation of Mendel. Thus, Galton's *Law of Ancestral Inheritance*, which was based upon the statistics of the inheritance of eye-colour in dogs, states that each ancestral generation participates in it in half the strength of its immediately following generation, with a total result of one:

$\frac{1}{2}$ (parents together) + $\frac{1}{4}$ (grandparents together) + $\frac{1}{8}$ . . . . + 1/N-1 = 1.00.

This method of analysis by mass-statistics led in the early days of human eugenics to entirely wrong conceptions. The idea that by elimination (artificial castration) of individuals with undesirable characters it would be easily possible to eliminate the undesirable hereditary units from a population is entirely wrong, as modern research has shown. Not only was the ecological influence of the social environment entirely neglected, but the recurrence of mutations (see later) makes this procedure as a rule futile. Unfortunately, these early opinions still linger, not only with the public at large, but also in many textbooks.

The careful studies of individual variation in beans led W. L. Johannsen (1857/1927) to conclusions which were quite different from Darwin's assumptions concerning fluctuating variations. He started with a bean population which produced a great range of variations as to the weight of the seeds. Selecting individual beans for his experiments and self-fertilizing their offspring, he obtained a range in seed-weight which was much narrower than the range for the entire population. The smallest and the biggest bean of the same motherplant produced under equal conditions exactly the same range of variation. This proved that selection cannot increase the weight of the beans. If such is nevertheless apparently the case in selections within bean populations, this must be the outcome of the exclusion by chance of those ' pure lines ' with the highest or lowest average. This result was apparently a death blow for the theory of natural selection.

Yet about the same time Hugo de Vries (1848/1935) showed another way for its explanation. In studying the genus *Oenothera*, he found on the dunes near Amsterdam a few individuals which differed very definitely in one or more characters from the normal *O. lamarckiana*. All these variants, when bred amongst themselves reproduced offspring entirely true to the new variation. Even more so: in mass-breedings of *O. lamarckiana* a certain small percentage of the same variants appeared in every generation. De Vries explained these sudden variants, which remained true by inheritance, as sudden changes of the basic (Mendelian) units of heredity. This type of inherited sudden change he called mutation. Small mutations are perfectly in agreement with Darwin's theory of Natural Selection. It is an often repeated irony of history that the changes in *Oenothera* noted by De Vries, were no mutations at all, but connected with complicated conditions of hybridization—known also from willows and roses—whilst the principle of mutation has been confirmed in an abundance of other cases.

Further theories of evolution which deserve mention are Wagner's theory of isolation, Rosa's Hologenesis, Eimer's Orthogenesis and Jollos' Orthomutation.

A most important step in the progress of genetics was the discovery of the sex-determining X- and Y-chromosomes by T. H. Montgomery (1901) and C. E. McClung (1901), and that of sex-linked heredity by Doncaster (1908), which lead to the most imposing mapping of the genes in the chromosomes of *Drosophila* by T. H. Morgan (1866/1945) and his school. Other progress includes the physiological theory of heredity by Richard Goldschmidt (1927), according to which, not absolute characters are inherited, but differences of intensity and speed, which reach the same level of manifestation at different stages of the development of the other characters; Müller's discovery of the regular recurrence of mutations in fixed percentages within a population; the discovery of the position effect; and especially modern population genetics with its neo-darwinistic trend. Sewell Wright established by the latter means (1931) his drift theory, according to which the greatest variability is manifested in low—the smallest, in dense—populations.

Neo-lamarckism has found its main promoters in Richard Semon's theory of mneme, which maintains that engrams of every impact suffered by the organism may undergo heredity. This point of view has been supported by Wood-Jones' discovery that in various aplacentalians the various regions of the direction of the hair in the embryos corresponds with that established in the adults by toilet movements.

While the evolution of all life on earth from a few simple forms is now a little contested hypothesis which explains more facts than any other theory and is contradicted by none, the scientists are as divided as ever with regard to the mechanism of this transformation. Neo-darwinistic and neo-lamarckistic views are vigorously maintained. Many mature biologists, Goldschmidt, Caullery, Jennings, and others have concluded that it is too early to decide about the mechanism of speciation, and probably more than one way is at work in nature.

References.

Meyer. Geschichte der Botanik. 4 vols. Königsberg, 1854/57.

J. V. Carus. Geschichte der Zoologie. München, 1872.

E. Radl. Geschichte der biologischen Theorien. 2 vols. Leipzig, 1907/08. Engl. ed., 1928.

J. Sachs. Geschichte der Botanik. München, 1875. Engl. ed., 1906.

W. A. Locy. Biology and its makers. London, 1908.

E. Nordenskiöld. Geschichte der Biologie. Jena, 1926.

F. S. Bodenheimer, Materialien zur Geschichte der Entomologie. 2 vols. Berlin, 1927/28.

C. Singer. A History of Biology. Oxford, 1931. Rev. ed., New York, 1951.

PART III

Sources from the History of Science

(In this brief analysis) " I have attempted to illustrate not only the importance of success, but the significance of failure. In criticising our predecessors we must not forget that we stand on their shoulders, and that we owe as much to their errors as to their wisdom."—F. J. Cole, *The History of Protozoology. London,* 1926. *p.* 57.

" Thus science gropes from error to error, and the truth of today is only the banality or the error of tomorrow."—F. S. Bodenheimer, *Arch. Gesch. Math. Nat. wiss. u. Technik.* 13. 1931. *p.* 416.

1

FROM THE HUMAN ANATOMY AND PHYSIOLOGY OF THE ANCIENT EGYPTIANS

The following quotations are based mainly upon translations of medical papyri dating from the period of 1900 to 1250 B.C., mainly from the so-called Papyrus Ebers. The following are English versions of German translations quoted in the literature, which should be referred to for further information. Literature:—(A) H. Grapow, Ueber die Anatomischen Kenntnisse der altaegyptischen Aerzte. *Hinrichs, Leipzig*, 1935. 30 pp. (B) H. Grapow, Grundriss der Medizin der alten Aegypter. I. Anatomie und Physiologie. *Berlin, Akad. Verl.* 1954. 102 pp.

(*a*) The apophysis of the lower jaw is the end of the lower jaw; the lower jaw ends against the zygomatic temporal bone, like the claw of a [certain bird] catching something. The lower jaw is [also] kept in position by the sinew which is ' bound ' [when the] *mt* (read: musculature) at the end of the lower jaw is stiff. (A, p. 12).

(*b*) The air [for respiration] enters the nose; it enters the heart and the lung; they [i.e. heart and lung] give [it] to the whole stomach [here, as always, sensu latissimo]; (Pap. Ebers 99, 12-13; B, p. 38).

(*c*) Hunger is in the stomach (s.l.), thirst in the lips (B, p. 40).

(*d*) [It is so that] heart and tongue have power over all organs, according to the theory that the heart is in every body and the tongue in every mouth, of all gods, men, and animals. The heart thinks everything it wants, and the tongue orders everything it wants. The seeing of the eyes, the listening of the ears, the inspiration of the nose, they [all] give information to the heart. It is the heart which produces all understanding, and it is the tongue which repeats what the heart has thought. Thus, every function is performed and every work: the creation of the hands, the going of the feet, the motion of all limbs, according to this order [of the heart]. (Pap. Ebers, 3, transl. H. Junker; B, p.67).

(*e*) The heart is in its place, i.e. the mass of fat (?) surrounding the heart is on the left side. Therefore it [the heart] cannot rise upwards, it cannot descend downwards, but remains in its place (A, p. 14; B, p. 68; Pap. Ebers 101, 15).

(*f*) There are vessels of it [the heart going] to every organ. And if the physician lays his hands and fingers on the head, on the occiput, on the hands, on the place of the heart, on the arms or on the legs, he feels something of the heart; because its vessels go into every organ and this is the reason that it ' speaks ' in the ends of the vessels of every limb (A, p. 15).

(*g*) [In the interior there is one single vessel], a vessel called the ' receiver ' [read: aorta], which gives water to the heart, and according to others, to the entire body.

Four vessels go to the liver, giving it water and air, and cause many diseases to appear in the liver because of the excess of blood.

Four vessels go to the lung and to the spleen, giving them water and air.

Two vessels go to the two testes, giving them the semen.

Two vessels go to the bladder, giving it urine.

Two vessels go to the buttocks, one to the one, the other to the other.

There are four vessels opening to the anus, letting it produce water and air. The anus communicates with every vessel, to the right and to the left side, in the arms and in the legs, and they are filled with excrements. (A, p.17).

It is quite obvious that some beginnings of anatomical knowledge were in the possession of the ancient Egyptians. We may assume, that the knowledge witnessed by these papyri is much older than the time of their writing. The weakest point, of course, is the interpretation of the vessels. The term *mt* refers to every hollowed tube, filled with liquid, air or excrement. The bowels are thus counted as vessels. Still less satisfactory is the physiological knowledge. The conception of the heart as the organ of perception and as giving the orders to all other organs may be the source of Aristotelian physiology. Remarkable, in contrast, is the excellent empirical knowledge of the pulse all over the body, apparently of importance for practical diagnosis. The taking out of the organs of the trunk during the ritual preparations for mummification facilitated, of course, a certain anatomical knowledge, based upon human organs.

2

THE HAR-RA=HUBULLU OF ASHUR
(9TH CENTURY B.C.).

The Har-ra=Hubullu is a bilingual Sumero-Accadian lexicon in cuneiform letters. The plates XI to XV contain in systematic enumeration the wild and domestic animals of the air, of the water and of the land. The prefixes of the Sumerian names, which were written but not spoken, each indicated a zoological group. Thus, we have in these plates a complete system of the zoology of the Sumerians. Plate XIV contains 409 names, mainly of wild terrestric animals of all classes. This part of the Har-ra=Hubullu can be regarded as the oldest book on zoology. It has been edited and translated into German by B. Landsberger in Abhdl. phil.-hist. Kl. Saechsische Akad. Wiss. 1934, no. 6. Our numbers 138-165 correspond to Landsberger pp. 10-13.

No.	*Sumerian Name.*	*Translation.*	*Akkadian Name.*	*Translation.*
138.	nu. um. ma	jackal	zi-i-bi	jackal
139.	ur. idim. ma	wild dog	zi-i-bi	jackal
140.	ur. bi. ku	the omnivorous	zi-i-bi	jackal
141.	ur. b i. ku	the omnivorous	a-ki-lu	the devourer (jackal)
142.	udu. idim	wild sheep	bi-ib-bu	wild sheep
143.	sheg	(great) wild sheep	a-tu-du	great wild sheep
144.	sheg. bar	foreign gr. w. sheep	sh/sap-pa-ri	foreign gr. wild sheep
145.	lu. lim	deer	lu-lim-mu	deer
146.	si. mul	star horned	a-a-ra	deer
147.	dara	ibex	tu-ra-khu	ibex
148.	dara. bar	foreign ibex	a-a-lu	deer
149.	dara. mash. da	gazelle-like ibex	na-a-lu	? roe-deer
150.	dara. Khal. khal. la	shy ibex	na-a-lu	? roe-deer
151.	mash	gazelle	ssa-bi-tu	gazelle
152.	mash-da	gazelle	ssa-bi-tu	gazelle
153.	mash-nita	male gazelle	da-ash-shu	male gazelle
154.	amar. mash. da	young gazelle	uz-za-lum	young gazelle
155.	gu. edin. na	young gazelle of the field	an-na-bu	hare
156.	dam. shakh	pig-like	da-bu-u	bear
157.	dim. shakh	pig-like	da-bu-u	bear
158.	ze. ih	pig	sha-hu-u	pig
159.	shakh	pig	sha-hu-u	pig
160.	shakh. tur	piglet	kur-ki-za-an-nu	piglet
161.	shakh. Gish. gi	pig of the reed	shakh-kha-pu	wild boar
162.	shakh. gish. gi. i. Ku.e	pig eating reed	bur-ma-mu	porcupine
163.	shakh. bar. gun. nu	multi-coloured pig	bur-ma-mu	porcupine
164.	shakh. ze. da. bar. shur. ra	?	bur-ma-mu	porcupine.
165.	shakh. ma. gan. na	pig from Magan	ma-ak-ka-nu-u	pig from Magan

3

THE OLD TESTAMENT ON LOCUSTS

The Bible is no book on natural history. Yet, the many mentions of the Desert Locust (*Schistocerca gregaria* Forsk.) form together a most valuable *corpus observationum* on its biology. The many names of it in the Bible have embarrassed many translators. The remarks on the role of eastern and western winds in Egypt (Exodus), the influence of sun radiation on its activity (Nahum) are still today remarkable. The following passages are quoted: Exodus X: 12-16; Leviticus XI: 21-23; Joel I: 2-4. 6-7, 10-12, 18; II: 2-11, 19-20, 24-25; Nahum III: 17.

Exodus X: 12-15, 19. And the Lord said unto Moses: Stretch out thine hand over the land of Egypt for the locusts, that they may come up upon the land of Egypt, and eat every herb of the land, even all that the hail has left. And Moses stretched forth his rod over the land of Egypt, and the Lord brought an east wind upon the land all that day, and all that night; and when it was morning, the east wind brought the locusts. And the locusts went up over all the land of Egypt, and rested in all the coasts of Egypt: very grievous were they; before them there were no such locusts as they, neither after them shall be such. For they covered the face of the whole earth, so that the land was darkened; and they did eat every herb of the land and all the fruit of the trees which the hail had left; and there remained not any green thing in the trees, or in the herbs of the field, through all the land of Egypt . . . And the Lord turned a mighty strong west wind, which took away the locusts, and cast them into the Red sea; there remained not one locust in all the coasts of Egypt.

Leviticus XI: 21-23. Yet these may ye eat of every flying creeping thing that goeth upon all four, which have legs above their feet, to leap withal upon the earth. Even these of them ye may eat: the arbeh after his kind, the ssal'am after his kind, the khargol after his kind and the khagav after his kind. [These four names refer to four stages of the Desert Locust *Schistocerca gregaria* Forsk.]. But all other flying creeping things, which have four feet, shall be an abomination to you.

Joel I: 2-4, 6-7, 10-12, 18, *II:* 2-11, 19-20, 24-25. Hear this, ye old men, and give ear, all ye inhabitants of the land. Hath this been in your days, or even in the days of your fathers? Tell ye your children of it, and let your children tell their children, and their children another generation. That which the gasam has left hath the arbeh eaten; and that which the arbeh hath left hath the yelek eaten; and that which the yelek hath left hath the khassil eaten. . . . For a nation is come up upon my land, strong, and without number, whose teeth are the teeth of a lion, and he hath the cheek teeth of a great lion. He has laid my vine waste, and barked my fig tree; he hath made it clean bare, and cast it away; the branches thereof are made white . . . The field is wasted, the land mourneth; for the corn is wasted; the new wine is dried up, the oil languisheth. Be ye ashamed, O ye husbandmen; howl, O ye vinedressers, for the wheat and for the barley; because the harvest of the field is perished. The vine is dried up, and the fig tree languisheth; the pomegranate tree, the palm tree also, and the apple tree, even all the trees of the field, are withered; because joy is withered away from the sons of men . . . Now do the beasts groan! the herds of cattle are perplexed, because they have no pasture; yea, the flocks of sheep are made desolate.

A day of darkness and of gloominess, a day of clouds and of thick darkness, as the morning spread upon the mountains; a great people and a strong; there hath not been ever the like, neither shall be any more after it, even to the years of many generations. A fire devoureth before them; and behind them a flame burneth; the land is as the garden of Eden before them, and behind them a desolate wilderness; yea, and nothing shall escape them. The appearance of them is as the appearance of horses; and as horsemen, so shall they run. Like the noise of chariots on the tops of mountains shall they leap, like the noise of a flame of fire that devoureth the stubble, as a strong people set in battle array. Before their face the people shall be much pained; all faces shall gather blackness. They shall run like mighty men; they shall climb the wall like men of war; and they shall march every one on his way, and they shall not break their ranks. Neither shall one thrust another; they shall walk every one in his path; and when they fall upon the sword, they shall not be wounded. They shall run to and fro in the city; they shall run upon the wall, they shall climb up upon the houses; they shall enter in at the windows like a thief. The earth shall quake before them ; the heavens shall tremble; the sun and the moon shall be dark, and the stars shall withdraw their shining.

And the Lord shall utter his voice before his army: for his camp is very great; for he is strong that executeth his word; for the day of the Lord is great and very terrible; and who can abide it? . . . But I will remove far off from you the northern army, and will drive him into a land barren and desolate, with his face toward the east sea, and his hinder part toward the utmost sea, and his stink shall come up, and his ill savour shall come up, because he hath done great things . . . And the floors shall be full of wheat, and the vats shall overflow with wine and oil. And I will restore to you the years that the arbeh has eaten, the yelek, and the khassil and the gasam, my great army which I sent among you.

Nahum III: 17. Thy crowned are as the locusts, and thy captains like the gov which camp in the hedges in the cold day, but when the sun ariseth they flee away, and their place is not known where they are.

4

HANNO OF CARTHAGE (500 B.C.)

The Carthaginian sailor Hanno describes in his *Periplus* his sailing through the straits of Gibraltar southwards along the west coast of Africa to a point which was probably the mouth of the Gabun River. There he discovered, before his return, the Gorilla.

The latter history of our knowledge on the Gorilla underlines strongly the greatness of this discovery of the *Periplus.*

"The third day after our depart we crossed the Tropic of the Cancer and arrived in a bay called the Southern Horn. In this port is an island with a lake on it; in this lake is another island full of wild men. Most of them were females with hairy body called gorillas by the interpreters. We pursued them, but did not succeeed to capture any male as these climbed along the precipes and defended themselves with rocks. But these female (gorillas) did bite and scratch those who captured them and did not want to follow us. So we killed them, took their hides and brought these to Carthage; as we had no further provisions we had to return home from there."

The next report is from a British sailor of the Elisabethan period, who was made prisoner in Angola by the Portuguese (published: Bartlett, 1625):

" The greatest of these monkeys is called by them Pongo, the smaller one Engecog. The Pongo resembles a man in all its proportions, a giant man with heavy eye-brows. His body is loosely covered by brown hair. It moves on its legs and holds its hands crossed over the neck when it moves on the soil. The Pongos sleep in the trees and cover them as protection against the rain. They feed upon fruits which they find in the forest and upon coco-nuts, but do not eat meat. They do not speak and understand more than animals. When men sleep in the forest they come to their fire places and sit there until the fire extinguishes, as they have not the intelligence to put in fresh wood. They move in big flocks and have killed more than one negro working in the forest. Sometimes they attack the elephants which come to feed on their habitats and beat them strongly with their closed fists and with sticks, that the elephants grumbling leave the place. These Pongos can never be taken alive, as ten men are unable to master the strength of one Pongo. But young animals are killed by the natives with poisoned arrows . . . When they die in company of others, these cover the corpse with branches and leaves."

The first authentic reports came from Paul B. du Chaillu in 1855. In 1876 the first living Gorilla was brought to Europe. Garner (1894/5) studied their ' language.' And Carl Akeley has devoted his life since 1922 to the study of the Gorilla.

5

HERODOTUS OF HALIKARNASSOS (485-425 B.C.)

Herodot, the Father of History, describes in his famous *Historia* many historical, geographical and natural history observations from his travels which led this keen observer throughout the Middle East. We give here his description of the crocodile in Egypt. The English translation is by H. Cary, London, 1891. pp. 104-105.

68. The following is the nature of the crocodile: During the four coldest months it eats nothing, and though it has four feet, it is amphibious. It lays its eggs on land, and there hatches them. It spends the greater part of the day on the dry ground, but the whole night in the river; for the water is then warmer than

the air and dew. Of all living things with which we are acquainted, this, from the least beginning, grows to be the largest. For it lays eggs little larger than those of a goose, and the young is at first in proportion to the egg; but when grown up it reaches a length of 17 cubits, and even more. It has the eyes of a pig, large teeth, and projecting tusks, in proportion to the body; it is the only animal that has no tongue; it does not move the lower jaw, but is the only animal that brings down its upper jaw to the under one. It has strong claws, and a skin covered with scales, that cannot be broken on the back. It is blind in the water, but very quick-sighted on land; and because it lives for the most part in the water, its mouth is filled with leeches. All other birds and beasts avoid him, but he is at peace with the trochilus, because he receives benefit from that bird. For when the crocodile gets out of the water on land, and then opens its jaws, which it does most commonly towards the west, the trochilus enters its mouth and swallows the leeches: the crocodile is so well pleased with this service that it never hurts the trochilus.

69. With some of the Egyptians crocodiles are sacred; with others not, but they treat them as enemies. The inhabitants of Thebes and Lake Moeris consider them as very sacred. Each of them train up a crocodile which is taught to be quite tame; and they put crystal and gold ear-rings into their ears, and bracelets on their fore paws; and they give them appointed and sacred food, and treat them as well as possible when alive, and when dead they embalm them, and bury them in sacred vaults. But the people of Elephantine eat them, not considering them sacred. They are not called crocodiles by the Egyptians, but 'champsae'; the Ionianes called them crocodiles, as they thought they resembled lizards, which also are called crocodiles and which are found in the hedges in their country.

70. Many and various are the modes of taking the crocodile, of which I shall report only the most noteworthy ones. When the fisherman has baited a hook with the chine of a pig, he lets it down into the middle of the river and holding a young live pig on the brink of the river, beats it; the crocodile, hearing the noise, goes in its direction, and meeting with the chine, swallows it; but the men draw it to land; when it is drawn out on shore, the hunters first plaster its eyes with mud and manage it afterwards quite easily; but until he has done this, he has a great deal of trouble.

6

HIPPOKRATES OF KOS (460-370 B.C.).

Hippokrates was the head of the most important Koan school of medicine. All the books from which we quote are either from him or from his school. Apart from many classical descriptions of diseases and their healing and prevention by diet and a proper life, we find many discussions on important biological problems. Our selections are:

(*a*) The famous oath of the physicians who graduated from this school.

(*b*) ' The Sacred Disease ' (a treatise on epilepsy, is called by Singer the *Magna Charta* of Science, as it puts observation above magic).

(*c*) In the ' Book of the airs, the waters and the places ' the first fundaments of ecology are laid, in application to health and disease.

(*d*) The production and the heredity of the longheaded nation gives opportunity to discuss Hippokrates' views on heredity.

(*e*) The observations on the development of the chicken in the egg were probably written by the son-in-law of Hippokrates.

We follow the English translation by G. F. Adams in: The Genuine Works of Hippokrates. 2 vol. London, 1849. Sydenham Soc. 872 pp. No. (d) seems to be regarded as pseudo-hippokratic. In D. J. F. K. Grimm, *Hippokrates Werke.* German translation Vol. I, p. 425-426. Altenburg, 1781.

The Medical Oath. I swear by Apollo Physician and Aesculapius and Hygeia and Panacea and all the gods and godesses, making them my witnesses, that I will fulfil according to my ability and judgement this oath and this covenant:

To hold him who has taught me this art as equal to my parents and to live my life in partnership with him, and if he is in need of money to give him a share of mine, and to regard his offspring as equal to my brothers in male lineage and to teach them this art—if they desire to learn it—without fee and covenant; to give a share of precepts and oral instruction and all the other learning to my sons and to the sons of him who has instructed me and to pupils who have signed the covenant and have taken an oath according to the medical law, but to none else.

I will apply dietetic measures for the benefit of the sick according to my ability and judgement; I will keep them from harm and injustice. I will neither give a deadly drug to anybody, if asked for it, nor will I make a suggestion to this effect. Similarly I will not give to a woman an abortive remedy. In purity and holiness I will guard my life and my art. I will not use the knife, not even on sufferers from stone, but will withdraw in favor of such men as are engaged in this work.

Whatever houses I may visit, I will come for the benefit of the sick, remaining free of all intentional injustice, of all mischief and in particular of sexual relations with both female and male persons, be they free or slaves. What I may see or hear in the course of the treatment or even outside of treatment in regard to the life of men, which on no account one must spread abroad, I will keep to myself, holding such things shameful to be spoken about.

If I fulfill this oath and do not violate it, may it be granted to me to enjoy life and art, being honored with fame among all men for all time to come; if I transgress it and swear falsely, may the opposite of all that be my lot.

(*b*) *The Sacred Disease.* It seems to me that the so-called Sacred Disease (epilepsy) is no more divine than any other. It has a natural cause, just as other diseases have. Men call it divine merely because they do not understand it. But if they called everything divine which they do not understand, there would be no end of divine things!

Those make a to-do about such things as being due to the gods appear to me, therefore like certain magicians who pretend to be very religious and to know what is hidden from others. If you watch these fellows treating the disease, you will see them use all kinds of incantations and magic, but they are also very careful in regulating diet. Now if food makes the disease better or worse, how can they say it is the gods who do this? Even in saying such a thing, they show impiety and suggest that there are no gods.

The fact is that this invoking of the gods to explain diseases and other natural events is all nonsense. It does not really matter whether you call such things divine or not. In nature all things are alike in this, that they can all be traced to preceding causes. Shall we say then that they are divine or not divine? Since they are all alike in this respect, it is really only a matter of words.

(*c*) *On airs, waters, and places.* 1. Whoever wishes to investigate medicine properly, should proceed thus: in the first place to consider the seasons of the year, and what effects each of them produces; for they are not all alike, but differ much from themselves in regard to their changes. Then the winds, the hot and the cold, especially such as are common to all countries, and then such as are peculiar to each locality. We must also consider the qualities of the waters, for as they differ from one another in taste and weight, so also do they differ much in their qualities. In the same manner, when one comes into a city to which he is a stranger, he ought to consider its situation, how it lies as to the winds and the rising of the sun; for its influence is not the same whether it lies to the north or to the south, to the rising or to the setting sun. These things one ought to consider most attentively, and concerning the waters which the inhabitants use, whether they be marshy and soft, or hard, and running from elevated and rocky situations, and if saltish and unfit for cooking; and the ground, whether its be naked and deficient in water, or wooded and well watered, and whether it lies in a hollow, confined situation, or is elevated and cold; and the mode in which the inhabitants live, and what are their pursuits, whether they are fond of drinking and eating to excess, and given to indolence, or fond of exercise and labour, and not given to excess in eating and drinking.

2. From these things he must proceed to investigate everything else. For if one knows all these things well, or at least the greater part of them, he cannot miss knowing, when he comes into a strange city, either the diseases peculiar to the place, or the particular nature of the common diseases, so that he will not be in doubt as to the treatment of the diseases, or commit mistakes, as is likely to be the case provided one had not previously considered these matters. And in particular, as the season and the year advances, he can tell what epidemic diseases will attack the city, either in summer or in winter, and what each individual will be in danger of experiencing from the change of the regimen. For knowing the changes of the seasons, the risings and settings of the stars, how each of them takes place, he will be able to know beforehand what sort of a year is going to ensue. Having made these investigations, and knowing beforehand the seasons, such a one must be acquainted with each particular, and must succeed in the preservation of health, and be by no means unsuccessful in the practice of his art. And if it shall be thought that these things belong rather to meteorology [then a part of astronomy],

it will be admitted on second thoughts, that astronomy contributes not a little, but a very great deal, indeed, to medicine. For with the seasons the digestive organs of men undergo a change.

3. But how each of the aforementioned things should be investigated and explained, I will now declare in a clear manner . . .

(*d*) *Concerning the people with elongated heads.* There is a people with pointed and elongated heads. In the beginning this head was prescribed by law. Later nature has helped, as those having such heads were regarded as the most noble ones. This habit has the following history: As soon as a child was born, they (the parents) formed the still soft head of the neonate with their hands and forced it by bandages and other suitable means to change its [natural] globular shape, and pressed it into an elongate form and growth. This law had the after effect of becoming nature, and in due time no force was anymore needed [to obtain this effect]. The semen concentrates from all parts of the body, from normal and abnormal, from healthy and diseased parts, according to the qualities of their parents. Thus bald parents have usually bald, blue-eyed offspring, crippled parents crippled offspring. What should interfere with this law in order to prevent that elongate-headed children should be born from elongate-headed parents?

(*e*) *Development of the bird egg and of the human embryo.* If you breed twenty or more eggs under two or more hens, take every day away one of these eggs from the second to the last day of their breeding, and open it for study, then you will find that all my [previous] observations [concerning the development of the human embryo] are true, and that therefore you are justified to compare the development of both. You will discover that what I have described before from the human embryo is true also in the bird egg, namely, that the envelopes spread out from the navel, etc. If somebody has not yet seen this, he will be astonished when he sees the navel in the bird egg.

7

XENOPHON OF ATHENS (435-354 B.C.).

Xenophon described in his *Anabasis* the march of a Greek army through Anatolia and N. Irak, giving an excellent description of animal life in the latter region. Much neglected is *Cynegetica*, the

Hunting with Dogs, where, however, we find a chapter on the hare which even today rouses attention and respect for this careful observer. Apart from the Greek text we have used the English translation of J. S. Watson of Xenophon's Minor Works. London, 1886. pp. 342-347. *Xenophon. Cynegetica*, ch. V, 9-33 (*abbreviated*).

When the hare wishes to settle, it makes a nest, usually in warm spots when it is cold, in shady spots when it is hot, in sunny places in spring and autumn . . . As the hare reclines, it draws the inner side of the thighs under its flanks, puts as a rule the outstretched forelegs together, letting rest the shin on the feet and spreads the ears over the shoulder-blades. Thus it covers the soft parts of the neck which are also protected by the thick and soft hair. When the hare is awake, it winks with the eyelids; but when it is asleep, the eyelids are raised and fixed, and the eyes continue to remain unmoved; also, while asleep, it moves its nostrils frequently, but when not asleep, less often. When the vegetation sprouts, cultivated crops attract the hare more than the hills . . . It is so prolific an animal, that when the female has brought forth (i.e. is pregnant), she is ready to bring forth again, and may at that same time conceive a third brood. The young hares give a stronger scent than the full-grown, as they drag their still weak limbs all along the ground . . . The hare when discovered and pursued by the dogs, may cross over brooks and double, or slink away into clefts and tortuous hiding places. They fear not only dogs, but also eagles, and when they pass over flat slopes and open landscape, they are frequently carried off by them, as long as they are less than a year old. The older ones the dogs pursue and carry off . . . There are two kinds of hares: some are large and blackish with much white in their face, while others are smaller and yellowish with little white. Some have the tail varied with rings of various colours, others have a white streak along its sides. Some have greyish, others bluish eyes. The black spot at the tips of the ears is big in some, small in others. The smaller hare is mostly found on islands where they abound . . . as on most islands no foxes live nor eagles nest . . . For many reasons the hare has a weak sight. It has prominent eyes with small eyelids which give no protection to the eyeballs. Hence, their sight is dim and dispersed. The much sleeping of the hare does not help vision, and its great swiftness tends greatly to trouble it . . . If the hare would run straightway, it would only seldom meet with such mishaps; but as it winds about and is attracted to its birth place, it is finally captured. It is rarely captured by the swiftness of the dogs, more often by chance, as no animal of equal size equals the hare in swiftness.

Such is the nature of the parts which compose its body: the head is light, small, facing downwards and narrow in the forepart; the neck is slender, round, not stiff and of a proper length; the shoulder-blades are straight, and not contracted at the top; the legs are joined to them, light and well attached; the breast is not heavy with flesh; the sides light and symmetrical; the loins agile; the hams fleshy; the flanks yielding and sufficiently loose; the hips round, full everywhere, and separated above by a proper interval; the thighs long, of due thickness, tense on the outside, and not turgid within; the hind-legs long and firm; the fore-feet extremely flexible, narrow and straight; the hind ones firm and broad; all the feet caring nothing for rough ground; the hind-legs much thicker than the forelegs, and bending a little outwards; the hair short and light. Accordingly to these parts the hare is strong, agile, and very nimble. For its nimbleness we have many proofs. When it goes along quietly, it proceeds by leaps, and nobody has seen it or will see it walking; but putting the hindfeet in advance of the forefeet, and on the outside of them, it jumps forward. This is plainly seen by the traces which it leaves in the snow. Its tail is not very helpful for speed, as it is short and cannot steer the body; it steers its body by the alternate movement of its ears, which it continues, even when heavily pursued by dogs; for, lowering one ear, and turning it obliquely on that side on which it is threatened with annoyance, it first sways itself in that direction, and then turns off suddenly in the other, and leaves its pursuers behind in a moment.

The hare is so pleasing an animal, that no one who sees it, whether when it is tracked and discovered, or when it is pursued and caught, would not forget whatever other object he admired.

8

PLATON OF ATHENS (427-347 B.C.).

Platon, with his desire for observation and his a *prioristic* acceptance of preconceived ideas was inimical to science. Where- and when-ever in later periods Platonism was strong, it created an atmosphere unfavourable for the development of biology. *Timaeus* is the only one of his books devoted to nature, trying to establish a mythological homology between the human micro-cosmos and the macro-cosmos of the universe. We bring here his theory of vision (Timaeus 45b-46a), using the English translation of T. Taylor (1870), R. D. Archer (1888) and F. M. Cornford (1948).

First of the organs they fabricated the eyes to bring us light, and fastened them there for the reason which I will now describe. Such fire as has the property, not of burning, but of yielding a gentle light, they contrived should become the proper body of each day. For the pure fire within us is akin to this, and they caused it to flow through the eyes, making the whole fabric of the eyeball, and especially the central part (the pupil), smooth and close in texture, so as to let nothing pass that is of coarser stuff, but only fire of this description to filter through pure by itself. Accordingly, whenever there is a daylight round about, the visual current issues forth, like to like, and coalesces with it and is formed into a single homogeneous body in a direct line with the eyes, in whatever quarter the stream issuing from within strikes upon any object it encounters outside. So the whole, because of its homogeneity, is similarly affected and passes on the motions of anything it comes in contact with or that comes into contact with it, throughout the whole body, to the soul, and thus causes the sensation we call seeing.

But when the kindred fire (of daylight) has departed at night-fall, the visual ray is cut off; for issuing out to encounter what is unlike it, it is itself changed and put out, no longer coalescing with the neighbouring air, since this contains no fire. Hence it sees no longer, and further induces sleep. For when the eyelids, the protection devised by the gods for vision, are closed, they confine the power of the fire inside, and this dispasses and smooths out the motions within, and then quietness ensues. If this quiet be profound, the sleep that comes on has few dreams; but when some stronger motions are left, they give rise to images answering in character and number to the motions and the regions in which they persist—images which are copies made inside and remembered when we awake in the world outside.

There will now be little difficulty to understand all that concerns the formation of images in mirrors . . . As a result of the combination of the two fires inside and outside and again as a consequence of the formation, on each occasion, at the smooth surface of a single fire which is in various ways changed in form; all such reflections necessarily occur, the fire belonging to the face (seen) coalescing, on the smooth and bright surface, with the fire belonging to the visual ray. Left appears right because reverse parts of the visual current come into contact with reverse parts (of the light from the face seen), contrary to the visual rule of impact.

9

ARISTOTELES OF ATHENS (384-322 B.C.).

Aristotle is called rightly the Father of Biology. His entire philosophy is built upon his vast biological experience. He is the founder of scientific biology. We doubt if any biologist of his valour has appeared unto our day. The following small selections contain descriptions, observation and analysis. We hope that the reading of these few tit-bits will raise the appetite of many readers to read more of his classical and fundamental books. Every rise of biology until the Renaissance was connected with a revival of Aristotle's biological works.

All the translations are according to the standard English translation of the Clarendon Press, Oxford (ed. Ross; the biological volumes being translated by d'Arcy Thompson, Ogle and Platt).

(*a*) On the anatomy of the sea-urchins. Hist. Anim. 530b1—531a7.

(*b*) On the migrations of the fishes. Hist. Anim. 598a1—599a19.

(*c*) On vole outbreaks. Hist. Anim. 580b15—29.

(*d*) The correlation of organs: neck, legs and tail in birds. Part. Animal. 694a18—694b23.

(*e*) On the generation of bees. Gener. Animal. 760a10—760b33. This part contains a most important methodological conclusion.

(*a*) *On sea-urchins.* The urchins are devoid of flesh, and this is a character peculiar to them; and while they are in all cases empty and devoid of any flesh within, they are in all cases furnished with the black formations. Of the several species of urchins one is made use of for food; this is the species in which the so-called eggs, large and edible, are found, both in the big and in the small specimens, even the very small ones containing them. Two other species, the spatangus and the bryssus, are pelagic and scarce. There are also the echinometrae or mother urchins, the biggest of all. A further, small kind has large hard spines, inhabiting the sea at the depth of some fathoms; this is a specific medicine against strangury. Around Torone live white sea-urchins, shells, spines and eggs, which are longer than the common urchin, with spines not large nor strong, but rather limp; and the black formations around the mouth are unusually numerous and communicate with the external duct, but not with one another, so that the animal is quasi divided up by them. The edible urchin moves with greatest freedom and most often, as is indicated by the fact, that they have always something or other upon their spines.

All urchins are supplied with eggs, but in some species the eggs are very small and unfit for food. It is strange, that the urchin has its so-called head and its mouth down below, and a place for the issue of the residuum up above. For the food on which they live lies down below; consequently the mouth has a position well adapted for getting at the food, and the excretion is above, near to the back of the shell. The urchin has, also, five hollow teeth inside, and in the middle of these teeth a fleshy substance serving as a tongue. Next comes the oesophagus, followed by the stomach, which is divided into five parts, and filled with excretion, all the five parts uniting at the anal vent, where the shell is perforated for an outlet. Underneath the stomach, in another membrane, are the so-called eggs, identical in number in all cases, and their number is always odd, to wit five. Up above, the black formations are attached to the starting-point of the teeth, and they are bitter to the taste, and unfit for food. A similar or at least an analogous formation is found in many animals, e.g. in the tortoise, toad, frog, the stromboids, and generally in the molluscs; but the formation varies here and there in colour, and in all cases is altogether uneatable, or more or less unpalatable. In reality the mouth-apparatus of the urchin is continuous from one end to the other, but to outward appearance it is not so, but looks like a horn-lantern with the panes of horn left out. The urchin uses its spines as feet; for it rests its weight on these, and when moving shifts from place to place.

(*b*) *On the migrations of fishes.* Of fishes, some are observed to migrate from the high sea towards the shores and from the shore to the high sea, to avoid the extremes of cold and heat. Fish living near the shore eat better than deep-sea fish, as they have more abundant and better feeding, because wherever the sun's heat can reach, vegetation is more abundant, better in quality, and more delicate, as we see in every ordinary garden . . . Some fishes are found near the shore; others are deep-sea fishes; and others again are found alike in shallow and deep waters. These fishes vary, however, at various localities; thus, the goby and all rock-fish are fat off the coast of Crete. Again, the tunny is out of season in summer, when it is being preyed on by its own peculiar louse-parasites, whilst after the rising of the Arcturus, when the parasites have left it, it comes into season again. Some fish also are found in sea-estuaries, such as most gregarious fishes. The coly-mackerel passes the summer in the Propontis, where it spawns, and winters in the Aegean. The tunny proper, the pelamys and the bonito penetrate into the Euxine in summer and pass the summer there; as do also most fish which swim in shoals with the currents, or congregate in shoals together. And most fish congregate in shoals, and shoal fishes always have leaders.

Fish penetrate into the Euxine for two reasons, and firstly for food; for food is more abundant and better in quality because of the many fresh river-waters that discharge into the sea; and the large fishes of this inland sea are smaller than the large fishes of the great sea. Actually there is no large fish in the Euxine besides the dolphin and the porpoise, and the dolphin is of a small variety; but as soon as you get into the high sea the big fishes are on the big scale. Yet fish also penetrate into this sea in order to breed in recesses which are favourable for spawning, and the fresh and exceptionally sweet water has an invigorating effect upon the spawn. After spawning, when the young fishes have attained some size, the parent fish swim out of the Euxine immediately after the rising of the Pleiads. If winter comes it begins with a south wind, they swim out with more or less deliberation; but when a north wind blows, they swim out with greater rapidity, as that breeze is favourable to their own course. The young fish are caught about that time near Byzantium very small in size, as might be expected from the shortness of their life in the Euxine. The shoals in general are visible both as they quit and enter the Euxine . . .

Tunny-fish swim into the Euxine keeping the shore on their right, and swim out of it with the shore upon their left. They do so apparently due to their weak sight, seeing better with the right eye. During the day shoal-fish continue on their way, but rest and feed during the night. But at moonlight they continue their journey without resting at all. Some people familiar with sea-life assert that shoal-fish at the winter solstice never move at all, but keep perfectly still wherever they may happen to have been overtaken by the solstice and this lasts until the equinox.

Hibernation is observed in fishes as well as in terrestric animals. During winter they conceal themselves in out-of-way places, which they quit in the warmer season. But animals conceal to avoid the extreme heat as well as the extreme cold. Sometimes, a whole genus will thus conceal, in other genera some species do and others not. Thus, the shell-fish all conceal: the purple murex, the ceryx, etc.; but whilst this phenomena is conspicuous in the mobile species, which all hide or are protected by an operculum like that of the land-snails, in the sessile molluscs the concealment is not so clearly observed. They do not go into hiding at one and the same season ; but the snails go in winter, the purple murex and the ceryx for about thirty days at the rising of the Dog-star, and the scallop at about the same period. But most go into concealment when the weather is either extremely cold or extremely hot.

(*c*) *On vole outbreaks.* The rate of propagation of voles in country places, and the destruction that they cause, are beyond all telling. In many places their number is so incalculable that but very little of the cereal-crop is left to the farmer; and so rapidly do they increase that sometimes a farmer will one day observe that the time has come for reaping, and on the next morning, when he comes with the reapers to the field, he finds his entire crop devoured. Their disappearance is unaccountable: in a few days not a vole will there remain. And yet a few days earlier men fail to keep down their numbers by fumigating and unearthing them, and by regularly hunting them and turning in swine upon them; for pigs, by the way, turn up the earths of the voles by rooting with their snouts. Foxes also hunt them, and the wild ferrets in particular destroy them, but they make no way against the prolific increase of the voles and the rapidity of their breeding. When they are on the peak of their abundance, nothing can diminish them except heavy rains; but after those they disappear suddenly.

(*d*) *Correlation of organs.* All this is the necessary consequence of the process of development. For the earthy matter in the body issuing from it is converted into parts that are useful as weapons. That which flows upwards gives hardness or size to the beak; and should any flow downwards, it either forms spurs upon the legs or gives size and strength to the claws upon the feet. But it does not at one and the same time produce both these results, one in the legs, the other in the claws; for such a dispersion of this residual matter would destroy all its efficiency. In other birds this earthy residue furnishes the legs with the material for their elongation; or sometimes, in place of this, fills up the interspaces between the toes. Thus it is simply a matter of necessity, that such birds as swim shall either be actually web-footed, or shall have a kind of broad blade-like margin running along the whole length of each distinct toe. The forms, then, of these feet are simply the necessary results of the causes that have been mentioned. Yet at the same time they are intended for the animal's advantage. For they are in harmony with the mode of life of these birds, who, living on the water, where their wings are useless, require that their feet shall be such as to serve in swimming. For these feet are so developed as to resemble the oars of a boat, or the fins of a fish; and the destruction of the foot-web has the same effect as the destruction of the fins, namely, to put an end to all power of swimming.

In some birds the legs are very long, as they inhabit marshes. Nature makes the organs for the function, and not the function for the organs. As these birds are not meant for swimming, their feet are without webs, and because they live on ground that gives way under the foot, their legs and toes are elongated, and these latter in most of them have an extra number of joints. Again, though all birds have the same material composition, they are not all made for flight; and in these, therefore, the nutriment that should go to their tail-feathers is spent on the legs and used to increase their size. This is the reason why these birds when they fly make use of their legs as a tail, stretching them out behind, and so rendering them serviceable, whereas in any other position they would be simply an impediment.

(*e*) *On the generation of the bees.* Bees are not generated like flies and their likes, but from a kind different but akin to them, for they are produced by the kings . . . The kings resemble the drones in size, and the workers by the possession of a sting . . . There must be some overlapping unless the same kind is always to be produced from each; but this is impossible as then all bees would be kings. The workers are assimilated to them in their power of generation, the drones in size; if the latter had a sting they would also be kings. The difficulty is solved, as the kings are like both kinds at once, having the sting of the workers, the size of the drones. But the kings also must be generated from something. Since it is neither from the workers nor from the drones, it must be from their own kind. The maggots of the kings are produced last and are not many in number.

Thus what happens is this: The kings generate their own kind but also another kind, that of the workers; the workers again generate the drones, but not their own kind . . . And since what is according to nature is always in due order, therefore it is denied to the drones even to generate another kind than themselves. This is just what happens, for though the drones are themselves generated, they generate nothing else, but the process reaches its limit in the third stage. And so beautifully is this arranged by nature that the three kinds always continue to exist and none of them fails, though they do not all generate. Another fact is also natural, that in fine seasons much honey is collected and many drones are produced, but in rainy seasons a large brood of workers. For the wet causes more residual matter to be formed in the bodies of the kings, the fine weather in that of the workers, for being smaller in size they need the fine weather more than the kings do. It is right also that the kings, being made to produce young, should remain within, freed from the labour of procuring necessities, and also that they should be of considerable size, their bodies being formed to bear young; and that the drones should be idle as they have no weapon to fight for the food and as their bodies are slow. But the workers are intermediate between them in size, as this is useful for their work, and they are workers as having to support not only their young but also their fathers. According to this view the workers attend upon their kings, as they are their offspring. Whilst they suffer the kings to do no work as being their parents, they punish the drones as their children, for it is nobler to punish one's children and those who have no work to perform. The fact that the kings being few generate the workers in large numbers seems to be similar to the generation of lions . . . So the kings at first produce a number of workers, afterwards a few of their own kind; thus the brood of the latter is smaller in number than that of the former, but where nature has taken away in number she has made it up again in size.

Such appears to be the truth about the generation of bees, judging from theory and from what are believed to be the facts about them; the facts however, have not yet been sufficiently grasped; if ever they are, then credit must be given rather to observation than to theories, and to theories only if what they affirm agrees with the observed facts.

10

THEOPHRASTOS OF ERESOS (380-287 B.C.).

Theophrast, the most gifted pupil and successor of Aristotle, is the Father of Botany. In his books 'History of Plants' and 'Causes of Plants' we find much good observation and an open eye for general problems. The following selections have been translated with the use of Loeb's Classical Library (edition of the History of Plants):

(*a*) The terebinth (*Pistacia*; Hist. III: 14: 3 and 4).
(*b*) On the importance of position and climate for plants. (Hist. IV: 1).

(*a*) *The terebinth.* The terebinth has a 'male' and a 'female' form. The male is barren, which is why it is called male; the fruit of one of the female forms is red from the first and as large as an unripe lentil; the other produces green fruit which subsequently turns red, and ripening at the same time as the grapes, becomes eventually black and is as large as a bean, but resinous and somewhat aromatic. About Ida and in Macedonia the tree is low, shrubby and twisted, but in the Syrian Damascus, where it abounds, it is tall and handsome; indeed they say that there is a certain hill which is covered with terebinths, though nothing else grows on it. It has tough wood and strong roots which run deep, and the tree as a whole is impossible to destroy. The flower is like that of the olive, but red in colour. The leaf is made up of a number of leaflets, like bay leaves, attached in pairs to a single leaf-stalk. So far it resembles the leaf of the sorb; there is also the extra leaflet at the tip: but the leaf is more angular than that of the sorb, and the edge resembles more the leaf of the bay; the leaf is glossy all over, as is the fruit. It bears also some hollow bag-like growths, like the elm, in which are found little creatures like gnats; and resinous sticky matter is found also in these bags; but the resin is gathered from the wood and not from these. The fruit does not discharge much resin, but it clings to the hands, and, if it is not washed after gathering, it all sticks together; if it is washed, the part which is white and unripe floats, but the black part sinks.

(*b*) *Of the importance of position and climate.* The differences between trees of the same kind have already been considered. Now all grow fairer and more vigorous in their proper positions; for wild, no less than cultivated trees, have each their own positions, some love wet and marshy ground, as black poplar, white willow and in general those that grow by rivers; some love exposed and sunny positions; some prefer a shady place. The fir is fairest and tallest in a sunny position, and does not grow at all in a shady one. The silver-fir on the contrary is fairest in a shady place, and not so vigorous in a sunny one.

Thus there is in Arcadia near Krane a low-lying district sheltered from wind, into which they say that the sun never strikes; and in this district the silver-fir excels greatly in height and stoutness, though they have not such close grain nor such comely wood, but quite the reverse,—like the fir when it grows in a shady place. Wherefore men do not use these for expensive work, such as doors or other choice articles, but rather for ship-building and house-building. For excellent rafters, beams and yard-arms are made from these, and also masts of great length which are not however equally strong; while masts made of trees grown in a sunny place are necessarily short but of closer grain and stronger than the others.

Yew pados and joint-fir rejoice exceedingly in shade. On mountain tops and in cold positions odorous cedar grows even to a height, while silver-fir and Phoenician cedar grow, but not to a height,—for instance on the top of Mt. Cyllene; and holly also grows in high and very wintry positions. These trees then we may reckon as cold-loving; all others, one may say in general, prefer a sunny position. However this too depends partly on the soil appropriate to each tree; thus they say that in Crete on the mountains of Ida and on the White Mts. the cypress is found on the peaks whence the snow never disappears; for this is the principal tree both in the island generally and in the mountains.

Again, as has been said already, both of wild and of cultivated trees some belong more to the mountains, some to the plains. And on the mountains themselves in proportion to the height some grow fairer and more vigorous in the lower regions, some about the peaks. It is true of all trees anywhere that with a north aspect the wood is closer and more compact and better generally; and, as a rule, more trees grow in positions facing the north. And trees which are close together grow and increase more in height, and so become unbranched straight and erect, and the best oar-spars are made from these, while those that grow far apart are of greater bulk and denser habit; wherefore they grow less straight and with more branches, and in general have harder wood and a closer grain.

Such trees exhibit nearly the same differences, whether the position be shady or sunny, windless or windy; for trees growing in a sunny or windy position are more branched, shorter and less straight. Further that each tree seeks an appropriate position and climate is plain from the fact that some districts bear some trees but not others (the latter do not grow there of their own accord, nor can they easily be made to grow), and that even if they obtain a hold, they do not bear fruit—as was said of the date palm, the sycamore and others; for there are many trees which in many places either do not grow at all, or, if they do, do not thrive nor bear fruit, but are in general of inferior quality. And perhaps we should discuss this matter, so far as our enquiries go.

11

MARCUS PORCIUS CATO OF ROME (234-149 B.C.)

Cato, the fanatic enemy of Carthage, wrote one of the oldest Roman books on Agriculture (*De re rustica*), from which we quote the description of the seedbeds of cypresses. English translation from Loeb's Classical Library, by Hooper and Ash, De re rust. cap. 151.

As to cypress seed, the best methods for its gathering, planting, and propagation, and for the planting of the cypress bed has been given as follows by Minius Percennius of Nola: The seed of the Tarentine cypress should be gathered in the spring, and the wood when the barley turns yellow; when you gather the seed, expose it to the sun, clean it, and store it dry so that it may be set out dry. Plant the seed in the spring, in soil which is very mellow, the so-called 'pulla' close to water. First cover the ground thick with goat or sheep dung, then turn it with the trenching spade and mix it well with the dung, cleaning out grass and weeds; break the ground fine. From the seed-beds four feet wide, with the surface concave, so that they will hold water, leaving a footway between the beds so that you may clean out the weeds. After the beds are formed, sow the seed as thickly as flax is usually sowed, sift dirt over it with a sieve to the depth of a half-finger, and smooth carefully with a board, or the hands or feet. In case the weather is dry so that the ground becomes thirsty, irrigate by letting a stream gently into the beds; or, failing a stream, have the water brought and poured gently; see that you add water whenever it is needed. If weeds spring up, see that you free the beds of them. Clean them when the weeds are very young, and as often as is necessary. This procedure should be continued as stated throughout the summer. The seed, after being planted, should be covered with straw which should be removed when they begin to sprout.

12

MARCUS TERENTIUS VARRO OF ROME (116-27 B.C.).

Varro's book *De re rustica* II : 2 : 2-5, 13-14 describes the properties of sheep and the breeding of sheep. English translation by Hooper and Ash in Loeb's Classical Library.

I shall speak of the earliest branch of animal husbandry, as you claim that sheep were the first of the wild animals to be caught and tamed by man. The first consideration is that these be in good condition when purchased; with respect to age that they be neither too old nor mere lambs, the latter being not yet and the former no longer profitable—though the age which is followed by hope is better than the one which is followed by death. As to form, sheep should be full-bodied, with abundant soft fleece, with fibres long and thick over the whole body, especially about the shoulders and neck, and should have a shaggy belly also. In fact, sheep which did not have this our ancestors called ' bald ' (apicas), and would have none of them. The legs should be short; and observe that the tail should be long in Italy but short in Syria. The most important point to watch is to have a flock from good stock. This can usually be judged by two points—the form and the progeny; by the form if the rams have a full coating of fleece on the forehead, have flat horns curving towards the muzzle, grey eyes, and ears overgrown with wool; if they are full bodied, with wide chest, shoulders, and hind-quarters, and a wide, long tail. A black or spotted tongue is also to be avoided, for rams with such a tongue usually beget black or spotted lambs. The stock is determined by the progeny if they beget handsome lambs . . . The rams which are to be used for breeding are to be removed from the flock two months ahead, and fed more generously. If barley is fed them on their return to the pens from the pasture, they are strengthened for the work before them. The best time for mating is from the setting of Arcturus to the setting of Aquila (13. V.—23. VII.); as lambs which are conceived after that time grow undersized and weak. As the period of pregnancy of the sheep is 150 days, the birth thus occurs at the close of the autumn, when the air is fairly temperate, and the grass which is called forth by the early rains is just growing. During the whole time of breeding they should drink the same water, as a change of water causes the wool to spot and is injurious to the womb. When all the ewes have conceived, the rams should again be removed, as they are troublesome in worrying the ewes which have now become pregnant. Ewes less than two years old should not be allowed to breed, for the offspring of these is not sturdy and the ewes themselves are injured; and no others are better than the three-year-olds for breeding. They may be protected from the male by binding behind them baskets made of rushes or other material; but they are protected more easily if they are fed apart.

13

GAJUS PLINIUS SECUNDUS MAJOR OF ROME (23-79).

Pliny compiled his famous encyclopaedia *Naturalis Historia* from 20,000 notes of 2,000 books of 200 writers. It remained for many centuries one of the main sources for the knowledge of natural history. We have selected here two chapters of his description of Egypt. For the translation we made use of the edition in Loeb's Classical Library.

(*a*) Of the Ichneumon, the Crocodile and the Trochilus Nat. Hist. Lib. VIII: 35-37.
(*b*) Of the Papyrus and the production of paper. Nat. Hist. Lib. XIII: 21-23.

(*a*) *Of the Ichneumon, the Crocodile, and the Trochilus.* It is impossible to know if nature has given evils or benefits more bountifully to the asp. It has bestowed dim eyes on this accursed creature, not in the forehead for it to look straight in front of it, but in the temples—and consequently it is more quickly excited by hearing than by sight. On the other hand it has given to the asp a dreadful war to the death with the ichneumon.

The ichneumon is also a native of Egypt. It is especially known by the following reason: The ichneumon repeatedly plunges into mud and dries itself in the sun. When it has equipped itself thus with a cuirass of several coatings, it proceeds to the fight. Then it raises its tail and renders the blows it receives ineffectual by turning away from the asp. Finally, watching for its opportunity while holding its head sideways, it attacks its enemy's throat.

Not content with this victim, it vanquishes also the ferocious crocodile. The latter belongs to the Nile, is a curse on four legs, and equally pernicious on land and in the river. It is the only land animal that has no tongue, and the only one that bites by moving down its mobile upper jaw. Formidable is also its row of teeth set close together like a comb. Its size surpasses usually 18 ells. It lays as many eggs as a goose, and by a kind of prophetic instinct incubates them always outside the line to which the Nile in that year is going to rise at its highest flood. No other animal grows from a small beginning to greater size. It is armed with talons and its skin is invincible against all blows. It passes the day on land and the night in the water, in both cases seeking the greater warmth. When this animal is sated with fish and sunk deep in sleep on the shore, its mouth is always full with (remainders) of its food. Thus it invites a small bird, called there trochilus, but the king-bird in Italy, into its open mouth to enable it to feed. First it hops in and cleans out the mouth, and then the teeth and the inner throat, which then it yawns as widely open as possible for the pleasure of its scratching. The ichneumon watches for it to be overcome by sleep in the middle of this pleasant entertainment and darts like a javelin through the throat so opened and gnaws out the crocodile's belly.

(*b*) *On the papyrus and the production of paper.* Before we take leave of Egypt we will talk about the papyrus plant, as human knowledge and memory are connected with the use of paper. At the beginning one wrote on palm leaves, then upon the inner bark of certain trees. Public memorabilities were engraved into plates of lead and wax tablets were used privately. The use of slates were already known before the Trojan war, as we read in Homer.

When Homer wrote Egypt did not even possess its complete extension of today. Just those regions where now the papyrus grows were formed as alluvial deposits of the Nile floods. The Land was then, according to Homer, one day's and one night's journey distant from the island of Pharos near Alexandria . . . When Ptolemaeus prohibited the export of paper, the parchment was discovered in Pergamon . . .

The papyrus plant grows in the swampy plains of Egypt, there, where the Nile does not stream any more and the water is about two ells deep. From an oblique arm-thick root sprouts a three-edged, up to ten ells long stem, growing pointed towards its end. On this end is a leaf-bushel, whose only use is the coronation of statues of the Gods. The roots are used as fuel and for the manufacture of household utensils. From the husk of the stem sails, mats, cloths, covers and ropes are prepared. The stem is consumed crude as well as cooked, but only the sap is swallowed.

Paper is made by cutting the stem by a needle into thin, as broad as possible stripes of the husk. The best part is the middle (of the stem). This paper was called hieratic in ancient times and only used for holy books. Paper which is unfit for writing is used for the wrapping of wares. For the preparation of the paper the stripes of husks are pasted together on a plank which is moistened by water. Then it is pressed and dried in the sun.

The breadth of the sheets of paper is very different. The best are those which are 13 fingers broad. Other factors determining the quality of the paper are its fineness, its density, its whiteness and its smoothness. Unevenness is smoothed by a tooth or by a mussel-shell.

14

DIOSKORIDES OF ANATOLIA (about 60).

Dioscorides is the Father of Pharmacology. His *Thesaurus Plantarum* collects all available knowledge on medicinal plants. It was translated into all languages and remained the standard book of pharmacology until the Renaissance. We render here the description of two herbs from the Hebrew translation of Asaph ha-Ropheh.

(*a*) Meion (*Meum athamanticum*).
(*b*) Kardamon (*Eletteria cardamon*).

(*a*) *The herb Meion* (*Meum athamanticum*). Its leaves resemble those of *Ferula*, and its roots are large. Its odour is pleasant, and the fragrance of its root resembles that of the cedar. Its potency is warm and strong, beneficient to pain of the kidney, relieving the pain and breaking the stone.

The root must be ground and boiled in water till it be reduced to one third. When cooked, if drunk by one whose urine has stopped, it will enable him to urinate. It is highly effective in treating internal afflictions, and all sorts of gases which lodge in the body and the intestine. It will benefit all types of female disorders, pains in the thighs and chest, a phlegmatic cough or a dry one, the lungs and the breathing. It should be ground and mixed with honey, and drunk with hot water till recovery. It will also cure him who spits blood, coughs blood, urinates blood, or the woman whose menstruation is over long, if a shekel of the herb is drunk daily with honey and hot water. It will liberate impeded urination, and for small children who cannot urinate, bind the herb to their bellies, and they will urinate. But he who drinks of it in excess will thereby instigate headaches.

(*b*) *The herb called Kardamon* (*Eletteria cardamon*). Its potency will break the stone which descends from the kidneys, and it will be beneficial if drunk by him whose either side has been immobilized by the affliction of paralysis or whose arms have been immobilized by a loss of strength, these it will cure. It will also cure the acute cough and will cure the fool's misery, shortness of the wind, and breathing, and all stomach aches, familiar or unknown. It will serve to unseat intestinal burning, if drunk with wine. It will benefit the kidneys and stoppage of urination and all types of lethal narcotics, every fatal pestilence, snake or lizard bites, and scorpion stings. If the patients are made to drink, they will be delivered with God's help.

And it will benefit malaria and tuberculosis if mixed with laurel root. Take of this herb and of laurel root in equal parts, and give to drink of it to the diseased in kidney, and it will relieve the pain and break the stone. And if you give of it to a pregnant woman, she will immediately abort. And if the child has died in the womb, it will emerge. And if you bind it to a flame beneath a woman till the fumes enter her womb, then it will abort anything that is within, living or dead. It will benefit any chronic scabies, if ground and kneaded with vinegar and applied to the flesh it will cure impetigo and scabies.

15

GALENOS OF PERGAMON (130-201).

Galen was the greatest anatomist and physiologist of antiquity and was generally recognised as such. A new tradition began only with Vesal and Harvey.

(*a*) On dissection. From A. J. Brook, Greek Medicine. London, 1929.

(*b*) On the forces involved in digestion. From: *De naturalibis facultatibus*, III: 1. Translation from Loeb's Classical Library.

(*a*) *On dissections.* What tent poles are to tents, and walls to houses, so to animals is their bony structure; the other parts adapt themselves to this and change with it. Thus, if an animal's cranium is round, its brain must be the same; or if it is oblong, its brain must also be oblong. If the jaws are small, and the face as a whole roundish, the muscles of these parts will also necessarily be small; and similarly, if the jaws are prominent, the animal's face as a whole will be long, as also the facial muscles. Consequently also the monkey is of all animals the likest to man in its viscera, muscles, arteries, veins and nerves, because it is so also in the form of its bones. From the nature of these it walks on two legs, uses its front limbs as hands, has the flattest breastbone of all quadrupeds, collarbones like those of a man, a round face and a short neck. And these being similar, the muscles cannot be different; for they are extended on the outside of the bones in such a manner that they resemble them in size and form. To the muscles, again, correspond the arteries, veins, and nerves; so these, being similar, must correspond to the bones . . .

First of all, I would ask you to make yourself well acquainted with the human bones, and not to look on this as a matter of secondary importance. Nor must you merely read the subject up in one of the books, such as: ' Osteology ', ' The bones ', or simply ' On bones ', as in my book, which is better, by the way, than any book written before in exactitude as well as in brevity and clearness. Not only read, but learn human osteology with your own eyes by actual observation. This is very easy at Alexandria, where teaching is accompanied by opportunities for personal inspection. Hence, try to go to Alexandria for this as well as for other reasons. If you are unable to go there, it is not impossible to look at human bones. I had often such a chance, when tombs or monuments were broken up. Once a rising river easily disintegrated a tomb, cleaned it of its putrid flesh and carried the skeleton a mile away, where it lay then ready on the shore like prepared for a medical student's inspection. At another time I examined the skeleton of a robber, lying on a mountain-side near the road, where he had been killed by a traveller whom he had attacked. The inhabitants were delighted to see his corpse being eaten by the birds of prey, which cleaned it in two days, leaving the skeleton for the inspection of those enjoying anatomical demonstrations.

Even if you do not have the luck to see anything like this, you still can dissect an ape and learn each of the bones from it, by carefully removing the flesh. For this purpose you must choose the apes which most resemble man . . . When you then meet later with a human skeleton, you will easily recognise and remember everything . . . When apes are not available, be prepared to dissect the bodies of other animals, distinguishing at once in what ways they differ from apes.

(*b*) *On assimilation.* It has been made clear in the preceding discussion that nutrition occurs by an alteration or assimilation of that which nourishes to that which receives nourishment, and that there exists in every part of the animal a faculty which in view of its activity we call, in general terms, alterative, or more specifically, assimilative and nutritive. It was also shown that a sufficient supply of the matter which the part being nourished makes into nutriment for itself is ensured by virtue of another faculty which naturally attracts its proper humour that that humour is proper to each part which is adapted for assimilation, and that the faculty which attracts the humour is called, by reason of its activity, attractive or epispastic. It has also been shown that assimilation is preceded by adhesion, and this again by presentation, the latter stage being, so to say, the end or goal of the activity corresponding to the attractive faculty. For the actual bringing up of the nutriment from the veins into each of the parts takes place through the activation of the attractive faculty, whilst to have been finally brought up and presented to the part is the actual end for which we desired such an activity; it is attracted in order that it may be presented. After this, considerable time is needed for the nutrition of the animal; whilst a thing may be even rapidly attracted, on the other hand to become adherent, altered, and entirely assimilated to the part which is being nourished and to become a part of it, cannot take place suddenly, but requires a considerable amount of time. But if the nutritive humour, so presented, does not remain in the part, but withdraws to another one, and keeps flowing away, and constantly changing and shifting its position, neither adhesion nor complete assimilation will take place in any of them. Here too, then the (animal's) nature has need of some other faculty for ensuring a prolonged stay of the present humour at the part, and this not a faculty which comes in from somewhere outside but one which is resident in the part which is to be nourished This faculty, again, in view of its activity our predecessors were obliged to call retentive.

Thus our argument has clearly shown the necessity for the genesis of such a faculty, and whoever has an appreciation of logical sequence must be firmly persuaded from what we have said that, if it be laid down and proved by previous demonstration that Nature is artistic and solicitous for the animal's welfare, it necessarily follows that she must also possess a faculty of this kind.

16

OPPIANUS OF CILICIA (about 200).

It is still doubtful if the two beautiful Nature eposes of the Middle East: the *Halieutika*, a description of life in the waters of the sea; and the *Cynegetika*, a great hunting epos, are the work of one or of two authors of the same name. We are inclined to the former opinion. Our selections are (see Loeb's Classics):

(*a*) On the fishery of sponges. Hal. II: 2: 3.
(*b*) The hunt of the bear. Cyn. IV: 320 ff.

(*a*) *The Fishing of Sponges.* Than the task of the sponge-cutters I declare that there is none worse nor any work more woeful for men. These, when they prepare themselves for their labour, use more meagre food and drink and indulge themselves with sleep unfitting fishermen. As when a man prepares himself for the tuneful contest and he studies all care and every way takes heed, nursing for the games the melody of his clear voice: so do they zealously take all watchful care that their breath may abide unscathed when they go down into the depths and that they may recover from past toil. But when they adventure to accomplish their mighty task, they make their vows to the blessed gods who rule the deep sea and pray that they ward from them all hurt from the monsters of the deep and that no harm may meet them in the sea. And if they see a Beauty-fish, then great courage comes into their hearts; for where these range there never yet has any dread sea-monster appeared nor noxious beast nor hurtful thing of the sea but always they delight in clean and harmless paths: wherefore also men have named it the Holy fish. Rejoicing in it they hasten to their labours. A man is girt with a long rope above his waist and, using both hands, in one he grasps a heavy mass of lead and in his right hand he holds a sharp bill, while in the jaws of his mouth he keeps white oil. Standing upon the prow he scans the waves of the sea, pondering his heavy task and the infinite water. His comrades incite and stir him to his work with encouraging words, even as a man skilled in foot-racing when he stands upon his mark. But when he takes heart of courage, he leaps into the eddying waves and as he springs the force of the heavy grey lead drags him down. Now when he arrives at the bottom, he spits out the oil, and it shines brightly and the gleam mingles with the water, even as a beacon showing its eye in the darkness of the night. Approaching the rocks he sees the sponges which grow on the ledges of the bottom, fixed fast to the rocks; and report tells that they have breath in them, even as other things that grow upon the sounding rocks. Straightway rushing upon them with the bill in his stout hand, like a mower, he cuts the body of the sponges, and he loiters not, but quickly shakes the rope, signalling to his comrades to pull him up swiftly. For hateful blood is sprinkled straightway from the sponges and rolls about the man, and many a times the grievous fluid, clinging to his nostrils, chokes the man with its noisome breath. Therefore swift as thought he is pulled to the surface; and beholding him escaped from the sea one would rejoice at once and grieve and pity: so much are his weak members relaxed and his limbs unstrung with fear and distressful labour. Often when the sponge-cutter has leapt into the deep waters of the sea and won his loathly and unkindly spoil, he comes up no more, unhappy man, having encountered some huge and hideous beast. Shaking repeatedly the rope he bids his comrades pull him up. And the mighty sea-monster and the companions of the fisher pull at his body rent in twain, a pitiful sight to see, still yearning for ship and shipmates. And they in sorrow speedily leave those waters and their mournful labour and return to land, weeping over the remains of their unhappy comrade.

(*b*) *The Hunt of the Bear.* For bears an exceeding glorious hunt is made by those who dwell on the Tigris and in Armenia famous for archery. A great crowd go to the shady depths of the thickets, skilful men with keen-scented dogs on leash, to seek the mazy tracks of the deadly beasts. But when the dogs descry the signs of footprints, they follow them up and guide the trackers with them, holding their long noses nigh the ground, and afterwards if they descry any fresher track, straightway they rush eagerly, giving tongue the while exultingly, forgetting the previous track. But when they reach the end of their devious tracking and come to the cunning lair of the beast, straightway the dog bounds from the hand of the hunter, pitifully barking, rejoicing in his heart exceedingly. As when a maiden in the season of milky spring roams with unsandalled feet over all the hills in search of flowers and while she is yet afar the fragrance tells her of the sweet violet ahead . . .; even so the stout heart of the dog is gladdened. But the hunter for all eagerness constrains him with straps and goes back exulting to the company of his comrades. He shows them the thicket and where himself and his helper ambushed and left the savage beast. And they hasten and set up strong stakes and spread hayes and cast nets around. On either hand in the two wings they put two men at the ends of the net to lie under piles of ashen boughs. From the wings themselves and the men who watch the entrance they stretch on the left hand a well-twined long rope of flax a little above the ground in such a way that the cord would reach to a man's waist.

Therefrom are hung many-coloured patterned ribbons, various and bright, a scare to wild beasts, and suspended therefrom are countless bright feathers, the beautiful wings of the fowls of the air, vultures, whilt swans and long storks. On the right hand they set ambushes in clefts of rock ; or with green leaves they swiftly roof huts a little apart from one another, and in each hide four men, covering all their bodies with branches. Now when all things are ready, the trumpet sounds its tremendous note, and the bear leaps forth from the thicket with a sharp cry, and looks sharply as she cries. And the young men rush on in a body and from either side come in battalions against the beast and drive her before them. And she, leaving the din and the men, rushes straight, where she sees an empty space of open plain. Thereupon in turn an ambush of men arises in her rear and make a clattering din, driving her to the brow of the rope and the many-coloured scare. The wretched beast is utterly in doubt and flees distraught, fearful of all alike—the ambush of men, the din, the flute, the shouting, the scaring rope; for with the roaring wind the ribands wave aloft in the air and the swinging feathers whistle shrill. So, glancing about her, the bear draws nigh the net and falls into the flaxen ambush. Then the watchers at the ends of the net near at hand spring forth and speedily draw tight above the skirting cord of broom. Net on net they pile; for at that moment bears greatly rage with jaws and terrible paws, and many a time they straightway evade the hunters and escape from the nets and make the hunting vain. But at that same moment some strong man fetters the right paw of the bear and widows her of all her force, and binds her skilfully and ties the beast to planks of wood and encloses her again in a cage of oak and pine, after she has exercised her body in many a twist and turn.

17

EARLY CHINESE TEXTS ON NATURAL HISTORY (290 B.C.—400 A.D.)

We place these few texts which we have selected from a greater choice which Professor J. Needham of Cambridge put at our disposal here, as they belong in chronological order to late antiquity (in the West). We are convinced, however, that most of them go back to much older traditions.

(*a*) Chouang Chou, Chuang Tzu. On the Unity of Nature. From Feng Yo-lan, Translations of Chuang Tzu. Shanghai, 1933.

(*b*) Yang Fu, Records of strange Things. Description of the Banana. From P. K. Reynolds and Fang Lien-Chih, Harv. J. Asiat. Stud. 1940. p. 165.

(*c*) Chi Han, Description of the Flora of the Southern Regions. On the Kan-Orange. From H. S. Reed, A short History of the Plant Sciences. 1942. Waltham, Mass. (refers to the ant *Oecophylla smaragdina*, which still today is in use for the biological control of pests in fruit-gardens in Lignan).

(*d*) Anonymous, Book of Strange Things and Spirits. On Mammouth rests. From B. Laufer, Field Mus. Nat. Hist. Chicago. Anthrop. Leaflets no. 21. 1925.

(*e*) Chu Hsi, Complete Works. On the nature of Fossils. From A. Forke, The World Conception of the Chinese. London, 1928.

(*a*) *Unity of Nature.*

"Tung Kuo Tzu asked Chuang Tzu: Where is the so-called Tao?

Chuang Tzu said: Everywhere.

Q: Specify an instance of it.

A: It is in the ant.

Q: How can the Tao be in anything so lowly?

A: It is also in the wild grasses.

Q: How can it still be lower?

A: It is in that earthenware tile.

Q: How can it still be lower?

A: It is in that dung.

To this Tung Kuo Tzu made no reply.

Chuang Tzu said: Your questioning does not touch the fundamentals of the Tao. You should not specify any particular thing. There is not a single thing without Tao. There are three terms: complete, all-embracing, and the whole. These three names are different, but denote the same reality—all refer to the one thing."

(*b*) *Description of the Banana.* "The Pa-Chiao plant has leaves as large as mats. Its stem is like a bamboo-shoot. After boiling, the stem breaks into fibres and can be used for weaving cloth. Women weavers make this fibre into fine or coarse linen which is now known as Chinchina-linen. The centre of the plant is shaped like a garlic-bulb, and is as large as a bowl. There the fruit grows and holds the stem. One stem bears several tens of fruits. The fruit has a yellowish-red skin coloured like flame, and when peeled the inside pulp is dark. The pulp is edible and very sweet, like sugar or honey. Four or five of these fruits are enough for a meal. After eating the flavour lingers on among the teeth."

(*c*) *The Kan Orange* (*Citrus nobilis*) belongs to the chū class (*C. sinensis*). It has a sweet delicious flavour which is especially remarkable. The deep red are called ' Pot mandarine oranges ' . . . The people of Cochinchina use mat bags in which they store ants and sell them in the market. The nests of these ants are like thin silken floss. The mat bags are attached to the branches and leaves and when the ants are inside, they are removed and sold in the market. These ants are of reddish-yellow colour and larger than ordinary ants. In the southern regions, if the mandarin trees are without these ants, their fruits will be injured by swarms of boring-insects and there will not be one perfect . . .

(*d*) *Mammoth rests.* Anonymous, Shen I Ching (Book of strange Things and Spirits), about 400 A.D. " In the northern regions, where the ice is piled up over a stretch of country 10,000 miles long and reaches a thickness of 1,000 feet, there is a rodent, called chi-shu, living beneath the ice in the interior of the earth. In shape it is like a rodent, and subsists on herbs and trees. Its flesh weighs 1000 pounds and may be used as dry meat for food; it is eaten to cool the body. Its hair is about eight feet in length, and is made into rugs, which are used as bedding and to keep out the cold. The hide of the animal yields a covering for Drums, the sound of which is audible over a distance of a thousand miles. Its hair is bound to attract rats; wherever it is, rats will flock together."

Tulishen, a Manchu envoy, writes of Yenisseisk in his memoirs (1715 A.D.): " In the coldest parts of this northern country is found a species of animal which burrows under the ground and which dies when exposed to sun and air. It is of enormous size and weighs 10,000 pounds. Its bones are very white and bright like ivory. It is not by nature a powerful animal and is therefore not very ferocious. It generally occurs on the banks of rivers. The Russians collect the bones of this animal in order to make cups, saucers, combs, and other small articles. The flesh of the animal is of a very cooling quality, and is eaten as a remedy in fevers. The foreign name of this animal is mo-men-to-wa; we call it chi-shu."

(*e*) *Nature of fossils.* "One frequently sees on high mountains conches and oyster-shells, sometimes embedded in rocks. These rocks in pristine times were earth, and the shellfish and oysters lived in water. Subsequently everyting was inverted; things from the bottom came to the top, and the soft became hard. Careful consideration of these facts will lead to far-reaching conclusions."

18

FROM AN OLD ARAMAEIC PHYSIOLOGUS OF SYRIA (? 5TH CENTURY).

The *Physiologi* (verbal translations: naturalists) originated in the 4th century in the Levant and were the prototypes of the *Bestiarii* of the Middle Ages in Europe. This is one of the oldest texts available. It was translated from the Syriac into German by K. Ahrens (Das Buch der Naturgegenstaende. Kiel, 1892). It is one of the few Physiologi which are restricted to the natural history text, without any theological or moralising addition. We bring here:

(*a*) On the Leopard. Chap. V.
(*b*) On the Cranes. Chap. XXVII.

(*a*) *The Leopard.* The leopard is the friend of all animals, but the dragon's enemy. It is multicoloured and predatory by nature. When it has eaten and is satiated, it sleeps in its hole and does not rise for three days. And when it leaves its hole, it cries loudly; and by the sound of its voice every good smell of the herbs of that place is spread and comes forth. Then, the agreeable voice of the leopard calls the animals and they assemble around it. When you intend to kill it, anoint yourself with the fat of the hyena, enter its hole and kill it; it will not wound you.—It satiates itself only, when it tears the prey; it is light of body and quick in its movements.

(*b*) *On the cranes.* The cranes watch in turn during the night, like sentinels: some sleep and other watch. And when the sentinel has finished the time of its watch, it utters a loud cry and then goes to sleep. After it another rises to repay the protection which it enjoyed from its comrade. The cranes, when they fly,

fly in order, and in their flight they are guided mutually; they wait one for the other, when they fly, and the time is determined, when one advances upon the other in its flight and when another replaces it, and at another time the first will become the last and the last the first one, so that one honours the other to precede it as the guide.—Certain birds remain in the country where they live, but others migrate far away, live during the summer in the cold countries of the north and migrate at the approach of the winter before the snow into southern countries, where they are warmer.

19

FROM ANOTHER OLD ARAMAEIC PHYSIOLOGUS OF SYRIA (? 6TH CENTURY).

This later *Physiologus* contains already long theological interpretations. We render Chap. LVII. The Tale of the Bees. This *Physiologus* was edited by Lund (The Syriac Physiologus. 1775).

57. *Tale of the bees.* Solomon, the sage, said about the bee (Prov. 6: 8 LXX): " It is wise and careful in its work, so that from the food which it collects for its sustenance, the king and the weak gain health."

The Physiologus teaches and says: They live in one group together, one is their exit, one their common labour, together they fly and together they hurry to their work. And, still more important, they are under one prince and prefect, who shares their nature. They do not begin their common work, and do not go out to the meadows and the roots and the flowers, until their king has made the beginning. We say that they have not elected spontaneously their king, but he is imposed upon them by the order and the natural law of the Creator. The prince has a sting, which he does not usually inflict. When, however, he stings, he atones by a punishment for his audacity: When he wounds a man by his sting, the sting remains in the wound, and the king dies in pain.—They collect the honey from the plants by their mouth, the wax from the flowers by their legs. These they bring into the hive and build with wonderful artifice cells of wax. In a skilled way they distribute it in grooves, in which they pour out the honey. The honey is thin at the beginning and diluted. Latter it becomes spicy and mature, after it has been kept for some time in the combs, and is then much better and more compact.—When heavy winds are blowing, whilst the bee is out of the hive, the bee goes down and sits on a stone or on a rock until the wind has abated, and then returns home.

Theory: Let us seize now all that we have learned, compare it and consider it in a spiritual way. What of this illustrates our human mind? The weak bees which live in groups together, leave together and industriously work together, are the souls of the sons of the catholic and apostolic church, who live together in the same church, leave together and care for the common work, which is the excellent divine and sacrosanct service. And as a prince or prefect is set over them by the Creator, who shares their nature, thus the bishops and the sages of human nature have been ordered by God in the church, in order to establish and maintain its existence. And the prince has a sting, which he does not use to inflict injury. This is the liberty which God gave to the bishop and to all his subjects. But the good bishop does not inflict injury with his sting. By his liberty he preserves sanctity and does not make abuse of it. He is set by God as a bishop, to be a good example and a model. But when somebody with stupid reasoning deviates from the orders of the bishop and errs, and inflicts injury on somebody in his liberty, he dies from the crime and injury which he has done. The bees collect the honey by their mouth from the plants, and the wax by the legs from the flowers, and the bee brings it (to the hive) and distributes the honey whilst it is thin. Thus, the saints of the holy church of God pasture on the holy plants and on the most holy flowers of the holy books and collect the nourishment of spiritual sweetness.

20

TIMOTHEOS OF GAZA (about 500).

Timothy was Professor at the University of Gaza. His *Peri Zoon*, On Animals, is a compendium of what the educated gentleman should know about animals. Only an extract is preserved, from which we translate Chapter V, On the Fox. (From F. S. Bodenheimer and A. Rabinowitz. Leiden, 1949. pp. 20-21).

On the fox. That (when) hungry it stretches itself as (if it were) dead in a lonely spot, and when the birds gather to eat (it) up, suddenly jumping up, it catches one of them and devours it. The sea-frog and the torpedo do the same . . . That it makes seven apertures to its earth, and being pursued by the hounds or by men, it moves from one (aperture) to the other and escapes.—That it is never tamed.—That it knows how the wolf fears the squill and it sleeps underneath it in order not to be hurt by the wolf.—That it fears the bile of the chameleon . . .—That it, rolling on seed-land, causes the place where it rolls not to produce (anything henceforth); therefore, ' alopecia ' is called (the illness of (people) losing their hairs or not growing (any).—That it passes over the frozen Ister, called Danube by the Romans; and when it perceives that (the Danube) is about to thaw, it flees and indicates to the (animals) living on the land to flee likewise, but to the sailors to get ready; if it does not flee, this indicates that the river remains frozen.—That it beguiles the hunting hounds through a wind of its stomach, leading (them) astray and wagging them into intercourse; for (the vizen) having mated with a dog bears the so-called alopos.—But if (the fox) is a male one, and the dog a female, (then) the Laconian dog is born, (just) as from the mating of a dog and a tiger the Indian dog is born.

21

THE GEOPONICA, THE AGRICULTURAL PURSUITS OF BYZANZE (about 6th Century).

This book is a mixture of Hellenistic superstitions and popular tradition, mixed with notes on Oriental amulets, etc., old empiry and good practical judgements of experienced agriculturists. We follow here the translation of T. Owen. London. 2 vols. 1805, 1806.

(*a*) Concerning Field Mice. XIII: 5.
(*b*) Concerning Peacocks. XIV: 18.
(*c*) Concerning the Propagation of Fish. XX: 1.

(*a*) *Concerning Field Mice.* Apuleius recommends to smear seeds with ox gall, and the mice will not touch them; but it is better to pound in the dog-days, the seed of hemlock with hellebory, and to mix it with barley meal; or seed of the wild cucumber, or of the hyoscyamus, or of bitter almond, with black hellebore, and to mix it with an equal quantity of barley-meal, and to mix it up with oil, and to lay it near the holes of the field-mice; for when they eat it, they die. But persons in Bithynia who have tried the experiment, stop the holes with rhododaphne (rhododendrum or nerium), so that they, endeavouring to get out, gnaw it, and thus they perish.

Take some paper and write these words on it: " I adjure the mice taken in this place, that you do me no injury yourselves nor suffer another to do it: for I give you this ground (and you mention which); but if I again take you on this spot, I take the mother of the Gods to witness, I will divide you into seven parts." Having written these words, fasten the paper in the place where the mice are, before the rising of the sun, to a stone of spontaneous production, and let the stone be turned externally. This is written by me, that I may not seem to omit any thing; but I do not receive all these things, far be it from me, and I advise all to do the same, so as not to have recourse to any ridiculous things of this kind.

(*b*) *Concerning Peacocks.* Peacocks are chiefly bred in factitious islands (Palladius: then they are secure from foxes): but let the place have abundant plenty of grass, and an orchard: and you are to separate those of a generous breed from those that are weak; for those that are strong oppress those that are feeble. The hens indeed when they are three years old, breed; but they that are younger, either do not hatch, or do not feed the young fowls. You are also to give peacocks for food, during the winter, beans parched on a coal fire, and before their other food, six cyathi (measures) to each bird; and you are to set clean water for them, for they will thus be more prolific; and you are to spread hay or straw in the house for them that lay, that the eggs, when they drop, may not be broken; for they drop their eggs standing, and they do this twice a year, but they have not more than 12 eggs in all. But it is proper to set the eggs when the moon is nine days old, nine in the whole, five of its own, and four of the domestic fowl: and you must take away those of the domestic fowl in the tenth day, and set others, that the hens eggs may be hatched on the thirtieth day with those of the pea fowl. It is not proper indeed to give the young brood, that is hatched, food the first two days; but on the third day we carry them barley-meal made up with wine and bran dressed and boiled, and the tenderest leaves of leeks pounded with green cheese. But let barley be given them after six months.

(*c*) *Concerning Propagation of Fish.* Fish ponds are to be made in an inland situation, the extent one wishes, and has the power to make them: and they are to be filled with fish that breed in river-water; or one may transfer fishes from brackish water from the sea into river water: and persons who are near the sea or a lake, what kind of fish soever the part of the sea produces, stock their artificial pond with them. One is also to adapt them to the nature of the place; if it is indeed marshy, he is to put in fish that live in swampy situations; and if it is rocky, he is to put into those fishes breeding in rocky situations. The tenderest herbage is also thrown in to feed them, and very small fish, and the gills and intestines of fish, and tender figs cut small, and a soft cheese, to sea and rock fish, and squillae, and bran, or anything of this kind one may be supplied with, or some coarse bread, or dry figs cut small. There will also be plenty of fish in any place, if you throw the herb polysporos, which greatly resembles polygonos, well sliced, into the water in which fish are bred.

22

ASAPH HA-ROPHEH, OF SYRIA
(6TH OR 7TH CENTURY).

The great Book of Medicine of Assaph, the oldest Hebrew book on medicine is very primitive in its anatomical and physiological details. The pharmacology is just a translation of Dioskorides. Yet much good observation on diseases is contained in it. We give here the Medical Oath of his pupils, as translated by Dr. S. Muntner, Jerusalem.

THE OATH OF ASSAPH

And this is the oath which *Assaph, the Son of Berachia and Jochanan, the Son of Zabda* administered to their disciples and they swore these words:

Ye shall not dare kill a person with the juice of a root, and ye shall not administer a potion for abortion to a woman impregnated by adultery and ye shall not lust women for their beauty to commit adultery with them and ye shall not reveal secrets confided unto you; and ye shall not take ransom to ruin or to destroy and ye shall not make your heart stubborn against pity with the poor and wretched to heal them, and ye shall not say the good is bad and the bad is good. Ye shall not walk in the way of the sorcerers who unite, exorcise, bewitch to separate the man from his beloved wife and the wife from the husband of her youth and ye shall not covet any riches and reward an abetted misdeed.

Ye shall not partake in idolatry to heal with it and ye shall not trust that the words of their service heal.

For they make abominable and abhorrent and hateful their servants and those who confide in them and have others confide in them.

For they are without form and void, they are nothing, dead spirits are their idols who cannot help the lifeless pictures, how could they deliver living men.

But ye shall trust in the Lord, the God of truth, the living God, for He kills and revives, He smites and cures.

He also endows man with reason to help; He smites in righteousness and justice and He heals in love and pity.

He will not fortify evil device and craftiness and nothing will be hidden before His eyes.

He makes healing herbs grow and in His mercy endows the perfect heart with the might to heal, to declare His miracles.

In the great assembly so that all living things may perceive that He is the Creator and there is no saviour but Him. For the peoples trust in their idols.

Who do not help them in their distress and do not save them from their troubles, for all their hopes and all their expectations are addressed to the dead.

Therefore, it befits you that ye set yourself apart, turn aside and keep aloof from their abhorred idols, to invoke the name of the Lord.

The God of the spirits of all flesh, in whose hands rests the soul of all life, to kill or to revive. None evades His power.

Ye shall be mindful of Him at any time, ye shall search for him in truth, in uprightness and perfection that you shall succeed in all your doings.

He will succour you when you help others and all men will praise you and the peoples will leave their idols.

They will yearn for God's service, because they will perceive that they trusted in vanity worthlessness and wore themselves out in the service of gods who do not know to help them.

Therefore, be strong and not neglectful, for reward awaits you. God is with you if you are with Him.

Ye shall keep His covenant, ye shall walk in His ways and ye shall cleave unto Him, and *Ye Will Be Accounted Saints in the Eyes Of Men* and they will say:

" Blessed is the man that is in such a case, happy is the people of God."

And the disciples will respond and say: All you have ordered and ordained we shall do, for thus it is prescribed in the Scriptures.

And we must do it with all our heart, with all our soul and with all our might, to do and to be obedient and not to flinch and depart neither to the right nor to the left and they blessed them in the name of God the Creator of Heaven and Earth.

And they continued to admonish their disciples and said unto them: See God, His Holiness and His Scriptures are witnesses.

That ye shall stand in awe of Him and ye shall not depart from His precepts and walk in the ways of uprightness without bending towards advantage.

To help the godless who waylays an innocent soul. Ye shall not mix poison for man or woman to kill his friend and not to tell everybody which plants are poisonous.

Ye shall not hand over them to every man and ye shall not be persuaded to bring sickness to man and Ye shall not seek to injure man with an iron instrument or burning unless ye examined twice and thrice, with care, and only then ye shall consider. Be not ruled by pride and haughtiness. Do not harbour a grudge against the sick, be righteous and truthful in your words then you will find favour in God's eyes if you keep His precepts and bidding and walk in His ways and be honest, trustworthy and righteous physicians.

Thus Assaph And Johanan Admonished And Adjured their Disciples.

23

ISIDORUS HISPALENSIS (about 570-636), SEVILLA.

From the encyclopaedia named *Originum sive etymologiarum libri XX*. Written from 622 to 633 we render four small paragraphs on natural history.

(*a*) The mouse. (Lib. 12 : 3: 1).
(*b*) The centipede. (Lib. 12: 5: 6).
(*c*) The mastix-tree. (Lib. 17: 8: 7).
(*d*) The asarum plant. (Lib. 17: 9: 7).

(*a*) The mouse (*mus*) is a small animal. Its name is Greek and the Latin name is thence derived. Others say they are called mus, because they are born from the moisture of the soil, as mus is soil, and humus is derived from this root. Its liver increases with every full moon, as (do the intestines of) some maritime animals, and they become smaller with the decrease of the moon.

(*b*) The centipede is a soilworm, called so after the multitude of its legs; they may roll themselves when touched into a ball. It is born beneath rocks from humidity and earth.

(*c*) Mastic is a secretion of the Lentiscus (Pistacia) tree. These secretions are called granomastic (gargire ha-mastik) from their grainy structure. The best mastic is produced in the island of Chios, being of good smell and shining like the wax of Carthage. Hence it increases the shining of the skin; and it is falsified by resin and incense.

(*d*) The Asarum grows on shady mountains. Its flowers are similar to those of the Cassia. The purple flower is set upon the root and contains seeds similar to grapes. It has several roots, which are very tough and of a good smell, like that of Oil of Nard.

24

ABU UTHMAN AMR IBN BACHR AL JAHIZ OF BASRA (died about 868).

The animal book, the *Kitab al Hayawan*, of Jahiz is written in a difficult language, and has not yet been translated into any other language from the Arabic original. Dr. L. Kopf has translated for us the chapter " On the Peculiarities of the Ants," which reveals the originality of many of the observations of Jahiz.

On the language of the ants. When you will see something wonderful and find circumspection, turn to the inconspicuous, the small, the weak, to show you how fine are its senses, how miraculous its actions, how it foresees the future, how it becomes like man and outdoes him, to whom belongs the universe and what is in it.

We know, that the ant prepares stores in summer for the winter, that it is active in the season of plenty, not losing any time, where prudent action is possible. Their prudence, experience and foresight goes so far, that when they have made their winter-stores in summer, they fear, that the stores may decay or become infested by worms, whilst they rest in the womb of the soil. The ants then return the stores to the surface of the soil, to dry them, restore their hardiness, and to let the wind pass over them to take away the bad smell and decay.

The grains are often, even usually, stored in a wet place. When the ant fears that they may germinate, it cuts their shell in the middle of the grain—you know that there the grain begins to germinate—and thus splits all grains into two halves. But kuzbura-grains are split into four parts, because they alone amongst all cereals still germinate from half grains. By such prudence the ant surpasses the intelligence of all other animals, and therein it is probably even more prudent than man.

In spite of its tiny body and its small weight, its sense of smelling and of tracking are better than those of any other creature. When a man eats a locust or something similar, sometimes a locust or part of its drops to the soil. Even if no ant is present at that moment, and even if no ant has ever been seen in that house, an ant will approach the dropped locust within a short time and try to turn and move it, to draw it away and to pull it. When it is unable, in spite of all its efforts, to transport it, it returns to its nest. And the observer will see, that it comes back after a short interval, followed like a black thread by a line of its comrades. Then they help one another to move the locust.

This behaviour is first based upon an unfailing sense of smell, but also upon the great initiative and boldness of the ant, which dares to undertake the transport of a body, which weighs an hundredfold and more its own weight. No other animal is able to transport objects which weigh many times its own weight, whilst the ant does not desist from such an attempt, until its breath fails it. If you object: How does a man know, that just that ant, which first tried unsuccessfully to move the locust, does notify its comrades and that it returns at their head? we respond: On the ground of our long experience. Always when we observed an ant, which desisted from an attempt to move a locust, we saw it return with such a following as described. And even if we are unable to distinguish it with our eyes from its comrades, our explanation is the only plausible one. We never saw an ant returning to its nest, either with a burden or without one, which did not stop to communicate with another ant, which it chanced to meet. This shows, that the first ant, when returning from its unsuccessful attempt, served as a true scout, which does not cheat its comrades . . . .

25

ABU HANIFA IBN AL DINAWARI (820-895).

The famous Book of Plants, the *Kitab al Nabat*, is lost, and only a few fragments are preserved. From these follows the paragraph " On the Division of the Plants " (from B. Silberberg, Zeitschr. Assyriol. 25. 1911. pp. 63-64).

Plants are divided into three groups: In one, root and stem survive the winter; in the second the winter kills the stem, but the root survives and the plant develops anew from this surviving rootstock; in the third group both root and stem are killed by the winter, and the new plant develops from seeds scattered in the earth. All plants may also be arranged in three other groups: some rise without help in one stem, others rise also but need the help of some object to climb, whilst the plants of the third group do not rise above the soil, but creep along its surface and spread upon it. All plants are called '*trees*', small and big, which rise without help, whether they survive the winter or not . . . Those plants which sprout from seeds and not from a rootstock are called ' herbs.' And every plant in its first stage is a herb. What, however, sprouts from the rootstock (of last year) but whose stem does not survive in winter is called ' shrub,' as it differs from the tree with perennial stem and root, and also from the herb, whose stem and root are killed in winter, and thus forms a link between both. What is hanging upon trees, creeps and embraces them, lives like the winding plants, and what spreads horizontally, not vertically lives like creeping plants.

26

IL IKHWAN EL SAFI, THE BRETHREN OF PURITY AT BASRA (10TH CENTURY).

This sect edited an encyclopaedia of all knowledge, which is well known by its pantheistic cosmology. It was of great influence upon the Arab, Latin and Hebrew writers of the Middle Ages. The encyclopaedia also contains some fables, of which we quote one of the lion from the Hebrew translation of Qalonymos ibn Qalonymos (1286-1328) of Arles.

(9) *In exposition of the virtues of the Lion, King of the hyenas, and of his excellence amongst the beasts.*

And it came to pass on the third day at morningtide that the chieftains of the castes came for judgement, and stood at their posts as of yore, and the King looked to his right and to his left, and behold, he saw a jackal standing beside his ass, his crooked, sidelong glance darting from right to left like that of a deceiver, an informer, and a traitor, and he was terrified of the dogs. Quoth the King unto him in the interpreter's tongue, "Who art thou?"

Quoth he, "Emissary of the hyenas am I."

Quoth he unto him, "Who hath sent thee hither?"

Quoth he, "Our King." "And who may be thy King?"

Quoth he, "The Lion."

Quoth the King, "And who be his populace?"

Quoth he, "The beasts of the wilderness."

Quoth he, "And who be his legions and his retainers?"

Quoth he, "The tigers and the bears and the wolves and the wildcats and all possessors of the fang and the claw."

Quoth the King, "Describe unto me his qualities, his attributes, and his habits."

Quoth he, "Thus will I do my Lord, the King. Know you, that the Lion is the greatest of the hyenas, grand in creation, most powerfully built, most terrible and most brave, broad of breast, narrow of ham, slender of hind-quarter, large of head, round of visage, bare of brow, broad of jowel, thick of arms, sharp of fang, heavy-voiced, bright-eyed, strong-spined, brave-hearted, his aspect is very terrible, he shall not fear, nor shall he rise for another of his stature, not for buffaloes, nor elephants, nor armoured humans or horses. He is forceful of thoughts, agile, and tremendous. When he has thought upon a thing, he will go himself, and will not be assisted by any one of his legions or his retainers. He is kind of soul: When he has hunted his prey and eaten of it, he leaves somewhat for his armies and his attendants. He is dear of soul: He will not be aroused to petty affairs such as harming women or children, or those who sleep or drouse, and when he sees a light in the distance he will go unto it in the darkness and will stand apart from it, and when he is angered he will allay his anger and his wrath will be silenced, and when he heareth a goodly melody from afar he will go unto it and be placated by it, and nothing will terrify him nor will he fear—

only the tiny ants, for they overmaster him and his cubs, even as the tiny gnat overmasters the elephant and the buffalo, and as the flies overmasters great heroes. But the Almighty hath afflicted him, in order to temper his perfections by his failing, with four-day fever all the days of his life."

Quoth unto him the King, "How treateth he his populace?"

Quoth he, "Far better than can be told, and far more seemly, and now our Lord the King hath heard of his nature."

27

HILDEGARD OF BINGEN (1098-1179).

The book called *Causa et curae*, vel *Liber compositae medicinae* (about 1157) contains apart from much medicine, some chapters on animals and plants. The contents of these is abstruse. Here is the chapter of the Creeping Animals (from H. Schulz, Der Aebtissin Hildegard von Bingen Ursachen und Behandlung der Krankheiten. Muenchen 1933. p. 38).

Of creeping animals: All animals, the creeping ones also, are created to serve man, and help and serve him by making holes in the soil so that it may be wettened by water and rain. Therefore, they rest always in the moist soil, to warm and wetten it by their exhalations and evaporations and sweat, which give the soil some coherence and strength. The worms are poisonous because of the evil smell and the decay of the soil. Rain and dew wash the surface of the soil, the sun gives it warmth, and hence the surface is clean and produces pure fruits. Dirt and decay flow into its interior, and from them grow in it the poisonous worms, just as worms originate in the pus of man which damage him. Thus also originate the worms in the earth from which they feed. Yet bones have almost no worms . . . Some worms are without hair, as they come out of the soil-humidity, live in the soil and avoid the upper regions, and thus neither contact air, dew nor the warmth of the sun, to which the other animals owe their hair. They are inimical to man and the higher creatures, as they have a nature opposite to theirs, and thus they kill them by their poison and cause damage to man and to all animals higher than themselves. In spite of being poisonous, some of them serve as medicine for man and animals, if not as a whole, yet in some of their parts, as they have also from the good soil liquor in themselves which produces the medicinal herbs. Thus also the deer is rejuvenated when it has swallowed a snake.

28

ABRAHAM BEN DAVID HALEVI OF TOLEDO (1110—1180).

In the encyclopaedia Elevated Faith, *Emunah Ramah* (1160) science is regarded as an entrance to philosophy and theology. We give here the chapter on the outer and inner senses, which illustrates the psychology of the Middle Ages. Ed. S. Weil. Berlin, 1919. pp. 28.30.

These five external senses are like the spies and scouts of the desiring soul. They bring the things to the soul, and the force at appetition reacts then towards these things by approach to or by flight from the sensated object. And they are the servants of the force of appetition.

The general sense (*usually: common sense*) is a (common) root which God, the Blessed, gave to the five (external) senses. The senses spring out from the general sense and their sensations are brought to it. By this (root) it can judge (an object) sensated by one sense with regard to characters which fall into the domain of another sense. This force from which the five (external) senses are sent out is called the general sense.

It occurs in animals, and certainly in man. By its help we judge, that this cake which is yellow at the outside and white within be bread, the best of our victuals. Or that this yellow, easily crumbling thing be a bitter opuntia fig (zabr). Thus the eye will judge upon bitter or sweet, which sensation actually can be perceived through the tongue.

The eye also pronounces judgement, that this white man be a musician with an agreeable voice. Without this general (sense) we would be obliged to taste every food which we need, when we have seen it, and it would be a great toil for us, if we would have to taste (many) bitter, disagreeable or repellent things, until we find after much toil what we need. This would make our life uncertain and full of unpleasant experiences. Also the hungry donkey runs speedily to a heap of its fodder, whilst it does not run so to a heap of sand. It (obviously) estimates, that this white grain have a good taste. The donkey likewise runs away from a whip which is raised against it, judging, that this tool of punishment be painful. And if not all senses would be united by one (common) root, they would be unable to pronounce judgements (relating to other sensory perceptions).

Retentive imagination (ha-metzayyer) is another grace of God, the Blessed, to the animals. This is another internal sense, which makes them remember the impressions of sensations which have ceased to act. By this retentive imagination they perceive shapes which are (at present) hidden before our eyes. It also occurs in animals. When the pigeon returns (to its cot), it retains the (vision of the) shape of the cot, and flies speedily to it, whenever it has the opportunity. This is not the same as the general sense. The general sense extends to present (objects) only, the retentive imagination to things which are not now present.

Compositive animal (medameh) and human (machshav) imagination is a further grace of God, the Blessed, as the third internal sense. In animals it is called ha-medameh, in man ha-mechshav. Man composes from sensated bodies (of his experience) other ones which were not in his sensory perception. Thus we may compose in comparative imagination a being half man, half horse, or a sesam seed of the size of a water-melon, inspite that we have seen and measured both, a sesam grain as well as a water-melon. This sense creates from what the eye has seen, but renders the objects not always in their true shapes, but sometimes creates deceptive shapes, such as those just mentioned.

When somebody considers, how it be possible that on the sphere a star may be cut off for some time (? disappear), and shortly afterwards it is seen running in the opposite direction of its (normal) path, then it seems to stand still and then to continue its straight path, we explain this by the force of the comprehensive sphere by easy corrections of the theories. This force is called human compositive imagination.

Yet in animals the compositive animal imagination is their aim and their purpose. From it spring all their actions. By its help the *Kermes* scale forms almond-shaped produces from its surpluses, and the honey-bee the combs from the nectar which is ready to turn into honey. And not from every material the bee makes its combs, but from nectar alone. Neither makes the *Kermes* its almond-shaped produces from every substance, but from its own surpluses alone. Both (these productions) are not the work of their reason. Man alone will create by his creative imagination many inventions from the most variegated materials. He produces, what has not been produced so far. The animal shows by its behaviour, that it does not produce these things by reason and purpose. This means: it does not understand the use and purpose obtained by its (actions), but (they are) in the service of a higher purpose, as we will explain later. This coerces it to this production for its use, and puts the wish and endeavour to perform them into the animals, without that they know the purpose which moves them to that activity. And it produces only with certain materials. But man, who acts with consciousness of the purpose and of his actions, and who acts only following a plan known to him, can—as long as his reason is healthy—form combs from wax, copper, iron, silver, gold, wood, stone or from any other natural material which he wants (to use) or finds suitable. Do not be astonished when an animal reaches a purpose in the service of higher existences! And do not think, that this be different (from what we have said), and that the animal understands the purpose of all its actions, as that is not true!

The plant which certainly does not know what it does and which is without the beginnings even of actions by reason, obeys without doubt the orders of reasonable existences. When it grows, it preserves the exact proportion of the leaf, the exact colour of the flower as well as its smell and taste, etc. All its individuals preserve these shapes, measures, colours, smells, tastes, etc., in close relations (to the others of their kind). This demonstrates the existence of higher existences who occupy themselves with the actions of the lower existences, who have no reason, yet act directively, as if they had insight (into the purpose of their actions).

Know that this sense of compositive imagination alone of all the senses of the animal soul continues to work in sleep. It works basing itself upon the shapes perceived before by imagination, preserves the shapes of the past and the stores of (past) sensations, and then composes them one with another, or separates them one from another, conveys these (compositions or separations) to the general sense, and it feels as if it really had perceived these things. These are the visions or deceptive dreams, which we have to mention later. This is the sense of compositive human and animal imagination. As its pictures are not true, this imagination of man is not always true, but he will sometimes think true and false thought on known things and deeds.

To this end imagination is good, but it is similar to reason only, namely it is primarily an animal sense. By a divine force can it become compositive human imagination or a separate existence, as we will see in the discussion on the imitative power. It lets appear absurdities as true shapes, it finds false explanations, wrong proofs and empty desires, but occasionally it may lead man to discoveries and to straightness.

Estimation (*ha-ra'ayoni*) is the fourth internal sense with which the grace of God (the Blessed and Exalted), has gifted the animals. By estimation the animal concludes from sensated phenomena partial conclusions which are not sensated. Thus estimates the sheep, that this wolf be an enemy to be avoided by flight, and that that shepherd be a friend to be approached. And this flight away of the animals from certain animals takes place, without that it understands the true implication of its flight. Likewise it seeks the company of other animals, without understanding the true implication of this aggregation, which is of great benefit to itself. When you raise your hand against the eye of a neonate suckling or animal young, to whom you have never done any damage, they will close their eyes before the hand raised against them. And if you take the child on your hand, and then let it drop, it will hang unto you. In these ways estimation is useful to the animals.

Memory (*ha-socher*) is the fifth internal sense with which the grace of God, the Exalted and Blessed, has gifted the animals. It preserves things which are not anymore sensated, yet it is something entirely different from retentive imagination. The latter preserves the shape which is at present not sensated, but the former preserves the objects which we had forgotten. The animals have apparently also this sense, and it is mixed with the shape, namely they remember some shape or some place, and they also remember what occurred to them from the same shapes or in the same places, pleasant or unpleasant. They avoid houses, where they suffered pain, and they approach houses, where they felt joy.

These ten senses, the five external and the five internal senses, occur in the animal soul. They are all activated by the force of instinct, which guides the animal to approach or to flight.

Locomotion (*meniah*) is another force of animals. It comes from the brain, it comes to the nerve and contracts it, so that that muscle contracts or expands, with which the nerve is connected;

the sinews are moved and they contract the muscles which move the limb to which they are connected. Thus the body is moved to the (object) of its desire or removed from that of its pain. This is one type of motion, the locomotion (ha-athaqah).

Involuntary motion, such as is observed in the pulse and in respiration, is another type of motion. Locomotion is directed by will, whilst the motion of the pulse is natural, and the will can neither stop nor produce it. The same is true for respiration. Yet God, the Exalted and Blessed, gave to some animals lungs into which cold air penetrates by this respiration, in order to temperate the heat of the heart. However, this inspiration of cold air is not so needed as the pulse, and therefore it is possible to pass through dust and smoke or to dive into water.

These are the twelve forces which the animals have in addition to those of the plants: five external and five internal senses and two forces of movement, in addition to the seven forces of the plants. Altogether nineteen forces of the soul are thus to be found in animals.

Now we will prove, that the forces are truly existing, that every act is the consequence of a force, and that the same act does not spring from two (different) forces. We maintain, that the differences of the external senses are based upon a (qualitative) difference of the forces. A demonstration is, that the sweetness of honey is not perceived through the eye, but through the tongue. This cannot be derived from any character of the (sense) organ or by the absence of that character: that the tongue be soft and very sensitive because of its very many nerves, whilst the eye be harder and hence does not produce the stimulus (of sweetness). Actually, the eye is not so much harder than the tongue, that this (difference) should prevent the sensation of certain stimuli. To the contrary: the eye is also rather soft and has also many nerves. It is affected by accidental troubles, such as smoke or dust much more than the tongue. The difference in the sensation of different materials cannot be ascribed to the presence or absence of a character of the organ, as the sweetness of the honey remains the same for the tongue, which sensates it, as for the eye, which does not sensate it. Therefore the different (qualities of sensation) must spring from something immaterial, which is found in the body of the tongue, but not in that of the eye. And again, in the body of the eye there is something immaterial, the prime mover of the sensations of the colours; and similarly is it with all the other sense organs. That the activating forces (causes) of the sensations be substances, we can prove, as follows. Every accident is, indeed, within the object, but it is not (an integral) part of it in the sense, that its removal from the objects makes the object disappear. But, if we remove the sensations from an animal, it ceases to be an animal. Therefore the senses are an (integral) part of the animal, and what is an (integral) part of an object is itself a substance.

Regarding the internal senses, it is paradox to assume that they be active and not active, i.e. imagination and not imagination, thought and not thought, memory and not memory. If one activator moves all of them it should either act or not act, as it is impossible that it be acting and not acting at the same time. And when it be active at one time, it should activate all senses together, i.e. thought, imagination estimation all together. And if it be resting, all the senses should rest together. Yet we find, that this be not so. Some senses are active sometimes whilst others are resting. It is clear, therefore, that (qualitatively) different forces bring about the various activities.

Now it is the time to discuss the *forces of reason* and of the *rational mind*. We maintain that when the mixture is not fitted for higher benefits, than those we have mentioned so far, the object will remain an animal. The range of the kinds of animals is great: from the monkeys, which are close to the nature of man, down to the coral-trees, where plant- and animal-life are touching. Yet if the mixture is fitted for greater benefits, it becomes man. Man possesses all the forces of the soul mentioned hitherto, and in addition a rational mind, which primarily is potential reason which later develops into actual reason.

29

MOSHEH BEN MAIMON OF CAIRO (1135—1204).

In all the philosophy of the Middle Ages the theory of man as the microcosmos in homology with the universe, the macro-cosmos, was maintained. Perhaps nowhere it is described as clearly as in the Guide of the Perplexed, the *Moreh Nebuchim* (I; 72) of the Rambam, Mosheh ben Maimon. We made use of the English translation of M. Friedlaender. London, 1925. pp. 113-119.

(1) The entire universe is like one individual organism: The outermost heavenly sphere together with all included therein is an

individual organism, just as that of Sayid or Omar. The variety of the substances of that sphere and of its parts is like the variety of the substances of a human individual with its solid substances (flesh, bones, sinews, humours) and with its spiritual elements. The universe is composed of the celestial orbs, the four elements and their combinations, the space being filled with matter, not leaving a vacuum. In its centre is the earth surrounded successively by water, air, fire and by the fifth substance (which is the quintessence). (Follow further details of the structure and functions of the universe).

(2) (The number of the spheres encompassing the universe cannot be less than 18, which is in contrast to the not more than 9 spheres of the Ihwan as-Safi).

(3) (The spherical bodies have life and possess a soul by which they move spontaneously. They have no properties allowing them to come to a rest.)

(4) As the human body consists both of principal organs and of other members which depend on them and cannot exist without the control of those organs, so does the universe consist both of principal parts, viz. the quintessence, which encompasses the four elements, and of other parts which are subordinated and require a leader, viz. the four elements and the things composed of them.

(5) Again, the principal part in the human body, namely the heart, is in constant motion, and is the source of every motion noticed in the body; it rules over the other members, and communicates to them through its own pulsations the force required for their functions. The outermost sphere by its motion rules in a similar way over all other parts of the universe, and supplies all things with their special properties. Every motion in the universe has thus its origin in the motion of that sphere; and the soul of every animated being derives its origin from the soul of that same sphere.

(6) The forces which according to this explanation are communicated by the spheres to our sublunary world are four in number: (*a*) the force which effects the mixture and the composition of the elements, and which undoubtedly suffices to form the minerals. (*b*) The force which supplies every growing thing with its vegetative functions. (*c*) The force which gives to each living being its animal vitality, and (*d*) The force which endows rational beings with intellect. All this is effected through

the action of light and darkness, which are regulated by the position and the motion of the spheres round the earth.

(7) When for one instant the beating of the heart is interrupted, man dies, and all his motions and powers come to an end. In a like manner would the whole universe perish, and everything therein cease to exist if the spheres were to come to a standstill.

(8) The living being as such is one through the action of its heart, although some parts of the body are devoid of motion and sensation, such as the bones, the cartilage, and similar parts. The same is the case with the entire universe; although it includes many beings without motion and without life, it is a single being living through the motion of the sphere, which may be compared to the heart of an animated being. You must, therefore, consider the entire globe as one individual being which is endowed with life, motion and a soul. This mode of considering the universe is, as will be explained, indispensable. It is very useful for demonstrating the unity of God: it also helps to elucidate the principle that He who is one has created only *one* being.

(9) Again, it is impossible that any of the members of a human body should exist by themselves, not connected with the body, and at the same time should actually be organic parts of that body; that is to say, that the liver should exist by itself, the heart by itself, or the flesh by itself. In like manner, it is impossible that one part of the universe should exist independently of the other parts in the existing order of the things as here considered, viz., that the fire should exist without the co-existence of the earth, or the earth without the heaven, or the heaven without the earth.

(10) In man there is a certain force which unites the members of the body, controls them, and gives to each of them what it requires for the conservation of its condition, and for the repulsion of injury—the physicians distinctly call it the leading force in the body of the living being; sometimes they call it 'nature.' The universe likewise possesses a force which unites the several parts with each other, protects the species from destruction, maintains the individuals of each species as long as possible, and endows some individual beings with permanent existence. Whether this force operates through the medium of the sphere or otherwise remains an open question.

(11) Again, in the body of each individual there are parts intended for a certain purpose: such as the organs of nutrition

serving the preservation of the individual; the organs of generation for the preservation of the species; hands and eyes for definite purposes, such as (the finding and taking) of food, etc. Yet other parts are by themselves not intended for any purpose, and are mere accessories and adjuncts to the constitution of the other parts. The structure and function of the organs essential for the maintenance of the body leads by their function and in accordance with the variability of their substances to the production of other things, such as hair or complexion of the body. These are, however, mere accessories and are not formed according to rigid rules. They may be absent in some individuals or vary considerably in others. This is not so with the essential organs. You never find that the liver of one person be ten times larger than that of another person. But you may find a person without a beard, without hair on certain parts of the body, or with a beard ten times longer than that of another man. Instances of great variation, for instance, of hair and colour are not rare. The same differences occur in the constitution of the universe. Some species are an essential part of the whole system. These are rigidly constant: varying as far as their nature permits, it remains insignificant in quantity and quality. Other species do not fulfil an essential purpose. They take origin merely by the general nature of transient things. Such are the many insects which are bred in dunghills, the animals originating in rottening fruits or in fetid liquids, worms generated in the intestines, etc. In general: everything devoid of the power of (sexual) generation belongs to this class (of accidental species). Hence, they do not follow rigid rules, although their entire absense is just as impossible as the absense of different complexions and of different kinds of hair amongst men.

(12) The individual existence of certain substances of man is permanent, whilst others are constant in the species, but not in the individual, such as the four humours. The same is the case in the universe. Certain substances are constant in individuals, such as the fifth element which is constant in all its formations. Yet other substances are constant only in the species, such as the four elements and all that is composed of them.

(13) The same forces which operate in the birth and the temporal existence of man operate also in his destruction and death. This is true for the entire transient world. The causes of generation are at the same time the causes of corruption. For

instance: If the four forces present in every feeling being, namely attraction, retention, digestion and secretion, were like intelligent forces able to confine themselves to what is necessary, and to act at the proper time and within the proper limits, man would be exempt from those great sufferings and from his numerous diseases. As this is not the case and these forces perform their natural functions without thought and intelligence, without consciousness of action, they necessarily cause dangerous and painful diseases, although they are the direct cause of the birth and the temporal existence of man. The explanation is: If the attractive force would absorb only the absolutely beneficial and that only in the quantity required, man would be free from many sufferings and diseases. But, quite differently, the attractive force absorbs any humour in the range of its action, however ill-suited it be in quality or in quantity. (This is the reason for the diseases and pains as well as for the disasters of the earth or of the macrocosmos, such as earthquakes, thunderstorms, meteoric showers, etc.).

(14) Nothing in all these observations about the similarity between universe and man would warrant us to claim man as a microcosmos. Although the comparison in all its parts applies to the universe and any living organism in its normal state, we never heard any of the ancient writers call the ass or the horse a microcosm. This attribute pertains to man only because of the faculty of thinking peculiar to him, i.e. of the intellect, namely the hylic intellect which is his alone amongst animals. No animal requires for its sustenance any plan, thought or scheme. Each animal moves and acts by its nature, eats as much as it can find of suitable things, rests wherever it happens to be, and, when in heat in the period of reproduction copulates with any mate it meets. Thus each individual conserves itself for a certain time, and perpetuates the existence of its species without requiring for its maintenance the assistance or support of any of its fellow creatures, attending itself to all its needs. But in man, if there were a solitary individual left like an animal without guidance, he would soon perish, would survive by chance only for the first day, unless it happened to find some food. For the food needed for his sustenance demands much work and preparation, requiring reflection and planning; many vessels are needed and many specialised individuals must co-operate. By necessity one person must organise the work and direct the proper co-operation of the men and their mutual assistance. None of the following things can be

properly done without design and thought: protection before heat in summer and before cold in winter, shelter from rain, snow and wind, all requiring many preparations. Therefore, man needs the intellectual faculties for thinking, considering and acting, for preparing and securing food, dwelling and clothing, and to control every organ of his body, causing both the principal and the secondary organs to perform their respective functions. Consequently a man, deprived of his intellectual faculties and left only with his vitality, he would be lost in a short time. The intellect is the highest of all faculties of living creatures. It is very difficult to comprehend, and its true character cannot be understood as easily as man's other faculties.

30

ADELARD (AETHELARD) OF BATH (12TH CENTURY).

Adelard is one of the great English personalities of the Middle Ages. In his youth he travelled through all the Mediterranean countries, including those of the Levant. In his *Quaestines Naturae* (also *De eodem et diverso*), discussions with his nephew, we find amongst others a discussion: Why do plants grow spontaneously? (From H. Gollancz, *Dodi venechdi.* Oxford, 1920. pp. 10-11).

Why do plants grow without any previous sowing of seed?

Nephew: How do plants spring from the earth? What reason exists or can be given, seeing that the surface of the earth is at first level and motionless, what is it, I say, that stirs there, springs up, grows, and puts forth branches? For though, if you like, you can collect quite dry soil, and after carefully riddling it put it in a pot of earth or brass, yet as time goes on you will see a shoot springing up, and to what are you to ascribe this save to the wondrous effect of the wondrous Divine Will?

Adelard: Will on the part of the Creator there certainly is in the springing of plants from the earth, but it is not divorced from reason. That this may be quite clear, I grant that plants spring from the earth, but not from pure soil: it is assuredly a mixture of such a sort as to contain in each of its particles, which indeed are subjacent to sense, all four elements with their qualities. Most certainly these four samples compose the one substance of the universe in such a way that the component parts exist in each compound, but yet never in such a way as to be apparent to the senses. The fact is, that we mistakenly call the compound by the name of its simple, for no one has ever touched earth or water, just as no one has ever seen air or fire. The things which we apprehend by our senses are compounds,—not themselves, but made out of themselves. Hence as the philosopher says, we ought to use not the terms earth, water, air, fire, but instead earthy, watery, airy, fiery. Therefore, since in your earthy matter, no matter how subtly pulverised, the four causes necessarily exist, there arises hence a sort of compound, mostly earthy, to a less extent watery, less still airy, and least of all fiery: it owes to earth its power of cohesion, to water that of spreading, to air and fire the tendency to rise, for unless fire were contained in it, it would have no power of upward movement, and unless water or air, no power of lateral expansion: while finally, were it not for earth, it would have no coherence.

Hence it is that the things combined with wondrous subtlety in your dust, come forth into the light. However, that you may understand the matter with entire clearness, I place the cause of this process in the exterior elements arousing and drawing forth their like, and by their qualities driving it out; hence the inferiors, by a perpetual process of dissolution, pass over to their likes.

31

FREDERICK II OF HOHENSTAUFEN (1194—1250), PALERMO.

The Art of Falconry (*De arte venandi cum avibus*. Engl. transl. Wood and Fyfe. London, Reprinted 1955), is based upon much observation, experience and common sense. It is one of the few great biological books of the European Middle Ages. The emperor himself and his son Manfred had clearly taken a very active part in the composing of the book. The main part of the manuscript deals with various aspects of falconry. The first book, however, is devoted to the morphology and biology of birds in general. The chapters on bird flight and on bird migration are really amazing.

(*a*) A reasoned judgement on Aristotle. (From the Introduction to liber I, from Wood and Fyfe. 1955. pp. 3-4).

(*b*) On bird migration. (An extract made by Beebe, The Book of Naturalists. New York, 1945. pp. 26-28).

(*a*) Despite the handicaps of our arduous duties as a ruler we did not lay aside our self-imposed task and were successful in committing to writing at the proper time the elements of the art. Amongst others, we discovered by hard-won experience that the deductions of Aristotle, whom we followed when they appealed to our reason, were not entirely to be relied upon, more particularly in his descriptions of the characters of certain birds. There is another reason why we do not follow implicitly the Prince of Philosophers: he was ignorant of the practice of falconry—an art which to us has ever been a pleasing occupation, and with the details of which we are well acquainted. In his work, the *Liber Animalium*, we find many quotations from other authors whose statements he did not verify and who, in their turn, were not speaking from experience. Entire conviction of the truth never follows mere hearsay . . .

The whole subject of falconry falls within the realm of natural science, for it deals with the nature of bird life. It will be apparent, however, that certain theories derived from written sources are modified by the experiences set forth in this book. The title of our work is: " The Book of the Divine Augustus, Frederick II, Emperor of the Romans, King of Jerusalem and Sicily, *De Arte Venandi cum Avibus*, an Analytical Inquiry into the Natural Phenomena Manifest in Hawking."

(*b*) With a prophetic instinct for the proper time to migrate, birds as a rule anticipate the storms that usually prevail on their way to and from a warmer climate. They are conscious of the fact that autumn follows summer (when they are strongest and their plumage is at its best) and that after these seasons comes the winter—the time they dread most. They are instinctively aware of the proper date of departure for avoiding the winds to which they may be exposed in their wanderings and for eluding the local rains and hailstorms. They usually are able to choose a period of mild and favouring winds. North winds, either lateral or from the rear, are favourable, and they wait for them with the same sagacity that sailors exhibit when at sea. With such helpful breezes progress and steering in the air are made easy. With these to help them on their way, they reach, with comparative comfort, the distant lands of their heart's desire. When they fly before the wind they can rest on an even keel, still maintaining progress, especially when propelled in a proper direction. When becalmed they do not fly so satisfactorily, for they must exert themselves all the more. With head winds there is a threefold difficulty in attempting to float, to fly forward, and to overcome direct aerial obstacles.

Among flight obstructions there are also to be considered not only contrary winds but local rains, hailstorms, and other forms of bad weather that may affect both air and sea, so that some birds fall into the ocean and others, when possible, fly on board a ship (where they are easily caught), preferring that fate to certain death or to continued exposure to the rigours and dangers of oceanic storms.

We notice also that when a favouring wind springs up, whether by day or night, migrating birds generally hasten to take advantage of it and even neglect food and sleep for this important purpose. We have observed that migrating birds of prey, that have begun to devour food we have thrown to them, will abandon it to fly off if a favourable wind begins to blow. They would rather endure and travel day and night than forego such an advantageous opportunity.

The calls of migrating cranes, herons, geese, and ducks may be recognised flying overhead even during the night, and not, as Aristotle claims, as a part of their efforts in flight; they are the call notes of one or more birds talking to their fellows. For example, they understand wind and weather so thoroughly that they know when meteorologic conditions are favourable and are likely to remain so long enough to enable them to reach their intended haven. Weak flyers postpone their journey until they are sure of a prolonged period of good weather sufficient for their migrating venture, but hardy aviators take advantage of the first propitious period to begin their flight.

The slower migrants begin their departure early. For example, the smaller birds, as well as storks and herons, remain until the end of summer, and leave the last of August so that they may not be embarrassed by changeable weather or early (autumn) storms. The more robust species and better flyers remain until the beginning of harvest (in mid-September). Among the latter are the larger and smaller cranes. At that date strong flyers can readily defy the early winds and rains. There are, moreover, still better and swifter flyers who postpone their departure until the end of autumn, say until November. These include certain ducks and geese who do not fear high winds and heavy rains because of their skill in flight and because their plumage protection against cold is adequate. This rule also applies to the smaller geese who may remain behind in the sixth and seventh climates (i.e. zones) the whole winter through, inasmuch as they can find there the herbage on which they feed. The larger geese also possess unusual weather instincts and birdy alertness. In years with short summers, i.e. when the winter threatens to set in early, they migrate much sooner than usual.

Certain birds, cranes for instance, who pass the summer in the far north (where winter comes on early) on account of the longer journey before them, migrate sooner than others of their species who, having nested farther south, prolong their northern visit, since their winter comes later and they have a shorter journey to make. When autumnal winds are favourable, these birds resume their southern flight and, travelling without intermission, quickly accomplish the voyage. Inclement weather, however, may delay the flight of species that have hatched their young in more southern localities until the storm has passed. Those nearest the equator begin their migrations last.

The order of migration may be summed up as follows. Not all shore birds depart, pell-mell, like the disorderly land birds; the latter do not seem to care what birds lead the van or which form the rear guard of the migrating flocks. Water birds, on the contrary, preserve the following order: one forms the apex of advance, and all the others in the flock follow successively in a double row, one to the left and one to the right. Sometimes there are more in one series than in the other, but the two rows, meeting at an angle, form a pyramidal figure. Occasionally there is a single line. This order they maintain not only when migrating to distant points and returning but, as has been explained, in going to and from their local feeding grounds.

One member of the flock usually acts as leader and, especially in the case of the cranes, does this not because he alone knows the goal they seek but that he may be ever on the lookout for danger, of which he warns his companions; he also notifies them of any change to be made in the direction of flight. The whole flock is thus entirely under control of their leader or guide. When the latter becomes fatigued from the performance of this important work, his place in front is taken and his duties are assumed by another experienced commander, and the former leader retires to a rear rank. It is not true, as Aristotle asserts, that the same leader heads the migrant column during the whole of their journey.

32

ALBERTUS MAGNUS (1193-1280), KOELN.

This great reviver of the natural history of Aristotle, to whose zoological books he added six books of his own, is with Frederick II of Palermo, the only good observer of natural history in the Middle Ages. The manuscript preserved at Cologne is apparently the original from Albertus' own hand. Latin print: H. Stadler, Albertus Magnus de animalibus libri XXVI. Münster, Aschendorff, 1916. 1664 pp.

(*a*) On the ventral cord of the arthropods (from *De Vegetabilibus* V: 18).

(*b*) On the dissection of the bee (from *De Animalibus* XVII: 2).

(*c*) On the behaviour of the larva of the ant-lion (from *De Animalibus* XXVI).

(*d*) On the development of the egg (from *De Animalibus*, VI: 1: 4; English translation from J. Needham, History of Embryology. Cambridge, 1934. pp. 71-72).

(*a*) We find in animals a cord extending from the brain or from its equivalent. This is called *nucha* and runs through the whole length of the body of the animal, either on the upper side, along its back, or on its lower side, along the breast and belly. The latter is known in the crayfish, the scorpion, and some others.

(*b*) I have tried to make an anatomy of the bee. A shining, translucent vesicle is found in the abdomen behind its narrow waist. If we taste it (with the tongue) it has a fine taste of honey. Apart from this (vesicle), we find in the abdomen only a thin and little curved intestine and threadlike cords to which the sting is fastened. Around them is a sticky sap and the legs are inserted in that part of the body which lies before the narrow waist.

(*c*) The (larva of the) ant-lion is not a transformed ant, as some say. I have often observed and demonstrated to friends that this animal has a structure similar to that of a tick. It hides in the sand, digs there a half-globular pit. Its mouth is situated at the apex of that pit. If an ant passes, looking for food, the ant-lion catches and devours it. This I have seen repeatedly. It is said that during winter the ant-lion plunders the provisions of the ants, as it does not collect provisions for itself in summer.

(*d*) From the drop of blood out of which the heart is formed, there proceed two vein-like and pulsatile passages and there is in them a purer blood which forms the chief organs such as the liver and lungs and these though very small at first grow and extend at last to the outer membranes which hold the whole material of the egg together. There they ramify in many divisions, but the greater of them appears on the membrane which holds the white of the egg within (the allantois). The albumen, at first quite white, is changed owing to the power of the vein almost to a pale yellow-green tint. Then the path of which we spoke proceeds to a place in which the head of the embryo is found carrying thither the virtue and purer material from which are formed the head and the brain, which is the marrow of the head. In the formation of the head also are found the eyes and because they are of an aqueous humidity which is with difficulty used up by the first heat they are very large, swelling out and bulging from the chick's head. A short time afterwards, however, they settle down a little and lose their swelling owing to the digestive action of the heat—and all this is brought about by the action of the formative virtue carried along the passage which is directed to the head, but before arriving there is separated and ramified by the great vein of the albumen-membrane, as may be clearly seen by anyone who breaks an egg at this time and notes the head appearing in the wet part of the egg and at the top of the other members. For what appears first in the making of a foetus are the upper parts because they are nobler and more spiritual, being compacted of the subtler part of the egg wherein the formative virtue is stronger. When this happened one of the aforementioned two passages which spring from the heart branches into two, one of them going to the spiritual part which contains the heart and divides there in it carrying to it the pulse and subtle blood from which the lungs and other spiritual parts are formed, and the other going through the diaphragm to enclose within it at the other end the yolk of the egg, around which it forms the liver and the stomach. It is accordingly said to take the place of the umbilicus in other animals and through it food is drawn in to supply the flesh of the chick's body, for the principle of generation of the radical members of the chick comes from the albumen but the food from which is made the flesh filling up all the hollows is from the yolk.

33

SAKARYA BEN MUHAMMED EL QASVINI (1203-1283), PERSIA.

This famous geographer mentions many plants and animals in his *Cosmographia*. Like all the great writers of that time he wrote in Arabic. The most accurate translation of his botanical descriptions has been made by E. Wiedemann (in Phys. Mediz. Soz. Erlangen, 48, 1916, of which we render: no. 25, p. 290; no. 47, p. 292: no. 58, p. 300; no. 110, p. 303):

Dulb, the plane-tree, is called tshanar in Persian. It belongs to the greatest, highest, and most hardy of trees. If it has stood for a long time, its interior decays and the trunk remains hollowed. Its leaf is like five fingers. Bats avoid the leaf, and therefore some birds put it into their nests, as they fear bats.

Filfil, the pepper-tree, grows in India in the region called Malibar. It is a high tree, which grows always above water. When the wind blows, what it bears falls into the water, and hence it is wrinkled. It is only collected from the water. The heat cannot injure this tree. It bears fruits in summer and winter, which are fruit-clusters. When the sun burns upon them, a number of leaves covers each cluster, to protect it against the sun. When the sun ceases to shine, the leaves move from the clusters, in order that the wind may have access to them. Somebody who saw the pepper-tree, said, that it is like the pomegranate tree. Between each two leaves is one peduncle of a bunch of fruit, on which the pepper-corns are arranged in rows. The peduncle of the fruit is one finger long.

The Razijanag, the fennel, is a well-known plant. It has a wild and a cultivated kind. Democritus reports that reptiles eat the fresh fennel in order to strengthen the power of their sight, and that the snakes rub their eyes in it, when they come out of their holes after winter, to make their eyes bright. Praise to him who ordered them to do so!

The Kammun, the cumin, is a well-known plant. It is said that doves like it. In order to accustom them to a new home, some cumin is thrown into it, before they leave the cot, which increases their love for it. Ants avoid its smell.

34

GERSHON BEN-SHLOMOH (13TH CENTURY), ARLES.

Gershon's *Sha'ar ha-Shamayim*, Gate of Heaven (Venezia 1547; English translation by F. S. Bodenheimer, The Gate of Heaven. Jerusalem, 1953.) was the Hebrew encyclopaedia of natural history during at least three centuries of the Hebrew reader. We render here the chapter On the Biology of the Bees (VI: 1, Venezia. pp. 3 ro. — 32a). The chapters of the *H A* (*Historia Animalium*) of Aristotle referred to are indicated in order to show the great influence of the master in this age.

VI. *To explain the nature of the bees, the ants and the spiders.*

1. The bees build their nests in the hive and build them from wax, to prevent the cold and enemies to enter there (HA 9: 27: 3). 2. And they have a king and leader: they build a big palace for him (HA 9 : 27: 3). 3. The bees take the wax from the flowers and collect it with the forelegs, pushing it afterwards to the midlegs and then to the thighs of the hindlegs, and fly heavily (loaden home) with it (HA 9: 27: 6). 4. And the honey comes from the flowers of the sweet fruits, when they feed on almonds and their like, and acorns (" g'lanz "), and of willows; and when they eat these flowers, they vomit them (alter) by their mouth (m.HA 9 : 27: 3, 5: 19: 3, 5: 19: 6). 5. All bees have partly the same occupation, partly a different one: Some carry water, some flowers, and some make wax; some knead honey, and others collect it in their houses (HA 9: 27: 14 & 23). 6. Some go out to their work at the beginning of the day, and some stay busy until their work is done. Then they fly out to hunt, and when they return, they first cry out and raise a noise. And then the noise ceases bit by bit, until one bee goes out of the hive and stands to give them a signal when the time has come to sleep. Afterwards they all rest. And the biggest noise is (heard), when they are hungry (HA 9: 27: 23 & 24). 7. All the time when the weather is mild and the stars shine, they do their work steadily and quietly (HA 9: 27: 14). 8. And it is seen, that they take pleasure in the sound of noise: Then the army of the bees leaves the hive, beating with stones one upon the back of its foreman. They all assemble around this noise, and they assemble only for pleasure or for fright (HA 9: 27: 23). 9. There is a beautiful kind of

bees which occupy themselves, as do other beautiful women, with their beauty (HA 9: 27: 23). But they do not busy themselves only with their toilet. Others, however, are lazy and do not know how to work. 10. And the other bees, which are working, drive away the lazy amongst them (HA 9: 27: 23). 11. Also bees have been seen to leave their own hive and to come to another hive. The bees of that hive fight the strangers, and conquer them and throw them out from amongst them. This I have seen myself once with my own eyes (HA 9: 27: 11 & 12). 12. And sometimes the small bees fight with the big ones a great and strong battle (HA 9: 27: 12). 13. Smoke is most injurious to them. And when the bee-keeper wants to take out the honey from the hive, he makes much smoke, so that the bees do not sting him (HA 9: 27 : 2). Somebody told me, that the smell of salt irritates them to bad stings, against which there is no remedy. 14. And they have a leader and they build him houses as royal palaces. And he leaves the house only in the company of all his community (HA 9: 27: 6). 15. And it is said, that if one of their " chicken " goes lost, the leader searches for it, until he finds it, and he recognises it by its odour (HA 9: 27: 6: When the leader goes lost!). 16. When the leader cannot fly (anymore), the bee community beats him and if he dies, he dies (HA 9: 27: 6). 17. They make their " chicken " in the wax houses, which they call " breskes " (HA 5: 18: 3; 5: 19: 3). 18. Afterwards they are busy to make much honey (HA 5: 19: 3). 19. It is said, that when the leader lives, some males stand singly before the (royal palace) in one place, and the females are within the palace. And when the leader dies, the males enter the palace together with the females. 20. The males have no stinging tool. Only the females have a sting which is called " aguillon " (HA 5: 18: 3). The males sometimes build houses for themselves to live in them (HA 9: 27: 5). 21. And of the leaders they are in two kinds of colours: one red, the other black (HA 5: 18: 2). The red is good, the black one bad. And the preferred (verbally: special) female bee of the leader is good, thick, and twice as fat as the other females (HA 5: 18: 2). The small bee is thick, great and short. 22. And if one dies amongst them in the hive, they immediately push it out of it, as they are very sensitive to bad smell (HA 9: 27: 18). 23. And when the eggs have been laid and the " chicken " hatched, the mothers close them the cell and shut the opening, so that they may rest there and may not leave until they have grown and become strong, and then they (themselves) open the holes and leave (HA 5: 19: 6 & 7; 14; 9: 27:). 24. Sometimes small animals enter the hive and do much damage, and when the bees are strong, they drive these away. And they are not afraid before other animals (HA 9: 27: 16 ! !). 25. When the bees have stung, they usually die, as their " spur " sticks to the intestines. And the stinging injures them much, when meat presses, into which the poison of the sting entered. It has been seen, that a horse died after the sting of a bee. The leader does not sting (HA 9: 27: 17). 26. (The bee) is the cleanest and purest of all animals. They defaecate whilst flying outside the hive, and bad smells injures them (HA 9: 27: 18). 27. The new " chicken " make the honey and do not sting as do the older ones (HA 9: 27: 19). The body of the " chicken " is soft, that of the older ones hard (HA 9: 27: 19!!). 28. Sometimes small worms are found in the hives, which do as the spiders do (webs of the wax moth) and they feed upon the honey (HA 9: 27: 20 & 10. The two wax moths). 29. In the summer a cool place is suitable for (the bees), and in winter a warm place (HA 9: 27: 30). 30. And when a great and strong wind suddenly rises, the watchman of the hive spreads over it (probably the entrance) a cloth so that the wind may not enter the hive, and they are very sensitive against cold and rain (HA 9: 27: 25 ! ! ! 23 !). 31. When they are close to the river, the bees think that the river surrounds them and they do not want to go out, until this cloth is taken away. They drink water only (HA 9: 27: 21 ! ! !). 32. The (great) activity of the bees is in spring and autumn (prima vera and autumne). 33. The honey made in spring is better than that made in autumn: clear, good and sweet. And honey is (made) from new wax. When it is made from new wax, it is red. The good and white honey is good for eye-diseases. And the lower part of the honey in the pot is good and excellent (HA 9: 27: 21). 34. And when the flowers are seen on the trees and herbs, the bees begin to make honey. And then they take the ready (i.e. stored) honey from the hive for food and they make new honey (HA 9: 27: 22). 35. As long as there is honey (available in the hive) for their food, they will not go busy (HA 9: 27: 24). 36. There is one kind of bees, which is called " wespes," which do not make honey. They sting just as do bees (HA 4: 7: 4). 37. They are said to live two years (HA 9: 27: 1 . . . the one-yeared wasps: but

9: 28: 2!). 38. And their nests are in the soil made from ash and clay (HA 5: 5: 20: 1). 39. They mainly live upon meat, and therefore they stand close to the places of meat in the slaughter houses, near the refuse heaps because of the corpses (to be found there). But they also eat sweet fruits and honey, yet they do not feed upon flowers (HA 9: 28: 6). 40. Bees and flies are silent, except when they beat with their wings, the air then entering between them (i.e. the wings). 41. They do not respire, except when they eat (HA 4: 9: 2).

35

KAMAL MUHAMMED AL DAMIRI (1349-1405), CAIRO.

From the big zoological lexicon of Arabic zoology, the *Hayat al Hayawan*, the greater part of which has been translated into English by A. S. G. Jayakar (London, 1907, 1908) the following two paragraphs:

(*a*) *Al Saratan*, the crab (from Jayakar II, pp. 43-45).

(*b*) *Faras al Bahr*, the hippopotamus (from Jayakar II, pp. 551-552).

(*a*) *As-Saratan, the crab.* A certain well-known animal; it is also called the water-scorpion. Its sobriquet is abu-bahr. It is one of the aquatic creatures, but also lives on land; it is very quick in walking and running, and has two jaws, claws, sharp nails, and several teeth, and is hard in its back; a person seeing it would think that it is an animal without a head or tail. Its two eyes are placed on its shoulders, its mouth is in its chest, and its two jaws are split on the two sides. It has eight legs, walks on one side. It draws in through the nostrils both water and air together. It casts off its skin six times a year, and builds for its hole two doors, one opening into water and the other on dry land; when it casts off its skin, it closes the door which is next to the water, out of fear for itself on account of the animals of prey of the fish kind, and leaves the one which is next to the dry land open, so that the wind may reach it, dry up the moisture in it, and strengthen it; when it becomes strong again, it opens the door next to the water and seeks its nourishment. . . . .

(*b*) *Faras al-bahr, the hippopotamus.* A certain animal found in the Nile in Egypt, having a mane like that of a horse and cloven feet like those of a cow. It is wide in the face and has a short tail resembling that of the pig. Its appearance resembles that of the horse, but its face is wide, and its skin excessively thick; it ascends on dry land and eats green plants, and sometimes kills man and other animals.

(Its lawfulness or unlawfulness). It is lawful to eat it, because it is like wild horses, which mostly cause injury.

(Properties). If its skin be burnt and mixed with the flour of pulse and then applied to cancer, it will cure it in three days. If its gall-bladder be left for thirty days in water, and then pounded and mixed with honey that has not been exposed over fire, and used as a collyrium for fourteen days or twenty-four days, it will remove the black humour (water) from the eye. Its tooth is beneficial for pain in the belly, if it be hung on the person of one who is on the point of death from pain in the belly due to indigestion or over-feeding; he will be cured by the order of God. If its skin be burned in the middle of a town or a village, no calamities will occur in it, and if it be burnt and applied on an (inflammatory) swelling, it will take it away, and relieve the pain due to it.

(Interpretation of it in a dream). A hippo in a dream indicates a lie and an affair that will not be completed . . . .

36

LEONARDO DA VINCI (1452-1519), MILANO.

Leonardo da Vinci, the great artist of the Renaissance, was a general genius. In his diaries mathematics, mechanics, fossils, and anatomy are largely treated. He did not make a school, as his diaries remained unpublished for some centuries. Here we bring Notes on the flight of birds as model for the flying machine of the bird-man (from E. MacCurdy's translation of the " Notebooks of Leonardo da Vinci." London, 1938).

A bird is an instrument working according to mathematical law . . . Such an instrument constructed by man is lacking in nothing except the life of the bird, and this life must needs be supplied from that of man. The life in the bird's limbs will doubtless better conform to their needs than will that of man. But man will be able to copy and comprehend the most simple of the very many varieties of movements. . . .

The slanting descent of birds against the wind will always be made beneath the wind, and their reflex movement will be made upon the wind. But if this falling movement is made to the east when the wind is blowing from the north, then the north wing will remain under the wind and it will do the same in the reflex movement, wherefore at the end of this reflex movement the bird will find itself with its front to the north. And if the bird descends to the south while the wind is blowing from the north it will make the descent upon the wind, and its reflex movement will be below the wind; but this is a vexed question which shall be discussed in its proper place, for here it would seem that it could not make the reflex movement.

When the bird makes its reflex movement facing and upon the wind it will rise much more than its natural impetus requires, seeing that it is also helped by the wind which enters underneath it and plays the part of a wedge. But when it is at the end of its ascent it will have used up its impetus and therefore will depend upon the help of the wind, which as it strikes it on the breast would throw it over if it were not that it lowers the right or left wing, for this will cause it to turn to the right or left, dropping down in a half circle.

The bird maintains itself in the air by imperceptible balancing when near to the mountains or lofty ocean crags; it does this by means of the curves of the winds which, as they strike against these projections, being forced to preserve their first impetus, bend their straight course towards the sky with diverse revolutions, at the beginnings of which the birds come to a stop with their wings open, receiving underneath themselves the continual bufferings of the reflex courses of the winds, and by the angle of their bodies acquiring as much weight against the wind as the wind makes force against this weight. And so by such a condition of equilibrium the bird proceeds to employ the smallest beginnings of every variety of power that can be produced.

37

GONZANO FERNANDEZ DE OVIEDO Y VALDES (16TH CENTURY).

Spanish traveller in all the territories of Spanish South America, to whose *Historia Natural y General de las Indias* (Sevilla, 1535). English translation from the chapters on natural history by R. Eden (1555, pp. 189, 193). We bring here parts of the descriptions of
(*a*) The Ant-bear.
(*b*) The Coco-palm.

(*a*) *The ant-bear of South America.* On the firm land another beast, the ant-bear is found. This beast is very similar in hair and colour to the Spanish bear, but it has a much longer snout. They are often taken with staves, without any other weapons. They are also taken by dogs, as their bites are not harmful. They are mostly found around the great ant hills. There lives a certain little, black ant, which flees the woods to avoid the ant-bear . . . Yet the ant-bear assaults them even in their hard hillocks. This beast licks the hard and narrow chemnies of their hills moist, until it can put his very long and thin tongue into it. He redraws the ants, draws it into his mouth, eats them and repeats this practice until he is sated. Its meat is filthy and unsavoury. The first immigrants did eat their meat, but when they found out their food, they started to hate them. . . .

(*b*) *The coco-palm of the Indies.* The Cocos tree of the continent and of the islands is a kind of date tree. Their leaves are as long as those of the date palm, but of different shape. The Cocos leaves grow of the trunk, as fingers grow out of a hand. These trees are high, and in some parts abundant. The fruit is almost as big as a man's head, in midst of which is the fruit, the outer skin of it being covered with the webs of ropes. The eatable fruit is round and as big as a man's fist. When the fruit is chewed, it feels like almond fruits . . . While the Cocos is yet fresh they draw a milk from it which is much better and sweeter than milk from animals, and without offence to the stomach. It leaves a very delicate taste. The outer meat of the fruit is at that time not less delicate to eat. Those who drink this milk say that it is a marvellous remedy against the stone and is urotropic, and helps against all the diseases which Pliny says that the date palm heals.

38

VALERIUS CORDUS (1514-1544), WITTENBERG.

In contrast to the so-called Fathers of Botany, to which Leonhard Fuchs belonged, it is the early deceased Valerius Cordus, Professor of Medicine, who gave the good original descriptions. We give here the description of the Chaste-tree (*Vitex agnus castus*), which he met on his travels in Italy (from *Stirpium descriptiones* (1541; ed. nov. Nuernberg, 1751. p. 7, no. 8).

No. 8. On the Chaste-tree (*Vitex agnus castus*).

The *Agnus* or *Vitex* is usually called *Agnus castus*. It is a bushy shrub, growing almost into a slender tree, if it is permitted to grow into adolescence. It divides from a fairly thick trunk into many branches. From the trunk it sends out, from near the root, many shoots, which can be easily torn out and, firmly pressed into soil, readily grow again. The young branches and the new twigs rise at regular intervals opposite to each other, as opposed forking twigs from the knots (of the stalk) . . . The pedicles arises from a single knot, on both sides, one opposite the other, much shorter than a digit, smooth and tender, from the end of which come out five, seven, or sometimes more leaves, all arising from one common organ like a human palm or like the leaves of hemp: long, narrow and sharp, with a straight and prominent upper longitudinal vein (*spina*) all along the leaf, similar to an olive-leaf; but they are much longer and more tender, greenish above, greyish and moderately woolly underneath. The leaves drop in winter. The biggest and longest leaf is the terminal one, the others being slightly smaller, on both sides, the more they are distant from the biggest one. The last two, sometimes the last one, are very small.

It flowers in June and July at the forked ends of the long and straight twigs; small, elongate, concave, with the extreme margin (*labrum*) longer and indented five times, bringing out small and medium stamens (*stamina*); attached to a small green or greyish cup (*caliculus*) as in the flowers of lavender or rosemary, and equal, on all sides arranged in a ring around the twiglets in some whirls, forming at intervals a long ear (*spica*). Inside the little flowers are blue-purple, outside (*foris*, nec: *feris*) whitish blue, one each in the indentation of the cup. There grow the seeds which are small, rounded, smaller than pepper, greyish, hard, woody. They remain throughout the winter in the forks. They germinate and sprout late. The bark of these twigs is soft, flexible and not easily broken, just as is the nature of the young twiglets. Leaves, flowers, seed and bark have a 'heavy' smell which oppresses the head. The taste of these parts is slightly bitter and sharp, delicate only in the less active bark than in the other parts. The leaves and the seeds have a stronger taste than the flowers. Hence, the sharpness of the leaves and of the unopened flowers cleaves long to the palate when tasted.

It grows in wild fields and along the sea, not far from inundations and rapidly flowing rivers. Most common are they in the county of Ancona. It is sown in various gardens in Italy, as at Venice, Travisi, Ferrara, Padua and Bologna.

39

LEONHARD FUCHS (1501-1566), TUEBINGEN.

Fuchs, together with Braunfels and Bock, are called the Fathers of Botany. This name is entirely unjustified by the poor descriptions of these botanists, who, however, gave the first woodcuts of plants, still primitive, but making possible the identification of many herbs. The poverty of their descriptions can easily be appreciated from the following text. The longest paragraph, on the properties of plants, is merely an excerpt from Dioscorides, Galen, Pliny, and Symeon Sethi, hence the repetitions.

We have used here the Latin edition *De Historia Stirpium Commentarii insignes* (Basel 1542, pp. 615-618, fig. p. 616), the French edition *Histoire Générale des Plantes et Herbes avec leur proprietez* (Rouen, 1675, pp. 83-87), and the German edition *New Kreuterbuch* (Basel, 1545). The following chapter is No. 235 of the Latin edition.

235. *On the rue* (*Ruta*).

Names: The *Raut* or *Weinraut* in German, is the *Peganon cepeuton* in Greek, *Ruta hortensis* in Latin. The Latin name is used by the druggists.

Shape: A very strong smelling herb, almost always green; with many small, rounded leaves of greyish colour. The flowers are black-yellow, like a beautiful star; they grow into four- or five-sided pods which contain black seeds. Stem and root are woody; inside, yellow like the wood of *Buxus*.

Place: It is always grown in medicinal gardens. It prefers dry places where the sun shines (*aprica loca*).

Season: It flowers in summer, the seed matures in autumn, when it should be collected.

Fig. 12. The herb Raut or Peganum (*Ruta*) drawn after nature in the Herbal of Leonhard Fuchs. (French edition. B.M. Nat. Hist. *Ruta*, No. 236). Paris, 1519.

Nature: The taste is not only sharp, but also bitter. The rue is warm in the third degree and very drying.

Properties: The rue makes a man piss and brings about the periods of women. When the leaves are boiled in wine and then drunk, they are good and useful against many deadly poisons.

When the leaves are eaten before a meal or pulverised with welsh nuts and dry figs, they are potent against many poisons and against pestilential air, also against snakes and vipers. When the rue is always eaten or drunk, it destroys the human semen. When boiled with dry dill and drunk, it relieves pain. It also relieves pain in the sides, the breast, etc., also difficult respiration and coughing . . . When boiled in oil and used as a clyster, it relieves pains in the bowels . . . Boiled with oil and drunk it laxates all worms. Mixed with figs and applied as a plaster, it removes the swelling of dropsy. Boiled in wine until half of the liquid has evaporated, and drunk, it relieves dropsy. When the green leaves are sprinkled with salt, they sharpen the vision. Mixed with rose-oil and vinegar, it is a good remedy against headaches. Applied as a powder it stills the bleeding of the nose . . . Crunched with green laurel leaves, and used as a poultice it relieves swellings. Mixed with wine, pepper and nitre, and rubbed on the bare skin, it heals all kinds of moles . . . The sap of its leaves, when taken with wine, is useful to those who have been stung or bitten by scorpions, spiders, bees, wasps, hornets, and mad dogs. If someone rubs himself with the sap of rue, he is safe from these animals, which cannot endanger him. Mixed with rose-oil and put into the ears, it restores hearing and drives away singing in the ears . . . Boiled in oil and used as a poultice, it is good for frozen limbs. If somebody cannot let urine, let him boil rue in oil and use this as a poultice upon the bladder . . . (This is about half of the part on the properties, the remainder being mainly repetitions).

40

ANDREAS VESALIUS (1514-1564), BRUXELLES AND PADUA.

Vesal is the founder of modern anatomy. He and Leonardo da Vinci were the first to draw the parts of the human body in their proper perspective (C. Singer), as they saw them. We render here :

(*a*) Vesal's development as an anatomist. From the dedication of the *De Humani Corporis Fabrica* (1543). English translation by B. Farrington, Proc. R. Soc. Med. 1932. pp. 1357-1366.

(*b*) Demonstration of the connection of the voice with the recurrent nerves. Also from the *Fabrica*. Translated by F. S. Taylor, Science, past and present. London, 1945. pp. 67-68.

(*a*) My effort to restore anatomy at least to the level of the old teachers of that science, would not have succeeded, when at Paris I had acquiesced in the casual and superficial display to the students by certain barbers of a few organs at one or two public dissections. In this perfunctory manner anatomy was then treated, where it now is happily reborn. I myself, having trained myself without guidance in the dissection of brute creatures, at the third dissection at which it was my fortune ever to be present, as usually concerned exclusively with the viscera, led on encouraged by my fellow students and teachers to perform in public a more thorough dissection than usual. Later I tried a second dissection, to exhibit the muscles of the hand together with a more accurate dissection of the viscera. For, and that is true, except for eight muscles of the abdomen, disgracefully mangled and in the wrong order, no one ever demonstrated to me any single muscle, or any single bone, much less the network of the nerves, veins and arteries.

Subsequently, at Louvain, where during 18 years the doctors had not even dreamed of anatomy, in order to help the students of the academy and to perfectionate myself in this important science, I expounded somewhat more accurately than at Paris the whole structure of the human body, so that now the younger teachers of that academy are eager to gain knowledge of the parts of man. Furthermore, at Padua, in the most famous gymnasium of the world, I was charged for five years with the teaching of surgical medicine. And since anatomical dissections are of importance for surgical medicine, I devoted much effort in its enquiry, and exploding the ridiculous fashion of the schools, taught the subject in a manner not falling short of the tradition of the ancients . . .

All later writers borrowed from Galen, in whose books they assumed that not the slightest error could be found, whilst Galen frequently corrected himself in his later books, after acquiring more experience, removed oversight and taught contradictory views. Deceived by his monkeys, Galen frequently wrongly controverted the ancient physicians who had trained themselves by dissecting human corpses. Astonishing is the fact that in the manifold and infinite divergences of the organs of the human body from those of the monkey Galen hardly noticed anything except in the fingers and in the bend of the knee, which he would certainly have passed over with the rest, if they had not been obvious to him without dissection. But here I do not propose to criticise the errors of Galen, who is easily the foremost amongst the teachers of anatomy, and I lack not in respect for his authority,

Fig. 13. Miniature initial from the *Fabrica* of Vesal, illustrating the experiment concerning the voice of the pig.

even if in a single course of anatomy I now exhibit over 200 errors of Galen.

(*b*) *Dissection of a living, squealing sow:* Before the animal is thus tied down, I usually address myself to any spectators who are not yet versed in the anatomy of dead animals and go over the things that should be seen in the dissection in question, in order that a wordy explanation in the middle of the dissection shall not interrupt the work, and so that it shall not come to be disturbed by talking. And so I immediately make a long incision in the neck with a sharp razor, so as to divide the skin and the muscles beneath it as far as the windpipe; taking care that the incision does not go too far to the side and injure an important vein. Then I take hold of the windpipe in my hands and stripping it from the attached muscles with my fingers only, I look for the carotid arteries at the side of it, and for the nerves of the sixth pair of cerebral nerves stretched out thereon. Then I observe the recurrent nerves also attached to the sides of the windpipe, and these I sometimes compress by ligatures and sometimes sever: this I do on one side so that when the nerve is compressed or severed it may be clearly observed that half the voice is lost, and that when both nerves are damaged the voice is completely lost, and that if I loosen the ligature it returns.

41

KONRAD VON GESNER (GESSNER) (1516-1565), ZUERICH.

Gessner was one of the founders of modern animal description. In a series of foliants he compiles all the knowledge of antiquity with his own knowledge and descriptions.

(*a*) The description of the Forest-Raven, the Waldrap, *Comatibis eremita*, which has long since disappeared from the Alps, and has been rediscovered in this century in the mountains of the Syrian desert and of the High Atlas. From the Vogelbuch (Frankfurt, 1582. p. 199) the German edition of the third volume of the *Historiae Animalium* (1555. p. 337).

(*b*) Letter to Avienus (Vogel) in Glarus, giving expression to a modern feeling of nature and landscape. *Libellus de lacte et operibus lactariis*. Tur. 1541. Praefatio.

(*a*) *On the forest raven* (*Corvus sylvaticus*). The bird which is illustrated here, is usually called forest raven, as it inhabits desolate forests. There it nests on high mountains in old deserted towers and castles, and hence it is called rock-raven, or in Bavaria and Styria a claus-raven, from the rocks and narrow gorges where it builds its nest. In Lorraine and on the Lake Verbenum it is called a sea-raven. Elsewhere, as in Italy, it is called the forest-raven. There a man lets himself down on a rope to take (? the eggs) from the nests, which are regarded as a delicacy. Also in our spa at Pfaefer some hunters have collected them in a similar way. From its voice it is called in Germany also the tinkler (Scheller). Some believe it to be the Phalacrocorax, as it resembles in size and colour very much the raven, and, as I have seen, it also has a bald head in its old age. Turner believes that

Fig. 14. The Waldrap of Carl Gesner. (From Gesner: *Von den Voegeln*).

the water raven of Aristotle and the Phalacrocorax of Pliny are the same bird as our forest raven. But our bird does not resemble the description of these birds: it is not broad-footed, it is no water-bird. Our forest raven has the size of a domestic fowl: entirely black when you look at it from far. When you look at it from near, against the sun, it appears to be mixed with green. The feet are also like those of the domestic fowl, but longer and cleft. The tail is not long. On its head it has a crest behind, but I am not certain that this occurs in all individuals and at all times. The beak is reddish, long and fit to burrow in the soil and to thrust into the crevices of walls and rocks to extract from them the hidden worms and beetles. It has long, dark-red legs. Their food is grasshoppers, crickets, small fishes and small frogs. Usually it nests on old and high walls of ruined castles, which abound in Switzerland. When I opened the stomach of this bird, I found amongst many other worms some insects which do damage to the roots of cereals, mainly of millet. They also devour the grubs from which the cockchafer develop. These birds fly very high. They lay two or three eggs. They leave us first of all the birds, early in June. Their young have become fledged, and have been trained to leave the nest for the fields and to return to it speedily. Their young are also highly esteemed for the table, as their meat is sweet and their bones soft. But when these are taken from the nest, one is always left in it, to encourage their return in the following year.

(*b*) Dear friend, I am resolved as long as, by Divine providence, I live, to climb every year mountains, or at least one mountain, and to do this at the season when the vegetation is most vigorous. I wish to do so, partly to enlarge my knowledge, partly to strengthen my body, and to give to my spirit the most noble recreation. What splendid enjoyment, what a rapture it is to see and to admire the endless ranges of the mountains and to raise one's head above the clouds! This unaccustomed height impresses upon the soul the stamp of sublimity and forces upon it adoring admiration of the omniscient Creator. Only men of an indolent soul admire nothing, remain in dull apathy at home, do not enter the beautiful arena of the world, lie dormant, hidden like marmots in a corner, and do not consider, that the human race has been put upon this earth, to learn from the contemplation of its wonders something greater, namely invisible God himself... Who loves wisdom shall proceed to contemplate the rich arena of this world with the eyes of the body and of the soul; he shall ascend high mountains, turn his eyes upon that infinite chain of the Alps, he shall wander through the shade of the woods, he shall stay on the lofty heights of the mountains and he shall embrace there the endless variety of the things which are spread before his eyes... And furthermore, how refreshing and invigorating is the breathing of pure mountain air, which surrounds us everywhere! The sight is delighted by the greatest variety and is strengthened by it: Close by, by plants which excel in the most vivid colours and the most tender shapes; in the distance, by the admirable forms of the mountains, the dazzling surface of the lakes, the meandering course of the rivers, the rich and cultivated plains with their towns, villages and hamlets, or the grassy pastures of the mountains with their huts and grazing flocks. The ear perceives here the pleasant song of the birds; there, the holy awfulness of silence, not broken by even the lowest sound. Everywhere we are surrounded by aromatic odours, as even those plants, which in the lower valleys have no smell, breathe on the heights of the Alps tender, aromatic perfumes. And in this pure air every sensual enjoyment is purer, finer, nobler. The cold water refreshes the entire body, the balsamic milk refreshes and gladdens, and the hunger raised by the effort of the ascent makes the most simple meal in the hut of the shepherds an ambrosia . . .

42

OLAUS MAGNUS (1490-1558), UPPSALA AND ROMA.

Olaus Magnus wrote an exciting book *Historia de Gentibus Septentrionalibus* (1555, English translation in 1658), in which appear the first descriptions of many animals of the Arctic lands and of the Nordic seas. Of these we bring:

(*a*) On the Volverine or glutton (XVIII: 5-7, p. 180).
(*b*) On the fight of the Whale and of the Orca (XXI: 7, p. 226).

(*a*) *On the Volverine or Glutton.* The Volverine of the northern parts of Sweden, usually called Gulo, Jerff in the local language, Vielfrass in German, Rassamaka in Slavic, is believed to be the most insatiable of all animals, and from this character has got its name. It has the size of a big dog, cat-like face and ears, with very sharp claws, a woolly rump with long brown hair. Its tail is fox-like, but shorter and with denser hair; of these they make good winter caps. The Wolverine is really most voracious. It devours so much of a cadaver that it has found, that from the plenty of food its body is stretched and expanded like a drum. Then it squeezes itself strongly through (the narrow passage between) two neighbouring trees, to void its body by vomiting. Having thus become lean (empty), it returns to the cadaver and fills itself again to top capacity. Again it squeezes itself through the same narrow passage, as before, and returns then to the cadaver until it has consumed it entirely, and then it looks eagerly out for a new prey. It was apparently created as a shame to those men who are gluttons and drunkards, who vomit in order to spend night and day on the table, as Mechovita believes in his Sarmatia. This animal's flesh is not fit for human consumption, while its fur is agreeable and precious. It has a brownish black shine like the brocade of Damascus . . . (28: 7.): How the Wolverine is hunted: A fresh cadaver is taken into a forest, where these beasts are often seen, especially when there is deep snow, as their summer fur has no value. They smell the cadaver, and begin to devour it until they are extended like a sac. Then they squeeze themselves, not without pain, between two neighbouring trees, where the hunter's arrow can reach them easily. The Wolverines are also caught in traps in which they are strangulated.

(*b*) *On the fight of the Whale* (*Balaena*) *and the Orca.* The whale is an exceedingly great fish, 100 or even 300 feet long, with an enormous body-mass. But it has a deadly enemy in the orca which is very much smaller, but much superior by its agility to jump and to attack. This orca is like an upturned keel, a beast with fierce teeth which are like the stern of a war-ship, by which it lacerates the whale's guts as well as the body of its young. Or it quickly runs, attacking the whale, until it is driven by its prickly back into a fjord or into the shallow water of the shore. But the whale cannot turn its enormous body there, nor knows it to resist the wily orca, flight being its only refuge. But during flight the whale becomes weaker, as this sluggish beast is burdened by its own weight and finds no one to guide it out of the dangers of the shallow waters.

43

PIERRE BELON (1517-1564), PARIS.

Belon visited the countries of the Eastern Mediterranean in order to identify the animal and plant names of the ancients. He wrote books on birds, fishes and trees. He treated the importance of anatomy, especially of the skeleton, for the systematics of birds.

In his *Portraits d'Oiseaux* (1557) he demonstrates the homology of a bird with that of man (fig. 2, pl. 3b and 4a).

> When you desire to follow nature in the composition of a body, you shall conveniently begin with the bones, which are, so to say, its primary fundaments. Just as we find characters to distinguish the different birds by their exterior, others are found in their inner structure. The great Aristotle was not above observing and

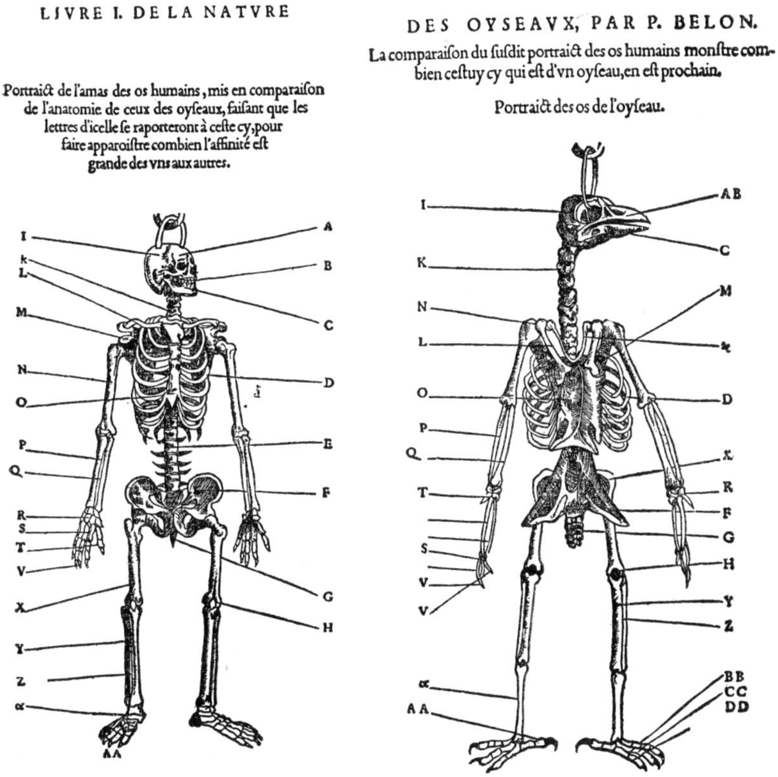

Fig. 15. Pierre Belon homologizes the bones of a bird with those of man in his book *Nature des Oyseaux* (1655). A view of the bones of the hands of man and of the wings of the bird show to what degree Belon has contributed to the beginnings of comparative anatomy.

dissecting them one by one, and found this very useful for the description of many hidden secrets of nature, which he could not discover in any other way. He states, that the parts and limbs of the various birds differ as much as their exterior. From the exterior we learn if a part be proportionate, big or small, and the bones, their inner fundaments, follow these indications of the exterior. When you take a wing or a leg of a bird to compare it with that of a mammal or of man, you will find that their bones are, as it were, corresponding in one and the other. To demonstrate this to the layman we shall not lose time by long explications, but will put the corresponding letters on the corresponding bones in the drawings of the skeleton of a bird and of man. In our commentary of Dioscorides we shall compare the bones of the birds with those of other vertebrates as well. The ancients have dissected birds and fish, snakes and mammals not for medicine, but for the better understanding of their actions.

44

THEOPHRASTUS BOMBAS VON HOHENHEIM called PARACELSUS (about 1490-1541), BASEL.

Paracelsus was one of the early reformers of chemistry and medicine. Here is a small paragraph from his book *Vom Schwebel oder erden*, Hartz, Basel, 1567. He describes there the narcotic effects of a product from vitriol and alcohol, which he applied to quieten nervous patients. In studying the effects of this substance Paracelsus writes:

> " All sulphurs derived from vitriol and salts are stupefactive, narcotic, analgetic and hypnotic. They act, however, in a mild and transient way unlike hyoscyamum, poppy and mandregora. There is no need to prepare its *quinta essentia* (as in opiates). In its natural condition the substance is sweet and liked by chickens, which it causes then to sleep for some time, and afterwards they rise without any harm."

We owe this passage to Dr. W. Pagel, London, to whom we express our sincere thanks for this and other courtesies.

45

VOLCHER COITER (1534-1576), NETHERLANDS.

Coiter is one of the first comparative anatomists. From his writings we render the comparative description of the skeleton of the squirrel from *Opuscula Selecta Neerlandicorum*, vol. 18, 1955, p. 195. (Ed. B. T. W. Nuyens and A. Schierbeek). It was written in 1575.

Chapter 6. *The Squirrel.* For most aspects the squirrel does not differ much in bone structure from the hare. Its head is similar to the hare's; from the extremity of the upper jaw two long teeth, sharp at the ends, make their appearance, which are not far apart at their starting-points and meet at the ends. Two also make their appearance from the lower jaw, which run parallel for their full length; their ends, becoming broader and sharper, extend towards the upper pair. The edges of these upper and lower teeth do not meet, but the lower pair fit in under the upper ones. Squirrels have powerful collar-bones and fairly strong shoulder-blades, not unlike those of the hedgehog. The process growing out of the middle of the shoulder-blade is not separated, however, from the base of the shoulder-blade, as in the hedgehog. It is provided with twelve ribs on either side. It has a rather short, broad chest, containing six long, slender parts of the breast-bone, the first of which, on account of its connection with the collar-bones, is broader and more powerful, especially at the top. The forelegs correspond to those of the hedgehog, and have also, as in the marten, five separate toes on the feet, namely four true toes and one big toe. The big toe, like the human thumb, stands out, for gripping. For they use their forelegs not only for running, but also for gripping; since quadrupeds have forepaws instead of hands. Like the animals dealt with above, the squirrel has seven cervical vertebrae, corresponding in every respect to the vertebrae of the marten. The loins have eight vertebrae, which are rather strong and powerful in proportion to the body.

The hip-bones are nearly the same as in the hare, marten and hedgehog. The femur resembles that of hare and fox. The tibia has two bones, of which the fibula, with a cartilaginous connection on the outside, proceeds from the tibia and separates fairly widely from it. Descending, it comes closer to it again and almost at the bottom grows together very strongly with it as far as the talus, with which it articulates, as in human beings.

The tail is very long. The hind-feet are split into five toes, with no big toe at all standing out from the others. Small animals with toes have five toes on the hind feet as well because they are of a creeping habitude; this is, of course, because they can more easily climb higher by gripping with a larger number of toes. The larger toed animals on the other hand, such as lions, wolves, foxes, dogs, leopards, etc., have five toes in the forefeet only.

The bones of the domestic mouse resemble exactly those of the squirrel, as do those of the dormouse.

46

OLIVIER DE SERRES (*ca.* 1539-1619), PRADEL, FRANCE.

Serres was one of the founders of agricultural science. We borrow here from his *Théatre d'Agriculture et Mesnage des Champs* (1600) two paragraphs concerning novelties in French agriculture:

(*a*) The Potatoe. Théatre, p. 82f.

(*b*) The Silkworm. Théatre, p. 72 f.

(*a*) *The potatoe.* This shrub, called cartoufle, produces fruits of the same name which are like truffles, and are even called so by some. It came into the Dauphiné from Switzerland, not long ago. It lives for one year only, and therefore must be replanted every year. It reproduces by seed, i.e. by the fruit itself, which in early spring is put into the soil, after the great cold, when the moon is waning, four finger deep. It requires a good soil, well manured, better light than heavy, a moderate climate . . . All through the winter they have to be kept in fine sand in a temperate cave, so that the rats cannot reach them, which are much attracted to such food and devour it quickly when they reach it. Nobody takes the trouble to layer this plant (*provigner*), but one leaves it to grow and to bring fruit spontaneously, taking the fruit out in the proper season. Yet the fruit does not mature so well in air as it does in the soil, being also therein like the true truffles, to which the cartoufles are similar in shape but not in colour, as they are clearer than truffles. The peel is not rugged, but smooth and fine. This is how these two fruits differ one from the other. With regard to taste, little difference can be found between them, when they are eaten after having been cooked.

(*b*) *The Silkworm.* The silk can grow well in any part of this kingdom, with few exceptions. Wherever the grape-vine grows silk can also be grown, as is frequently demonstrated by experience. Even where only the mulberry-tree will grow, even when not accompanied by grape-vine, the silkworm will prosper, as was shown in Leiden in Holland in 1593/95, where the Duchess of Ascot succeeded to breed silkworms, and from the silk which was produced from them and which our ladies wore, to the admiration of all who saw them; and this in spite of the cold of that region. History proves that in the time of the ancient Gauls France did not produce wine, with which noble drink it is amply provided today, thanks to the skill of those who started its cultivation as a profitable curiosity . . . To breed them is not difficult. The mulberry trees in the open do not require any work. The worms which suffer from cold are kept in the houses, not in the open, even in spring and during part of the summer. The silk-crop is later than in more southerly countries; but who cares for that so long as the silk is good and beautiful, if it is collected not in May and June as in Languedoc and Provence, but in July and August? That is the same with the wine. Mulberry-trees existed before the science of silkworm breeding . . . Whoever takes care of them in all seasons makes a good profit. He will collect many mulberry leaves at the silkworm season, as is done at Nimes, e.g. . . .

47

ULYSSE ALDROVANDI (1522-1605), BOLOGNE.

Aldrovandi was a contempory of Gesner. He also compiled in many foliants the knowledge of the ancients with that of his own time and observations. We bring here the systematic division of his *Insecta.* (Bologna, 1602, in *De Animalibus Insectis Libri VII.* Foreword.), which include most terrestric Evertebrates.

Definition: Insects are small segmented animals, which are distinguished, as follows :

I. *Terrestrial Insects.*
  A. With legs.
    AA. with wings.
      1. without elytra.
        a. with four wings.
          aa. with hyaline wings.
            α. producing wax (bees, bumble-bees, wasps, hornets).
            β. not wax-producing (cicada, Perla, bugs, Orsodacne).
          bb. with scaly wings (butterflies and moths).
        b. with two wings (flies, gnats, ephemerids).
      2. with elytra (grasshoppers, crickets, beetles, cockroaches).
    BB. without wings.
      1. with few legs.
        a. with 6 legs (ant, bed-bug, louse, tick, flea, mole-cricket, earwigs).
        b. with 8 legs (scorpion, spider).
        c. with 12 or 14 legs (caterpillars).
      2. with many feet (myriopods, woodlice, scolopender, julus).
  B. without legs. (The worms which originate in men, animals, plants, stones and metals, Teredo, earthworm, snails, moths, Otips).

II. *Aquatic insects.*
  A. with legs.
    1. with few legs: (larvae of phryganids, water fly, water beetles and various water bugs).
    2. with many legs: (marine flea, marine louse, the marine oestrus, the marine scolopender, worms in tubes (annelids and crustaceans).
  B. without legs (worms, echinoderms, hippocamps).

I distinguish the insects first by the place where they are born and live, then according to colour, size, food, origin, motion and coitus . . .

The insects are perfect animals, as nothing more can be desired in perfection of their body and soul . . .

Colours are important for the distinction of species especially in butterflies, moths and caterpillars, whilst ants and worms are each of one colour only, more or less : Some are big, others small;

some round, angular or oval; smooth, rough or hairy. The upper half differs from the lower, but all insects have three parts: head, thorax and abdomen. First in importance is the separation of terrestrial from aquatic groups. The number of legs and wings follows in importance.

48

FRANCIS BACON OF VERULAM (1561-1626), LONDON.

Francis Bacon underlined strongly the importance of observation of nature, which he still called *experimenta.* Ecological observations on the changes in plants due to their environment are found in his *Sylva Sylvarum*, or A Natural History (7th edition. London, 1658. pp. 111-113).

On the transmutation of plants.

It is certain, that in sterile years corn sown will grow to another kind.

> Grandia saepe quibus mandavimus hordea sulcis
> Infelix lolium, et steriles dominantur avenae.

And generally it is a rule, that plants that are brought forth by culture, as corn, will sooner change into other species, than those that come of themselves. For that culture gives but an adventitious nature which is more easily put off.

This work of the transmutations of plants, one into another, is inter Maglia Naturae: For the transmutation of species is, in the vulgar philosophy, pronounced impossible. And certainly it is a thing of difficulty, and requires deep search into nature. But seeing there appear some manifest instances of it, the opinion of impossibility is to be rejected; and the means thereof to be found out. We see, that in living creatures, that come of putrefaction, there is much transmutation, of one into another. For it is the seed, and the nature of it which locks and bounds in the creature, that it does not expatiate. So as we may well conclude, that seeing the earth, of itself, does put forth plants without seed, therefore plants may well have a transmigration of species. Wherefore wanting instances, which do occur, we shall give directions of the most likely trials. And generally, we would not have those, that read this work of *Sylva Sylvarum*, account it strange, or think that it is an over-haste, that we have set down particulars untried; for contrarywise, in our own estimation, we account such particulars, more worthy than those that are already tried and known. For these latter must be taken as you find them; but the other do level point blank at the inventing of causes and axioms.

First, therefore, you must make account, that if you will have one plant change into another, you must have the nourishment over-rule the seed. And therefore you are to practise it by nourishments, as contrary as may be, to the nature of the herb; so nevertheless as the herb may grow. And likewise with seeds that are of the weakest sort, and have least vigour. You shall do well, therefore, to take marsh-herbs, and plant them upon tops of hills, and fields; and such plants require much moisture upon sandy and very dry grounds. As for example, marsh-mallows and sedge upon hills, cucumber and lettuce-seeds and coleworts upon a sandy plot. So contrarywise plant bushes, heath, ling and brakes upon a wet or marsh ground. This I conceive also, that all esculent and garden herbs, set upon the tops of hills will prove more medicinal, though less esculent than they were before. And it may be likewise, some wild herbs you may make salad herbs. This is the first rule for transmutation of plants.

The second rule shall be to bury some few seeds of the herb you would change, amongst other seeds; and then you shall see, whether the juice of those other seeds do not so qualify the earth, as it will alter the seed, whereupon you work. As for example, put parsley-seed amongst onion-seed, or lettuce seed amongst parsley seed, or basil-seed amongst thyme-seed, and see the change of taste or otherwise. But you shall do well to put the seed you would change into a little linen cloth, that it mingle not with the foreign seed.

The third rule shall be the making of some medley or mixture of earth with some other plants bruised or shaven, either in leaf or root: as, for example, make earth with a mixture of colewort leaves stamped and set in it artichokes or parsnips. So take earth made with marjorn or origanum or wild thyme, bruised or stamped, and set in it fennell-seed, etc. In which operation the process of nature still will be (as I conceive) not that the herb you work upon should draw the juice of the foreign herb (for that opinion we have formerly rejected); but there will be a new confection of mould, which perhaps will alter the seed, and yet not to the kind of the former herb.

The fourth rule shall be, to mark what herbs some earths do put forth of themselves. And to take that earth and to pot it or to vessel it. And into that set the seed you would change: as for example, take from under walls or the like, where nettles put forth in abundance, the earth which you shall find there, without any string or root of the nettles; and pot that earth, and set in it stock-gilly-flowers or wall-flowers, etc. Or sow in the seeds of them, and see what the event will be; or take earth, that you have prepared to put forth mushrooms or itself (whereof you shall find some instances following), and sow it in purslane seed or lettuce seed, for in these experiments it is likely enough, that the earth being accustomed to send forth one kind of nourishment, will alter the new seed.

The fifth rule shall be to make the herb grow contrary to his nature; as to make ground-herbs rise in height: as for example, carry camomile or wild thyme or the green strawberry upon sticks, as you do hops upon poles, and see what the event will be.

The sixth rule shall be to make plants grow out of the sun or open air; for that is a great mutation in nature; and may induce a change in the seed: as barrel up earth, and sow some seed in it, and put it in the bottom of a pond; or put it in some great hollowed tree; trie also the sowing of seeds in the bottoms of caves; and pots with seeds sown, hanged up in wells, some distance from the water, and see what the event will be.

49

WILLIAM HARVEY (1578-1657), LONDON.

William Harvey is known in the history of physiology as the discoverer of the blood circulation within the body by calcul and by experiment. His *Exercitatio Anatomica de Motu Cordis et Sanguinibus in Animalibus* (London, 1628) was translated by R. Willis (London, 1847), from whom we render:

(*a*) How Harvey conceived the notion of the circulation of blood (Chapter VIII, pp. 45-47).

(*b*) Description of the motion of the heart (Chapter I, p.19; II, p. 21).

(*c*) Harvey's computation of the blood volume passing through the heart (Chapter IX, p. 48).

(*a*) *Of the quantity of blood passing through the heart from the veins to the arteries; and of the circular motion of the blood.* Thus far I have spoken of the passage of the blood from the veins into the arteries, and of the manner in which it is transmitted and distributed by the action of the heart; points to which some, moved either by the authority of Galen or Columbus, or the reasonings of others, will give in their adhesion. But what remains to be said upon the quantity and source of the blood which thus passes, is of so novel and unheard-of character, that I not only fear injury to myself from the envy of a few, but I tremble lest I have mankind at large for my enemies so much doth wont and custom, that become as another nature, and doctrine once sown and that hath struck deep root, and respect for antiquity influence all men: still the die is cast, and my trust is in love of truth, and the candour that inheres in cultivated minds. And sooth to say, when I surveyed my mass of evidence whether derived from vivisections and my various reflections on them, or from the ventricles of the heart and the vessels that enter into, and issue from them, the symmetry and size of these conduits,—for nature doing nothing in vain, would never have given them so large a relative size without a purpose,—or from the arrangement and intimate structure of the valves in particular, and of the other parts of the heart in general, with many things, besides, I frequently and seriously bethought me, and long revolved in my mind, what might be the quantity of blood which was transmitted, in how short a time its passage might be effected, and the like; and not finding it possible that this could be supplied by the juices of the ingested aliment without the veins on the one hand becoming drained, and the arteries on the other getting ruptured through the excessive charge of blood, unless the blood should somehow find its way from the arteries into the veins, and to return to the right side of the heart. I began to think whether these might not be:

A Motion, As it were, In a Circle.

Now this I afterwards found to be true; and I finally saw that the blood, forced by the action of the left ventricle into the arteries, was distributed to the body at large, and its several parts in the same manner as it is sent through the lungs, impelled by the right ventricle in the manner I indicated already. Which motion we may be allowed to call circular in the same way as Aristotle says that the air and the rain emulate the circular motion

WH conſtat per fabricam cordis ſanguinem
per pulmones in Aortam perpetuo
tranſferri, as by two clacks of a
water bellows to rayſe water
conſtat per ligaturam tranſitum ſanguinis
ab arterijs ad venas
vnde Δ perpetuum ſanguinis motum
in circulo fieri pulſu cordis
An ? hoc gratia Nutritionis
an magis Conſervationis ſanguinis
et Membrorum per Infuſionem calidam
viciſſimque ſanguis Calefaciens
membra frigifactum a Corde
Calefit

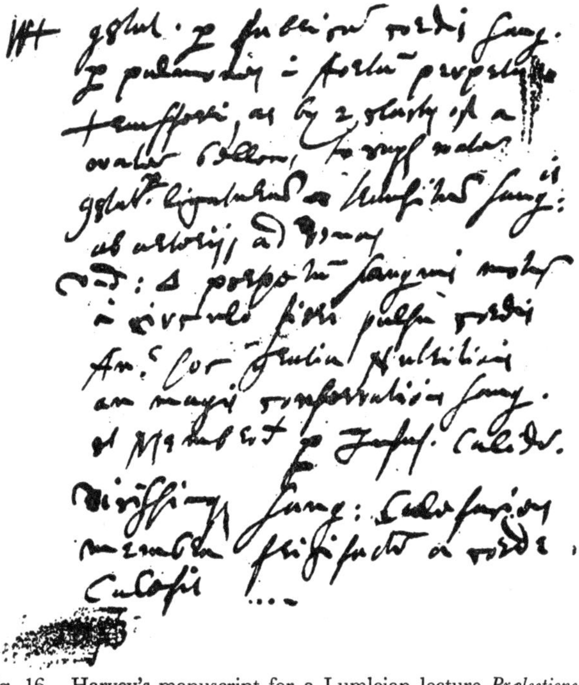

Fig. 16. Harvey's manuscript for a Lumleian lecture *Prælectiones anatomiæ universalis*
Above: the printed text (London, 1886)
Below: the original manuscript (1616)

of the superior bodies; for the moist earth, warmed by the sun, evaporates; the vapours drawn upwards are condensed, and descending in the form of rain, moisten the earth again; and by this arrangement are generations of living things produced; and in like manner, too, are tempests and meteors engendered by the circular motion, and by the approach and recession of the sun.

And so, in all likelihood, does it come to pass in the body, through the motion of the blood, the various parts are nourished, cherished, quickened by the warmer, more perfect, vaporous, spirituous, and, as I may say, alimentative blood; which, on the contrary, in contact with these parts become cooled, coagulated, and, so to speak, effete; whence it returns to its sovereign the heart, as if to its source, or to the inmost home of the body, there to recover its state of excellence or perfection. Here it resumes its due fluidity and receives an infusion of natural heat—powerful, fervid, a kind of treasury of life, and is impregnated with spirits, and it might be said with balsam; and thence it is again dispersed; and all this depends on the motion and action of the heart.

The heart, consequently, is the beginning of life, the sun of the microcosm even as the sun in his turn might well be designated the heart of the world; for it is the heart by whose virtue and pulse the blood is moved, perfected, made apt to nourish, and is preserved from corruption and coagulation; it is the household divinity which discharging its functions, nourishes, cherishes, quickens the whole body, and is indeed the foundation of life, the source of all action . . .

A vein and an artery, both styled vein by the ancients, and that not undeservedly, as Galen has remarked, because the one, the artery to wit, is the vessel which carries the blood from the heart to the body at large, the other or vein of the present day bringing it back from the general system to the heart; the former is the conduit from, the latter the channel to, the heart; the latter contains the cruder, effete blood, rendered unfit for nutrition; the former transmits the digested, perfect, peculiarly nutritive fluid . . ."

(*b*) *On the motion of the heart.* When I first gave my mind to vivisections, as a means of discovering the motions and uses of the heart, and sought to discover these from actual inspection, and not from the writings of others, I found the task so truly arduous, so full of difficulties, that I was almost tempted to think, that the motion of the heart was only to be comprehended by God . . .

In the first place, when the chest of a living animal is laid open and the capsule that immediately surrounds the heart is slit up

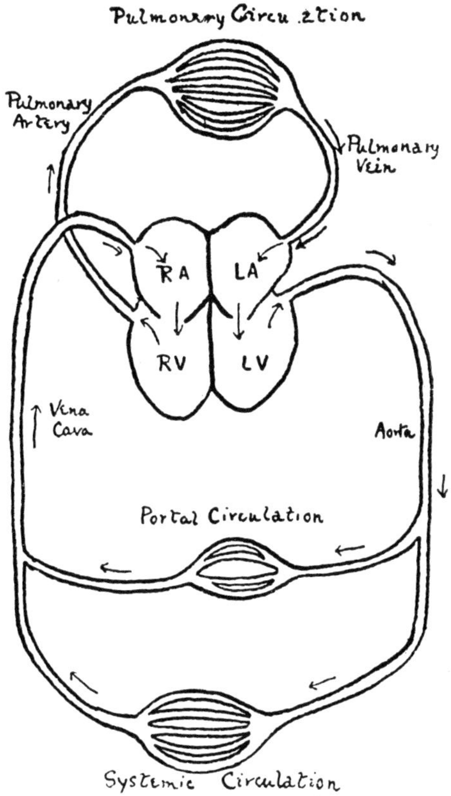

Fig. 17. Schematic drawing of the blood circulation of man after Harvey (1628), in which he describes the big and the small circulation.

or removed, the organ is seen now to move, now to be at rest; there is a time when it moves, and a time when it is motionless. These things are more obvious in the colder animals, such as toads, frogs, serpents, small fishes, crabs, shrimps, snails and shellfish. They also become more distinct in warm-blooded animals, such as the dog and hog, if they be attentively noted when the heart begins to flag, to move more slowly, and to die: the movements then become slower and rarer, the pauses longer, by which it is made much easier to perceive and unravel what the motions really are, and how they are performed. In the pause, as in death, the heart is soft, flaccid, exhausted, lying, as it were, at rest.

In the motion, and interval in which this accomplished, three principal circumstances are to be noted:

1. That the heart is erected, and rises upward to a point, so that at this time it strikes against the breast and the pulse is felt externally.

2. That it is everywhere contracted, but more especially toward the sides, so that it looks narrower, relatively longer, more drawn together. The heart of the eel taken out of the body of the animal and placed upon the table or the hand shows these particulars; but the same things are manifest in the hearts of small fishes and of those colder animals where the organ is more conical or elongated.

3. The heart, being grasped in the hand, is felt to become harder during its action. Now this hardness proceeds from tension, precisely as when the forearm is grasped, its tendons are perceived to become tense and resilient when the fingers are moved.

4. It may further be observed in fishes and the colder-blooded animals, such as frogs, serpents, etc., that the heart when it moves becomes of a paler colour; when quiescent, of a deeper blood-red colour.

From these particulars it appeared evident to me that the motion of the heart consists in a certain universal tension: both contraction in the line of fibres and constriction in every sense. It becomes erect, hard, and of diminished size during its action; the motion is plainly of the same nature as that of the muscles when they contract in the line of their sinews and fibres; for the muscles, when in action, acquire vigour and tenseness, and from soft become hard, prominent, and thickened: in the same manner the heart.

We are therefore authorised to conclude that the heart, at the moment of its action, is at once constricted on all sides, rendered thicker in the walls and smaller in its ventricles, and so made apt to project or expel its charge of blood. This, indeed, is made sufficiently manifest by the fourth observation preceding, in which we have seen that the heart, by squeezing out the blood it contains, becomes paler, and then, when it sinks into repose and the ventricle is filled anew with blood, that the deeper crimson colour returns. But no one need remain in doubt of the fact, for if the ventricle be pierced the blood will be seen to be forcibly projected outward upon each motion or pulsation when the heart is tense. These things, therefore, happen together or at the same instant: the tension of the heart, the pulse of its apex, which is felt externally by its striking against the chest, the thickening of its walls, and the forcible expulsion of the blood it contains by the construction of its ventricles.

Hence the very opposite of the opinions commonly received appears to be true; inasmuch as it is generally believed that, when the heart strikes the breast and the pulse is felt without, the heart is dilated in its ventricles and is filled with blood; but the contrary of this is the fact, and the heart, when it contracts (and the shock is given), is emptied. Whence the motion which is generally regarded as the diastole of the heart is in truth its systole. And in like manner the intrinsic motion of the heart is not the diastole but the systole; neither is it in the diastole that the heart grows firm and tense, but in the systole, for then only, when tense, is it moved and made vigorous . . .

Thus I am persuaded that the motion of the heart is as follows: First of all, the auricle contracts, and in the course of the contraction throws the blood (which it contains in ample quantity as the head of the veins, the storehouse and cistern of the blood) into the ventricle, which, being filled, the heart raises itself straightway, makes all its fibres tense, contracts the ventricles, and performs a beat, by which beat it immediately sends the blood supplied to it by the auricle into the arteries; the right ventricle sending its charge into the lungs by the vessel which is called vena arteriosa, but which in structure and function and all things else, is an artery; the left ventricle sending its charge into the aorta, and through this by the arteries to the body at large. These two motions, one of the ventricles, another of the auricles, take place consecutively, but in such a manner that there is a kind of harmony or rhythm preserved between them, the two concurring in such wise that but one motion is apparent, especially in the warmer-blooded animals, in which the movements in question are rapid. All the wheels seem to move simultaneously . . .

(*c*) But so that no one may say we are giving only words, and making specious assertions without foundation, and making innovations without just cause; there come up for confirmation three things from which, if granted, this truth necessarily follows and the matter is quite clear.

First: that the blood is continually and without interruption being transmitted from the vena cava into the arteries by the beating of the heart and in such quantity that it could not be supplied by what was taken in, so much so that the whole mass (of blood) passes out of the heart in a short time . . . Let us make a supposition (either by reasoning or experiment) as to the amount of blood the left ventricle holds when dilated (when it is full), whether two or three, or one-and-a-half ounces: I have found two in the dead body. Let us in the same way assume that the heart (when contracted), holds that amount less, by which it is smaller under those conditions, and by which the ventricle is then less capacious; and let us suppose that the above amount of blood is forced out of the ventricle into the aorta: . . . then one may reasonably conjecture that a fourth or fifth or sixth or at least an eighth part is sent into the artery. So we may assume that in man there is put forth from the heart half an ounce or three drachms or two drachms; which because of the closing of the valves, cannot flow back to the heart.

In one half-hour the heart makes more than a thousand beats; and in some people and at some times, two, three or four thousand. Now multiply the drachms, and you will see that in one half-hour there is poured through the heart into the arteries, a thousand times either three (or two) drachms, that is five hundred ounces, or some other proportionate quantity of blood, which is a greater quantity than is found in the whole body.

50

THE DUTCH EXPLORER PELSAERT DISCOVERED IN 1629 THE FIRST KANGAROO.

Geographical explorations enlarged the number of animal and plant forms known considerably. More striking even than the discovery of tapirs, armadillos, opossums and many others in the 16th century, was, perhaps, the discovery of the kangaroos of Australia. It is commonly held that Captain Cook was the first to describe and to draw kangaroos around 1770. But the Dutch explorer Pelsaert gave almost 150 years earlier (in 1629) a recognisable description of the Tammar Wallaby (*Thylogale eugenii Desm.*), as follows (Pelsaert, Batavia, 1629, here quoted from C. J. Hartmann, Possums, Univ. of Texas Press 1952, p. 15).

" We found on these strands large numbers of a species of cat, which are very strange creatures; they are about the size of a hare, the head resembling that of a civet cat; the forepaws are very short, about the length of a finger, on which the animal has five small nails or fingers, resembling those of a monkey's forepaw. Its two hindlegs, on the contrary, are upwards of half an ell in length, and it walks on these only, on the flat of the heavy part of the leg, so that it does not run fast. Its tail is very long, like that of a long-legged monkey; if it eats, it sits on its hindlegs and clutches its food with its forepaws, just like a squirrel.

Their manner of generation is exceedingly strange and highly worth observing. Below the belly the female carries a pouch, into which you may put your hand; inside this pouch are her nipples, and we have found that the young ones grow up in this pouch with the nipples in their mouths. We have seen some young ones lying there, which were only the size of a bean, though at the same time perfectly proportioned, so that it seems certain that they grow there out of the nipples of the mammae, from which they draw their food, until they are grown up and are able to walk. Still they keep creeping into the pouch with them when they are hunted."

51

GALILEO GALILEI (1564-1642), PADUA.

Galilei is not only the founder of modern physics, combining mathematical deductions with inductive and experimental methods, breaking away from the complete, rationalised, *a prioristically* deduced system of physics of the Middle Ages. Here we give a brief extract of his epistomology (from E. A. Burtt, Metaphysical foundations of modern Science. London, 1925. p. 75).

I feel myself impelled by necessity, as soon as I conceive a piece of matter or corporal substance, of conceiving that in its own nature it is bounded and figured by such and such a figure, that in relation to others it is large and small, that it is in this or that place, in this or that time, that it is in motion or remains at rest, that it touches or does not touch another body, that it is single, few or many; in short by no imagination can a body be separated from such conditions. But that it must be white or red, bitter or sweet, sounding or mute, of a pleasant or unpleasant odour, I do not perceive my mind forced to acknowledge it accompanied by such conditions; so, if the senses were not the escorts, perhaps the reason or the imagination by itself would never have arrived at them. Hence I think that those tastes, odours, colours, etc., on the side of the object in which they seem to exist, are nothing else but mere names, but hold their residence solely in the sensitive body; so that if the animal were removed, every such quality would be abolished and annihilated.

52

RÉNÉ DESCARTES (1596-1650), NETHERLANDS.

Descartes became by his *Discours de la Méthode* (Leyden, 1637) one of the founders of modern thought. We bring here a small paragraph concerning the laws of observation of nature. English translation by J. Veitch (1887. 2nd edition. p. 19, ff. 58).

And as a multitude of laws often only hampers justice, so that a state is best governed, when with few laws, these are rigidly administered; in like manner, instead of the great number of precepts of which Logic is composed, I believed that the four following would prove perfectly sufficient for me, provided I took the firm and unwavering resolution never in a single instance to fail in observing them.

The *first* was never to accept anything for true which I did not clearly know to be such; that is to say, carefully to avoid precipitancy and prejudice, and to comprise nothing more in my judgement than what was presented to my mind so clearly and distinctly as to exclude all ground of doubts.

The *second*, to divide each of the difficulties under examination into as many parts as possible, and as might be necessary for its adequate solution.

The *third*, to conduct my thoughts in such order that, by commencing with objects the simplest and easiest to know, I might ascend by little and little, and, as it were, step by step, to the knowledge of the more complex; assigning in thought a certain order even to those objects which in their own nature do not stand in a relation of antecedence and sequence.

And the *last*, in every case to make enumerations so complete, and reviews so general, that I might be assured that nothing was omitted . . .

Now, in conclusion, the Method which teaches adherence to the true order, and an exact enumeration of all the conditions of the thing sought includes all that gives certitude of the rules of Arithmetic.

I had after this described the Reasonable soul, and shown that it could by no means be educed from the power of matter, as the other things of which I had spoken, but that it must be expressly created; and that it is not sufficient that it be lodged in the human body exactly like a pilot in a ship, unless perhaps to move its members, but that it is necessary for it to be joined and united more closely to the body, in order to have sensations and appetites similar to ours, and thus constitute a real man. I here entered, in conclusion, upon the subject of the soul at considerable length, because it is of the greatest moment: for after the error of those who deny the existence of God, an error which I think I have already sufficiently refuted, there is none that is more powerful in leading feeble minds astray from the straight path of virtue than the supposition that the soul of the brutes is of the same nature with our own; and consequently that after this life we have nothing to hope for or fear, more than flies and ants; in place of which, when we know how far they differ we much better comprehend the reasons which establish that the soul is of a nature wholly independent of the body, and that consequently it is not liable to die with the latter; and, finally, because no other causes are observed capable of destroying it, we are naturally led thence to judge that it is immortal.

53

JAN BAPTIST VAN HELMONT (1577-1644), BRUXELLES.

This famous chemist and physiologist published a small note on the nutrition of the plant from the water of the soil. Deceptive as this method was, it was one of the first experiments. A similar experiment was already described (but not published) by Leonardo da Vinci. It is translated here from a German version in W. v. Buddenbrock, Bilder Gesch. Biol. Grundprobleme. Berlin ,1930. pp. 107-108).

I learned by experiment, that everything vegetable is derived directly and materially from the element of water. I took a vessel of earthenware and put into it 200 lot of dried earth. I watered this with rain water and planted into it a young willow, which had a weight of 16 lot. Five years later the same plant had a weight of 169 lot and some ounces . . . I dried the earth in the vessel and determined its weight at the end of the experiment as 200 lot minus two ounces. The total weight difference between the old and the young plant was thus clearly derived from the water.

54

ROBERT BOYLE (1627-1691), OXFORD.

Robert Boyle, the famous physicist and one of the founders of the Royal Society, demonstrated in one of the meetings of that Society that the air is necessary for respiration, and that the air fit for respiration is soon exhausted in a small closed vessel. (First published as a book in 1660. Here from the Works of Robert Boyle. Vol. I. London, 1772. pp. 97-113).

Experiment 40. It may seem well worth trying, whether or no in our exhausted glass the want of an ambient body or the wanted thickness of air would disable even light and little animals, as bees, and other winged insects to fly. At two times we put a large flesh-fly into a small and a humming bee into a greater receiver. The fly, after some exsuctions of the air dropped down from the side of the glass whereon she was walking. But with the bee we conveyed a bundle of flowers suspended by a string. We excited the bee to fly up and down the capacity of the vessel, till at length she lighted upon the flowers; whereupon we presently began to draw out the air. For some time the bee seemed to take no notice of it, yet within a while after she did not

Fig. 18. Robert Boyle experiments with the vacuum, demonstrating in the Royal Society the influence of the vacuum upon living animals. (From ICI).

fly, but fell down from the flowers, without appearing to make any use of her wings to keep herself. Later we procured a white butterfly, and enclosed it into a small receiver, where though at first he fluttered up and down, yet presently, upon the exsuction of the air, he fell down as in a swoon, retaining no other motion than some little trembling of the wings. Whether the fall of these insects proceeded from the mediums being too thin for them to fly in, or barely from the weakness, you will easily gather from the following experiment.

Experiment 41. To satisfy ourselves about the account upon which respiration is so necessary to the animals that nature has furnished with lungs, we took a lark with one broken wing. It was lively and, put into the receiver, sprang diverse times up in it to a good height. When the pump was applied, the bird for a while appeared lively enough; but upon a greater exsuction of the air, she began manifestly to droop, and appear sick, and very soon after was taken with as violent and irregular convulsions as in poultry when their heads are wrung off. And though upon the appearing of these convulsions we turned the stop-cock, and let in the air upon her, yet it came too late. Whereupon casting our eyes upon one of these accurate dials that go with a pendulum, and were of late geniously invented by the noble and learned Huygens, we found that the whole tragedy had been concluded within ten minutes, part of which time was employed in cementing the cover of the receiver . . .

A while after we put a mouse, newly taken, into a receiver, where it sprang high. Though for a while after the pump was set to work, he continued leaping up as before, yet it was not long ere he began to appear sick and giddy, and to stagger; after which he fell down as dead, but without such violent convulsions as the bird died with. Letting hastily in some fresh air, he recovered after a while his senses and his feet, and at length grew able to skip as formerly. Then the pump was plied again for eight minutes, about the middle of which space a little air by mischance got in and about two minutes after that the mouse diverse times leaped up lively enough, though after two minutes more he fell down quite dead, yet with convulsions far milder than those of the bird . . . I caused the bodies of these dead to be opened, but in such small bodies could discover little of what we sought for and what we may possibly find in larger animals . . .

A disgression containing some doubts touching respiration . . . (p. 111-113). The necessity of air to most animals unaccustomed to its want, may best be judged by the following experiments in our engine to discover whether insects have not either respiration or some other use of the air equivalent thereunto. We took a humble bee, a common flesh-fly and a hairy caterpillar into one of our small receivers and observed to the great wonder of the beholders, that not only the bee and the fly fell down and lay with their bellies upwards, but the caterpillar also seemed to be suddenly struck dead, all of them laying without motion and lifeless within less than one minute, in spite of the smallness of the animals in proportion to the capacity of the vessels. No sooner had we re-admitted air all three insects began to show signs of life and to recover. But when we again had drawn out the air, they fell down seemingly dead as before, as long as the vessel was kept exhausted. The creator by giving the air a spring has made it so very difficult to exclude a thing so necessary to animals. It gave us occasion to suspect, that if insects have no lungs or their analogy, the ambient air affects them and relieves them at the pores of their skin; like the moister parts of the air readily insinuate themselves into, and recede from the pores of the beards of wild oats and other plants, which almost continually wreath and unwreath themselves to even the light variations of the temperature of the ambient air . . . We scarce ever saw anything that seemed so much as this experiment to manifest that even living creatures (man always excepted) are a kind of curious engines, framed and contrived by nature much more skilfully than our gross tools and imperfect wits can reach to . . . Whereas it is known, that bees and flies will not only walk, but fly for a great while, after their heads are off; and sometimes one half of the body will, for diverse hours, walk up and down, when it is severed from the other; yet upon the exsuction of the air not only the progressive motion of the whole body, but those of the limbs cease, as if the presence of air were more necessary to these animals than the presence of their own heads.

But in these insects that fluid body, whether it be a juice or flame, wherein life chiefly resides, is nothing near so easy dissipable as in perfect animals. Whilst the birds were within two minutes brought to be past recovery, we were unable to kill our insects by the exsuction of air: as long as the pump was kept moving, they continued immovable, yet when we desisted from pumping, the slowly entering air restored them to the free exercise of life . . . Without denying that the inspired and expired air may be sometimes very useful, by condensing and cooling the blood that passes through the lungs, I hold that the depuration of the blood in that passage is one of the principal uses of respiration. But

I also suspect, that the air does something else in respiration, which has not yet been sufficiently explained . . .

55

MARCELLO MALPIGHI (1628-1694), BOLOGNA.

This great man in the history of anatomy, embryology, plant and insect anatomy is by far not yet sufficiently estimated. We bring here:

(*a*) and (*b*) The discovery of the capillaries in the lung. Two letters *De Pulmonibus* to Borelii (1661). M. Malpighi. Opera, Vol. II. Leiden, 1587, p. 320-330. (See also: E. Ebstein, Ärzte-Briefe. Berlin, 1920. p. 15-17).

(*c*) From the development of the chicken in the egg. From the *Observationes de ovo incubato* (1672; here from Paris 1686. pp. 348-352).

(*a*) During the post-mortems, to which I devoted myself with an eagerness increasing from day to day, I studied recently especially the structure and function of the lungs; much about them seemed to me to be still obscure. I will now communicate to you the results of my studies, in order that you with your eyes which are so expert in anatomy, may separate the true from the false and thus make really useful my discoveries. Now, I am of another opinion, as I have found by devoted researches, that the entire mass of the lungs which is connected with the vessels which spring from the heart, consists of very fine and tender membranes, and that these membranes, which are sometimes tense, and sometimes folded, form a great number of small vesicles, similar to the cells of a bee's comb. Their position and connection is such that the vesicles are connected with one another as well as with the trachea directly, and that they are all bounded by a continuous membrane. This is best seen when the lungs are taken from a living animal; then one sees distinctly, especially at the lower end, many small vesicles which are swollen with air. But this is also though less distinctly, recognisable in a lung which is cut through the midst and emptied of air . . .

On the surface of the lungs a wonderful net, extended, is visible when light falls upon them, which is intimately connected with the single vesicles; the same can be observed in an opened lung,

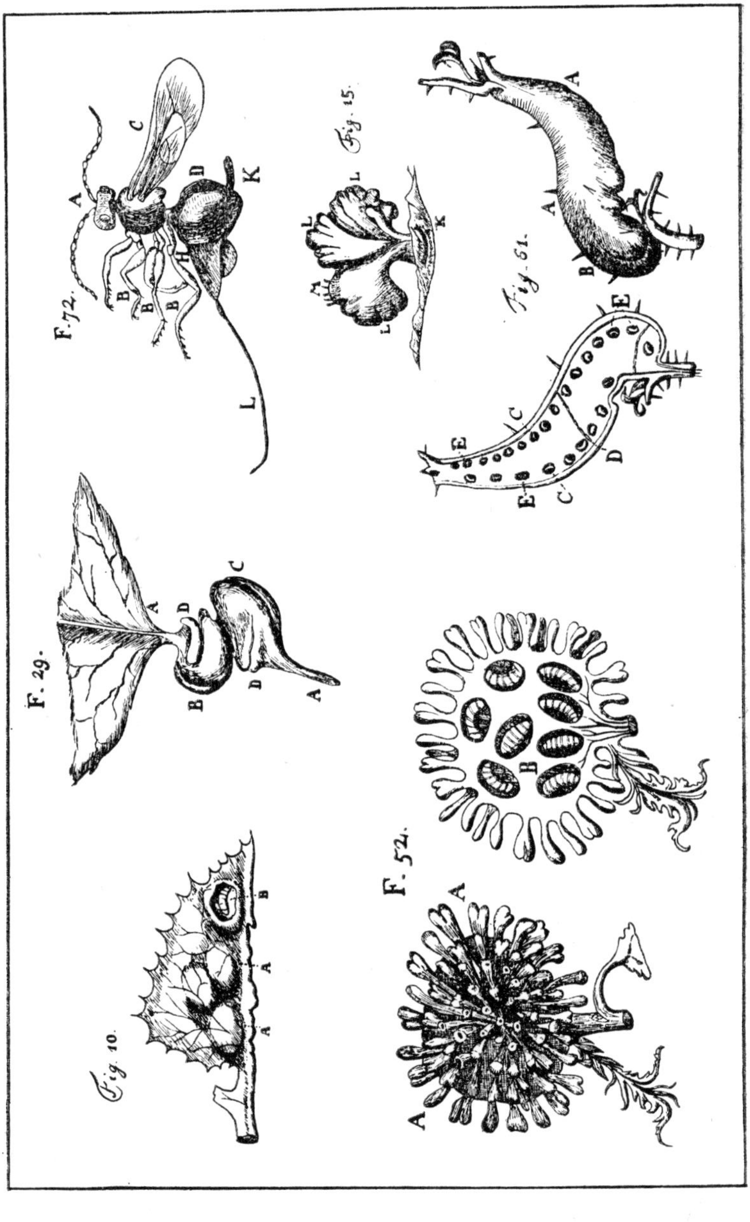

Fig. 19. Cross-section through a plant-gall with insect larva from Malpighi. De gallis plantarum, pl. XI, fig. 3.

but less distinctly . . . Generally the lungs are divided according to shape and position. Two main parts are distinguished, and between them the mediastinum; in man each of these parts has two, in the animals more subdivisions. I have found a more miraculous and more complicated division. The entire mass of the lungs consists of very small lobules, which are surrounded by a special membrane, which have special vessels and are formed from the processes of the trachea.

In order to make visible the single lobules, one must hold the half-inflated lung against the light, and then the intervals grow conspicuous, and by blowing air through the trachea, one must separate by small cuts the lobules, which are enveloped in a special membrane from the associated vessels. By very careful preparation one succeeds in this way . . . Concerning the functions of the lungs, I know, that much which was regarded as certain by the Ancients, is still very doubtful, especially the cooling of the lungs, which in the traditional opinion forms the main function of the lungs. This opinion is based upon the assumption of a warmth which rises from the heart, which requires an exit. By reasons, which I will give later, I think it probable, that the lungs are determined by nature to mix the blood. Concerning the blood, I do not believe that it is composed from the usually assumed four humours: the two gall-matters, the blood proper and the saliva, but I assume that the entire mass of the blood, which flows without interruption through the veins and arteries, and which is composed of very small particles, is composed only from two liquids, which resemble one another very much: a whitish one, usually called serum, and a reddish one . . .

Apart from the hitherto quoted functions of the lung, I could add as very necessary still, that the lungs are determined by nature to be a reservoir of blood, from which continuously blood flows to the heart; from there it is then driven by the motion of the heart through the entire body, and brings thus life and motion to all its parts. Yet this has already been described by others, and therefore I wish only briefly to mention that when you insert a tube into the already collapsed lungs, after the opening of the thorax of a still living animal, and when you blow air through the tube, the pulsation of the heart is restored, even when it had almost ceased before, since then, by the pressure of the air, blood enters into the left ventricle. Also experience at the sick-bed teaches the same: when the pulmonary vessels are obstructed you find first irregularities of the pulse near the ears, followed by death. The lungs are in all animals of such importance, that most diseases either begin in the lungs or end in them.

All these observations I have made in my anatomical researches, and I would have given a more detailed description, were it not that it would concern the smallest things, which the eye scarcely can discern. I ask you to preserve to me your friendship in the future, also, may a long and happy life be your lot!

(*b*) To eyes otherwise attentive solely to structure and composition (of the lung), microscopic observation reveals things even more remarkable. For, provided the heart still beats, contrary movements of the blood in the veins are (admittedly with difficulty) to be seen. By this is manifestly revealed the blood's circulation, which is also, and even more happily, to be discerned in the mesentery and other major veins contained in the abdomen. In this way the blood pours floodlike into the smallest openings, through the arteries, and by one or another of the branches crossing through or terminating here, into each little compartment. And, by being so much subdivided, it loses its red colour, and, conducted roundaboutly, is everywhere distributed until it reaches the compartmental walls and angles, and the reabsorbing branches of the veins. The power of the unaided eye could not be extended further in the dissected, animate living being, whence I had been led to believe that the blood's substance was emptied into a vacant space, and, by some other pathway, was recollected by the peculiar structure of the compartmental walls, the possibility of which was attested by the blood's motion and is coming together . . . In the elongated and equally membraneous and transparent lung of the tortoise the subdivided blood, by this impetus, ran through twisted vessels and was not poured out into empty spaces, but was driven through little tubes and dispersed by the frequent turnings of the vessels.

(*c*) When the egg was incubated for three days I found the chicken's body to be in a curved position. In the head, apart from the two eyes, five vesicles could be distinguished, filled with humour, and they compose the brain. Also the first stumps of the legs (D) and of the wings (E) appeared. The vesicles composing the brain are arranged as follows: The widest one is at the apex of the head, hemispherical, and receives blood from various vessels. This divides into two vesicles some days later, which made me doubt, if at the beginning this vesicle be really one or perhaps already two. Behind it on the head is another vesicle (F), almost triangular in shape. The part before it on the head is an oval vesicle (H), close to which two more vesicles (I) are located. The body is covered by flesh in such a way, that the path of the (circulation of the) blood cannot be easily observed. The eyes (B) appear: their lenses are formed by a small perforated (? déchiquetée) membrane which covers the entire surface except at the base. The cristalline (body) is enclosed in the vitreous liquor, and occupies the centre (of the eye) close to the place where the vessels leave it. The vesicle (K)

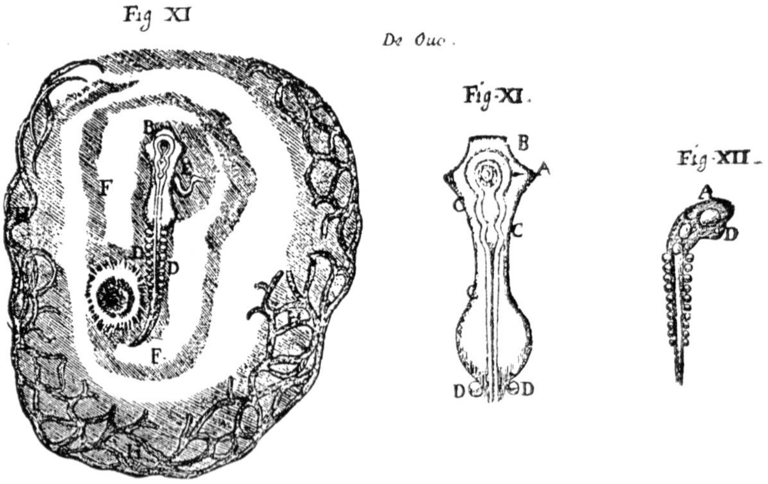

Fig. 20. From the development of the chicken embryo in the egg in Marcello Malpighi's *De formatione pulli in ovo* (1672), describing for the first time the development during the first hours of the breeding.

hangs outward, being 'watered' by some vessels. I took it to be a fleshy pouch. The structure of the heart is like that described before, and nature developed in those days the mysteries about which I talked: namely the auricle (L) which receives the blood of the veins had a double pulsation, like two separated ventricles, and thus the blood was driven into the heart by a route which should be followed more closely. The right ventricle of the heart (N) continues to pulsate as since its beginning, yet the left one had a different movement and grew bigger from day to day, until it was united with its accompanying ventricle. It appeared in the place of the left one, which became clearer a few days later.

56

FRANCESCO REDI (1626-1698), FIRENZE.

Redi, a physician at the time of the Medici was one of the courageous men in the history of science. His many observations and experiments which often widely diverged from the common opinion, could have been dangerous for him. We bring here from his *Esperienze intorno alla generazione degli Insetti* (1668, p. 1 ff.) a small experiment demonstrating that maggots are not born from the decay of meat or fruits, but by oviposition of insects.

Although content to be corrected by one wiser than myself, if I should make erroneous statements, I shall express my belief that the earth, after having brought forth the first plants and animals at the beginning by order of the omnipotent Creator, has never since produced any kinds of plants or animals, either perfect or imperfect, and everything which we know in past or present times that she has produced, came solely from the true seeds of the plants and animals themselves, which thus, through means of their own, preserve their species. And, although it be a matter of daily observation that infinite numbers of worms are produced in dead bodies and decayed plants, I feel, I say, inclined to believe that these worms are all generated by insemination and that the putrefied matter in which they are found has no other office than that of serving as a place, or suitable nest where animals deposit their eggs at the breeding season, and in which they also find nourishment; otherwise, I assert that nothing is ever generated therein. And in order, Signor Carlo, to demonstrate to you the truth of what I say, I will describe to you some of these insects, which being most common, are best known to us . . .

Having considered these things, I began to believe that all worms found in meat were derived directly from the droppings of flies, and not from the putrefaction of the meat, and I was still more confirmed in this belief by having observed that, before the meat grew wormy, flies had hovered over it, of the same kind as those that later bred in it. Belief would be vain without the confirmation of experiment. Hence in the middle of July I put a snake, some fish, some eels of the Arno, and a slice of milk-fed

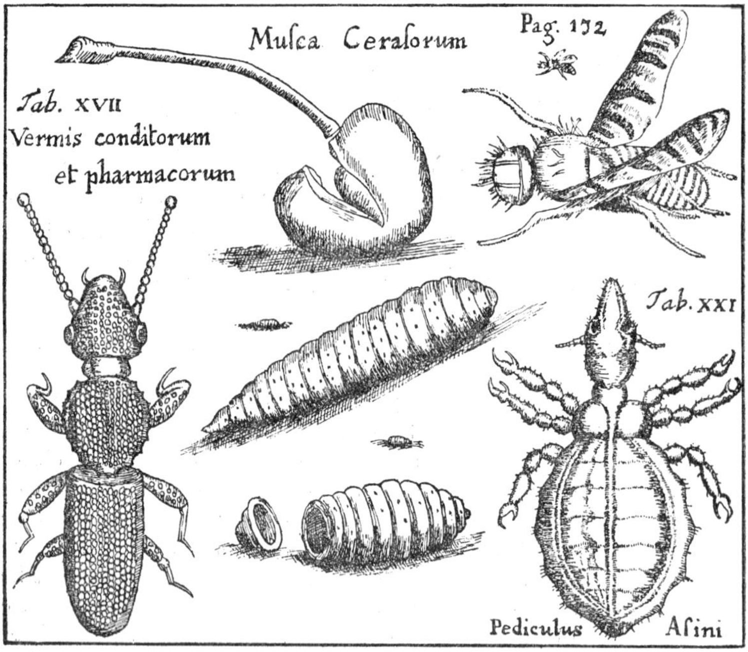

Fig. 21. In Redi's *Experimenta circa Insecta* we find the illustration of the louse of an ass (*Pediculus asini*), the presence of which was denied by Aristotle. This was a courageous deed at that time. (See Bodenheimer: *Mat. Gesch. Ent.* Vol. 1).

veal in four large, wide-mouthed flasks; having well closed and sealed them, I then filled the same number of flasks in the same way, only leaving these open. It was not long before the meat and the fish, in these second vessels, became wormy and flies were seen entering and leaving at will; but in the closed flasks, I did not see a worm, though many days had passed since the dead flesh had been put in them. Outside on the paper cover there was now and then a deposit, or a maggot that eagerly sought some crevice by which to enter and obtain nourishment. Meanwhile the different things placed in the flasks had become putrid . . .

Not content with these experiments, I tried many others at different seasons, using different vessels. In order to leave nothing undone, I even had pieces of meat put under ground, but though remaining buried for weeks, they never bred worms, as was always the case when flies had been allowed to alight on the meat. One day a large number of worms, which had bred in buffalo-meat, were killed by my order; having placed part in a closed dish, and part in an open one, nothing appeared in the first dish, but in the second worms had hatched, which pupated and became flies of the common kind. In the same experiment tried with dead flies, I never saw anything breed in the closed vessel.

57

REIGNIER VAN GRAAF (1641-1673), DELFT.

This famous Dutch anatomist believed to have discovered the mammalian egg. It was, in fact, the later called Graafian follicle, which contains the egg, which he described. We give here the important paragraph: On the ovaries as the producers of the eggs of the female. (From *De mulierum organis generationi inservientibus*, Cap. XII. Leyden, 1672. English translation G. H. Corner. 1943. Opera Omnia, 1705, p. 232).

Thus, the general function of the female testicles is to generate the ova, to nourish them, and to bring them to maturity, so that they serve the same purpose in women as the ovaries of birds. Hence, they should rather be called ovaries than testes because they show no similarity, either in form or contents, with the male testes properly so called. On this account, many have considered these bodies useless, but this is incorrect, because they are indispensable for reproduction. This is proved by the remarkable convolutions of the nutritive vessels about them, and is confirmed by the castration of females, which is invariably accompanied by sterility. Varro writes that spayed (= castrated) cows will conceive, if they copulate immediately; a thing which is no doubt true of males in which the seminal vesicles are still filled with spermatic fluid (after castration), but not of females, in which such vesicles are not present. How the ova are fertilised and proceed to the uterus will be explained in the following chapters.

58

NEHEMIAH GREW (1641-1712), LONDON.

Nehemiah Grew was together with Malpighi the founder of plant anatomy in his *The Anatomy of Vegetables* (1672) and *The Anatomy of Plants* (1682, p. 67 ff.). From the latter we give a paragraph on the anatomy of the trunk.

Of the parts of the *trunk*, the first occurring is its skin: the formation whereof is not from the air, but in the seed, being the production of the cuticle, there inverting the two lobes and plume. The next part is the cortical body, which here in the trunk is no new substantial formation, but as is that of the root, originated from the Parenchyma of the seed, of which it is only the increase and augmentation. The skin, this cortical body and some fibres of the lignous mixed herewith, all together make the barque.

Next, the lignous body which, whether it be visibly divided into many softer fibres, as fennel and most plants; or that its parts stand more compact and close, showing one hard, firm and solid piece, as in trees; it is in all one and the same body; and that not formed originally in the trunk, but in the seed; being nothing else but the prolongation of the inner body distributed in the lobes and plume thereof.

Lastly, the insertions and pith are here originated likewise from the plume, as the same in the root from the radicles. So that as to their substantial parts, the lobes of the seed, the radicle and plume, the root and trunk are all one.

Yet some things are more fairly observable in the trunk. First, the latitudinal shootings of the lignous body, which in trunks of several years growth, are visible in so many rings, as is commonly known: for several young fibres of the lignous body, as in the root, so here, shooting into the cortical one year, and the spaces betwixt them being after filled up with more (I think not till) the next, at length they become altogether a firm compact ring;

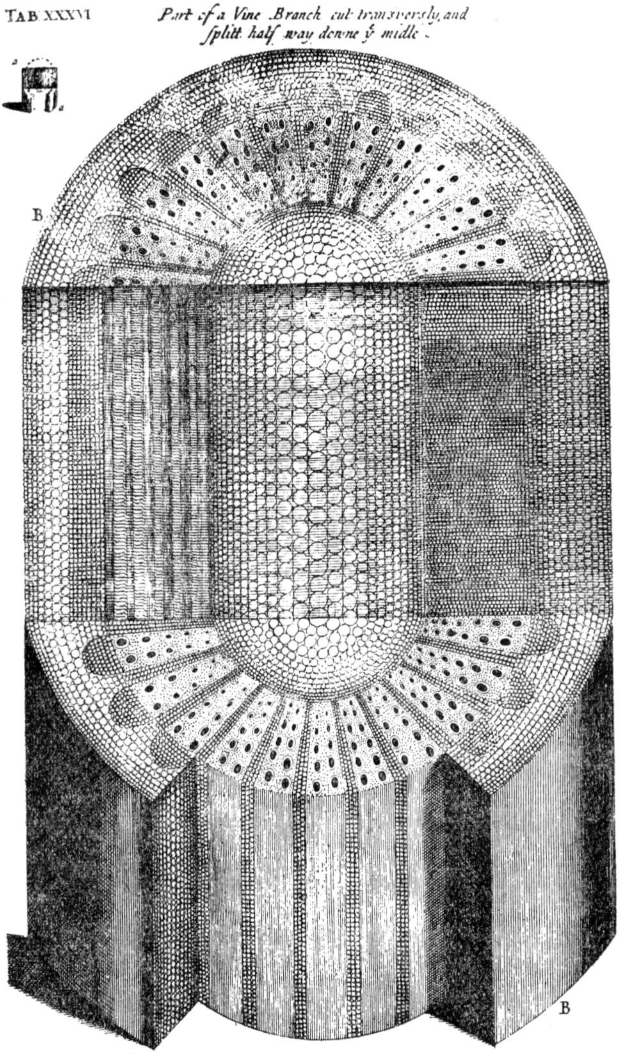

TAB XXXVI
Part of a Vine Branch cut transversly, and splitt half way downe ye midle.
B
B

Fig. 22. Section of a vine branch cut transversally and vertically to show microscopic structure. N. Grew, *Anatomy of Plants.* 1682. p. 36.

the perfection of one ring, and the ground-work of another being thus made concomitantly.

From these annual younger fibres it is, that although the cortical body and pith are both of the same substantial nature, and their pores little different; yet whereas the pith, which the first year is green, and of all the parts the fullest of sap, becomes afterwards white and dry; the cortical body, on the contrary, so long as the tree grows, ever keepeth green and moist, so, because the said fibres annually shoot into, and so communicate with it.

The pores likewise of the lignous body, many of them in well-grown timber, as in oak boards, are very conspicuous, is cutting both lengthwise and traverse; they very seldom run into one another, but keep, like so many several vessels, all along distinct; as by cutting, and so following any one of them as far as you please, for a foot or half a yard, or more together, may be observed.

These greater pores, though in wainscot, tables, and the like, where they have lain long open, they are but mere vacuities, and so would be thought to contain only sap in the tree, and afterwards only air. Yet upon a fresh cut, each of them may be seen, filled up with a light and spongy body, which by glasses, and even by the bare eye, appears to be a perfect pith, sometimes entire and sometimes more or less broken.

Besides these, there are a lesser sort, which by the help of a microscope also appear, if not to be filled up with a pith, yet to contain certain light and filmy parts, more or fewer, of a pithy nature within them. And these are all the pores the best glasses which we had at hand, would show us. But the learned and most ingenious naturalist Mr. Hook(e) sheweth us moreover besides these a third and yet smaller sort, the description whereof I find he hath given us amongst his *Microscopical Observations* . . .

Upon further enquiry, I likewise found lignous body in the trunk of plants, which at first we only supposed, by the help of good glasses are very fairly visible; each fibre being perforated by 30, 50, 100 or hundreds of pores. Or what I think is the truest notion of them, that each fibre, though it seems to the bare eye to be but one, yet is indeed a great number of fibres together; every pore being merely a space betwixt the several pores of the wood, but the concave of a fibre: so that if it be asked, what all that part of a vegetable, either plant or tree, which is properly called the wood part, is nothing else but a cluster of innumerable and most extraordinary small vessels or concave fibres.

59

ANTONY VAN LEEUWENHOEK (1632-1723), DELFT.

Leeuwenhoek, secretary of the township of Delft, was one of the early microscopists whose researches penetrated into everything, from sperms, tissues, protozoans, bacteria, etc. We have selected here:

(*a*) A letter concerning little animals observed in rain-, well-, sea-, and snow-water, as also in water wherein pepper had lain infused. (Philos. Trans. London. 10. No. 133. 1676. p. 821).

(*b*) The observations of Mr. Antony Leeuwenhoek, on the animals engendered in the semen. (Ibidem, 1677. English translation by Dobell, from F. J. Cole, Early Theories of Sexual Generation. Oxford, 1930, pp. 9-12).

(*a*) *Little animals in the Rain Water.* In 1675 I discovered very small living creatures in rain water which had stood but few days in a new earthen pot glazen blue within. This invited me to view this water with great attention, especially those little animals appearing to me ten thousand times less than those represented by Monsieur Swammerdam, and by him called water fleas or water lice, which may be perceived in the water with the naked eye.

The first sort I several times observed to consist of 5, 6, 7 or 8 clear globules, without being able to discern any film that held them together, or contained them. When these animalcula or living atoms moved, they put forth two little horns, continually moving. The space between these two horns was flat, though the rest of the body was roundish, sharpening a little toward the end, where they had a tail, near four times the length of the whole body, of the thickness, by my microscope, of a spider's web; at the end of which appeared a globule of the size of one of those which made up the body. These little creatures, if they chanced to light on the least filament or string, or other particle, were entangled therein, extending their body in a long round and endeavouring to disentangle their tail. Their motion of extension and contraction continued awhile; and I have seen several thousands of these poor little creatures, within the space of a grain of gross sand, lie fast clustered together in a few filaments.

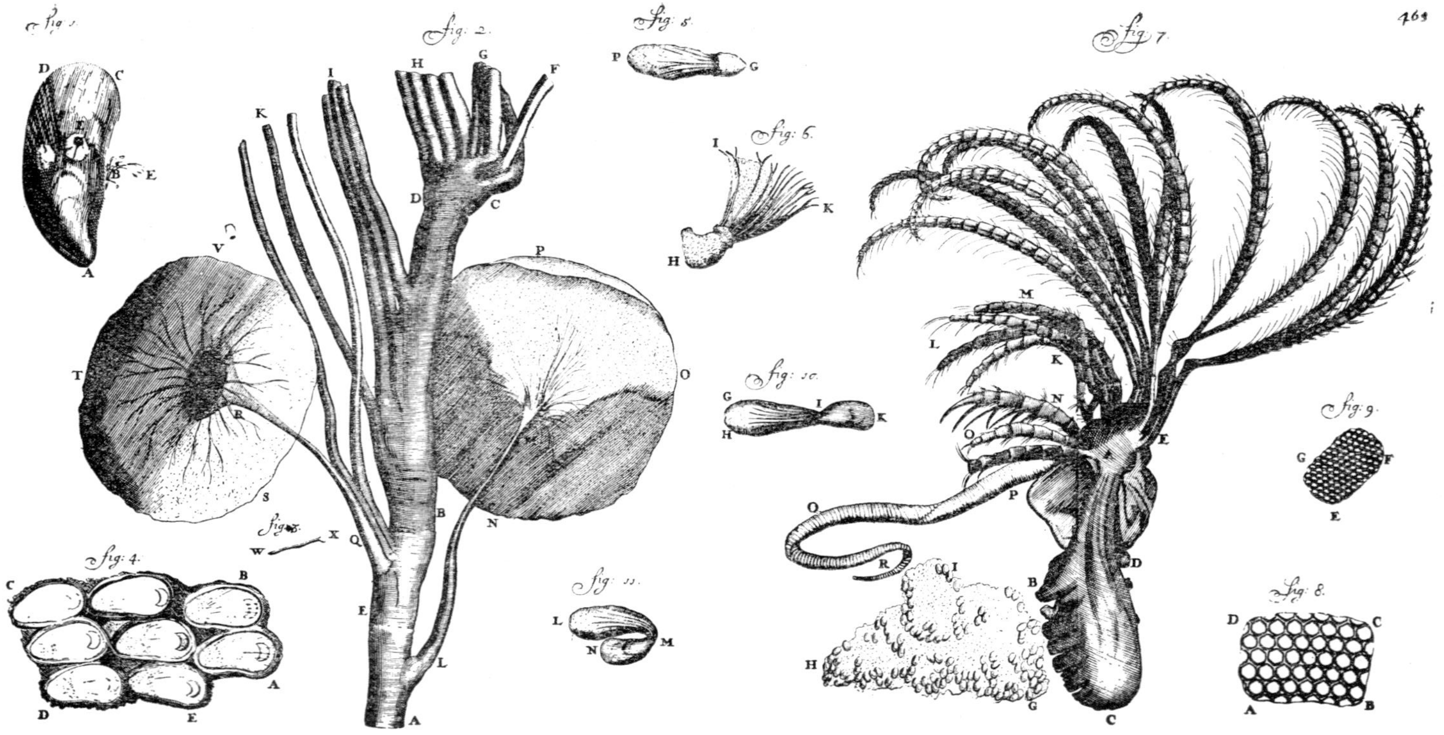

Fig. 23. Some drawings from the *Arcana Naturae Detectae* of Leeuwenhoek. 1695. Vol. I, p. 465, figs. 7-10.

I also discovered a second sort, of an oval figure; and I imagined their head to stand on a sharp end. These were a little longer than the former. The inferior part of their body is flat, furnished with several extremely thin feet, which move very nimbly. The upper part of the body was round, and had within 8, 10 or 12 globules, where they were very clear. These little animals sometimes changed their figure into a perfect round, especially when they came to lie on a dry place. Their body was also very flexible; for as soon as they struck against the smallest fibre or string their body was bent in, which bending presently jerked out again. When I put any of them on a dry place I observed that, changing themselves into a round, their body was raised pyramidal-wise, with an extant point in the middle; and having lain thus a little while, with a motion of their feet, they burst asunder, and the globules were presently diffused and dissipated, so that I could not discern the least thing of any film, in which the globules had doubtless been enclosed; and at this time of their bursting asunder I was able to discover more globules than when they were alive.

A third sort of animalcula was twice as long as broad and eight times smaller than the first. Yet, I thought, I discerned little feet, whereby they moved very briskly, both in round and straight line. A fourth sort was so small that I was not able to give them any figure at all. These were a thousand times smaller than the eye of a large louse and exceeded all the former in celerity. I have often observed them to stand still as it were on a point, and then turn themselves about with swiftness and then extending themselves straight forward. I also discovered by and by several other sorts of animalcula . . . (Follow observations on protozoa in fresh rainwater and in the waters of the river Maese).

Having put about one third ounce of whole pepper in water and it lain about three weeks in the water, to which I had twice added some snow water, the other water being in great part exhaled, I discerned in it with great surprise an incredible number of little animals, of divers kinds, and among the rest, some that were 3 or 4 times as long as broad; but their whole thickness did not much exceed the hair of a louse. They had a very pretty motion, often tumbling about and sideways; and when the water was let to run off from them they turned round like a top; at first their body changed into an oval, and afterwards, when the circular motion ceased, they returned to their former length. The second sort of creatures discovered in this water were of a perfect oval figure, and they had no less pleasing or nimble a motion than the former; and these were in far greater numbers. There was a third sort, which exceeded the two former in number, and these had tails like those I had formerly observed in rain water. The fourth sort, which moved through the three former sorts, were incredibly small so that I judged that if one hundred of them lay one by another they would not equal the length of a grain of coarse sand; and according to this estimate one million of them could not equal the dimensions of a grain of such coarse sand . . .

In snow water, which had been about three years in a glass bottle well stopped, I could discover no living creatures; and having poured some of it into a porcelain teacup, and put therein half an ounce of whole pepper, after some days I observed some animalcula, and those, exceeding small ones, whose body seemed to me twice as long as broad, but they moved very slowly, and often circularly. I observed also a vast multitude of oval-figured animalcula, to the number of 8000 in a single drop.

(*b*) *On the Animalcules engendered in the Semen* . . . When Mr. Ham visited me (in August 1677), he brought with him, in a small phial, the spontaneously discharged semen of a man . . . After a very few minutes, he had seen living animalcules in it which he believed to have arisen by some sort of putrefaction. He judged these animalcules to possess tails, and not to remain alive above twenty-four hours . . . But I observed that they were dead after the lapse of two or three hours.

I have divers times examined the same matter (human semen) from a healthy man . . . and I have seen so great a number of living creatures in it, that sometimes more than a thousand were moving about in an amount of material the size of a grain of sand. I saw this vast number of living animalcules not all through the semen, but only in the liquid matter which seemed adhering to the surface of the thicker part. In the thicker matter of the semen, however, the animalcules lay apparently motionless . . . These animalcules were smaller than the corpuscles which impart a red colour to the blood; so that I judge a million of them would not equal in size a large grain of sand. Their bodies were rounded, but blunt in front and running to a point behind, and furnished with a long thin tail, about five or six times as long as the body, and very transparent, and with the thickness of about one twenty-fifth that of the body; so that I can best liken them in form to a small earth-nut (*Bunium flexuosum*) with a long tail. The animalcules moved forward with a snake-like motion of the tail, as eels do when swimming in water: and in the somewhat thicker matter, they lashed their tails some eight or ten times in advancing a hair's breadth. I have sometimes fancied that I could even discern different parts in the bodies of these animalcules: but forasmuch as I have not always been able to do so, I will say not more . . .

I remember that some three or four years ago I examined seminal fluid at the request of the late Mr. Oldenburg, Secretary of the Royal Society. Looking into the matter I found that he wrote asking me to do so from London, on the 24th of April, 1674: and among other things, he besought me also to examine saliva, chyle, sweat, etc.: but at that time I took the animalcules just described for globules. Yet as I felt averse from making further inquiries, and still more so from writing about them, I did nothing more at that time. What I here describe was not obtained by any sinful contrivance on my part, but the observations were made upon the excess with which Nature provided me in my conjugal relations. And if your Lordship should consider such matters either disgusting, or likely to seem offensive to the learned, I earnestly beg that they be regarded as private . . . (L. describes the other matters contained in the semen) . . . Moreover, when this matter had stood a little while, there appeared therein some three-sided bodies terminating at either end in a point (crystals of the human spermine phosphate) . . .

60

ATHANASIUS KIRCHER (1602-1680), ROMA.

A. Kircher was one of the most learned men of his time. His experiments in magnetism, with the camera obscura, and others deserve praise. In between we find utter nonsense, thus when he pretends to create by mixtures of various composition different orders of insects. On the other hand, he proposed in his *Scrutinium physico-medicum contagione luis, quae pestis dicuntur* a well presented germ theory of diseases (1658). Here we quote his famous experiment on the hypnosis of a hen (from *Physiologia*, 1674; here quoted from the edition by J. S. Kestlerus. Amsterdam, 1680. p. 90, where also the fig.).

Lay a hen bound on its feet on any floor. It will, at first feel as a captive and will try by all means—shaking of its wings and movements of its entire body—to free itself from its fetters. But finally it despairs of obtaining its freedom, grows quiet and puts itself at the disposal of the victor's judgement. When now the

Fig. 24. The so-called hypnosis of a hen as described in the famous experiment of Athanasius Kircher. (*Physiol.* 1680. B.M. p. 90).

hen remains quiet, draw with chalk a straight line on the pavement, beginning from the hen's eyes, or draw there with any colour something on the pavement which resembles a rope, and then release the fetters of the hen. I say that whenever the hen is liberated from its fetters, it remains almost motionless, even if you goad it to fly away. There is no reason for this behaviour other than the very great imagination of the bird which takes the line drawn on the pavement for fetters by which it is bound. There is no doubt, that this occurs also in other animals.

61

ALFONSO BORELLI (1608-1679), ROMA.

Borelli, a pupil of Galilei, is one of the outstanding iatrophysicists. His *De motu animalium* (1680, here quoted from the 2nd edition, 1734) is an analysis of the movements of the external limbs and of the internal organs of animals.

(*a*) One paragraph from the analysis of the flight of birds (from *Aeronaut. Classics* No. 6. Aeronaut. Soc. Gr. Brit. 1911. pp. 30-32. Translated by T. O'B. Hubbard and J. H. Ledeboer).

(*b*) Introduction to the *De Motu*.

(*a*) *It does not appear possible that birds, flying horizontally, can depart quickly from their course by the transverse flexions of their head and neck.* Let us consider the two ways by which a ship, moving through the water, is able to turn to the right and to the left. Firstly, if the oars on one side impel the water more strongly towards the stern than the oars on the other side. Secondly, if, while the ship is moving, the rudder, either in the stern or in the prow, is turned laterally, perpendicularly to the horizon. These two operations largely differ from one another; for the turning of the ship is effected on the one hand, by a considerable exertion of motive force on the part of the rowers, and on the other by the inappreciable power of the steersman grasping the tiller, who, though directing the movement, does not affect it by his own strength, but from the impetus acquired by the ship and from the rudder resisting the impact of the water; moreover, the turning movement made by the oars on one side is performed very quickly, but very slow by the rudder.

From these facts we can judge, in the similar action of birds flying, whether the bending of the neck cannot fulfil the function of a rudder. Firstly, if the neck inclined laterally had the strength of a rudder and thereby the bird were able to alter its course to the right and left, therefore, in the same way, by raising or lowering its neck, a bird in flight were able to direct its course upwards or downwards. But, as it cannot be said that the large tail, which, acting as a rudder, so evidently produces the up and down movements, was made by nature for no reason, it must be confessed that the bending of the neck does not fulfil the function of a rudder. Secondly: eagles, hawks, and swallows have a very short neck and a small head of but little weight; therefore, the centre of gravity could only deviate a very small way from the direction of the axis of the bird and for this reason it would only turn with much difficulty and very slowly in a lateral direction; but so false is this, that in truth they whirl round almost in a twinkling of an eye; on the other hand: geese, ducks, swans, and other birds of this kind, possessed of a long neck and a very heavy head and neck, turn most slowly when they are flying. It must therefore be confessed that bending the head and neck laterally in no wise produces a horizontal turning movement.

Thirdly, if the centre of gravity of the entire bird should depart considerably from its axis by a lateral bending of the neck, the bird could not maintain its horizontal equilibrium, and therefore the side depressed would have to be righted by a violent exertion of the wing on that side. From which it would follow that a contrary action to the first would be made in the interests of gravity, which would interfere with the turning movement; and such an action would be vain and useless; moreover being foolish and sorely at variance with nature's shrewdness.

Nor may you say that the speediest turns of birds are made by the strong flapping of one wing towards the tail, and that slow turns can be made by bending the neck sideways without any special effort by the wing in the same way as ships are put about by means of a rudder and without using the oars, for I am of the opinion that the slow circling of a bird is accomplished not by a stronger movement of one wing than either of them exert in straight flight; for it is sufficient that the wing, to make the turn, should incline for a little while towards the tail and strike the air there, so that without any fresh exertion the slow lateral turn of the bird may be accomplished in the quickest way.

(*b*) *General Introduction:* The motions of animals are of various kinds, and first the transport by motion of the entire weight from one spot to another one. Such locomotion along the surface of the soil is called walking or running, in the water swimming, and in the air flying. Animal movements are partly performed by external limbs, such as hands, legs, head, etc., but partly they are internal, such as the motions of the intestines, heart, arteries, veins or of the muscles, bones, etc.; other motions are the streaming and coursing motion of liquids in cavities or vessels, which concerns blood and other humours.

To inquire into the faculties, instruments and mechanics by which nature performs these primary motions we have to make some assumptions which, however, are proven by the evidence of our senses. Thus, the soul is the principle and the effective cause of animal motion, as nobody ignores that animals live through the soul which moves the animal during its life, whilst after its death, when the soul has left the body, the animal machine turns inert and immobile.

All agree that the manifold and various animal motions are produced by choice or by natural appetite of the animal. It is also obvious, that ideas and appetites by themselves cannot move and incite the animal's limbs, but these are in need of certain instruments, without which they are unable to move. We usually distinguish the active instruments of motion from the organic or passive ones. The active instrument of the animal's property of motion is the soul, which generally is believed to reside in the animal spirits. Aristotle assumes, that the moving faculty of the soul is transmitted through the blood-vessels and the nerves to their corresponding bones, which move the joints. This doctrine is rejected by Galen and by all others, as well as by sensorial evidence which demonstrates that the muscles are the organs and machines by which the moving faculty of the soul sets the joints and limbs of the animals in motion. This is well confirmed by the following experiment: When we cut transversly through a muscle the joint near which it is inserted cannot be any more retracted, whilst the other motions of the joint which depend upon the insertions of of other muscles remain intact. It is also known, that the muscle by itself is an inert and dead machine, without the access of the moving faculty which brings orders and which awakens them from sleep and rest to action. The intact muscle of the elbow, for instance, will not move the arm, when not stimulated to action by the appetite (of motion).

Sensory experience and observation show easily how and by what means the orders of the soul and the moving faculty reach the muscle. Whilst arteries, veins and nerves lead into the muscle, the two former cannot fulfil this task, as after their being cut through, the muscle still moves as if it were intact. But when the nerve which leads into a muscle is cut or strongly ligatured, this muscle ceases to move and turns inert as in a cadaver. It is therefore the nerve which transmits the moving faculty and which communicates the orders of the appetite to the muscle, which stimulates its motion. Whether what is transmitted through the nerve is some incorporeal force, some gas, wind, humour, motion, impulse or something else, and how it can overcome the resistence of considerable weights, we will discuss later. Here let it suffice to say in summary, that the order of the moving faculty of the soul, without which no voluntary motion is possible, is transmitted through the nerves.

62

JAN SWAMMERDAM (1637-1680), AMSTERDAM.

Swammerdam was a religious enthusiast who tried to discover the greatness of God in His creatures. His *Biblia Naturae* appeared only long after his death. (Leyden, 1737/8. 2 vols.).

(*a*) Describes the discovery of the female character of the ' bee-king.' Vol. II, p. 369 f.

(*b*) A table showing the similarity of the praeformative development of animals and plants. Vol. II, p. 874.

(*a*) On the 22nd of August, 1673, I opened a bee-hive after the bees had swarmed. I counted some thousands of common bees in it (worker bees), some hundreds of breeding bees (drones) and one king (queen), applying the commonly used names. Actually there have never been kings or breeding bees in the hives. It is a great and inexcusable error to give such names to these animals. I will try to avoid them in future and will start to explain that the king is actually the female, as I will prove further on . . .

Most of the ovary is situated in the upper part of the abdomen, towards the narrow waist between breast and abdomen. The other viscera, such as stomach, intestines, yellow vessels, (read: Malpighian vessels) and others are placed much lower in the body. The ovary is double, just as it is in man, quadrupeds, fishes, frogs and many other insects. Yet in one animal the separation

PLATE XVI

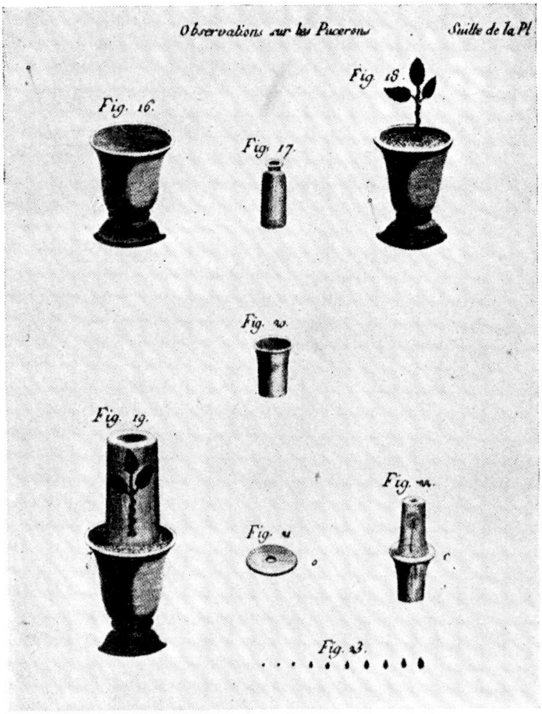

THE INSTRUMENTARY WITH WHICH CHARLES BONNET ASSURED THE COMPLETE ISOLATION OF THE APHIDS WHICH HE BRED IN ORDER TO DEMONSTRATE PARTHENOGENESIS.

PLATE XVII

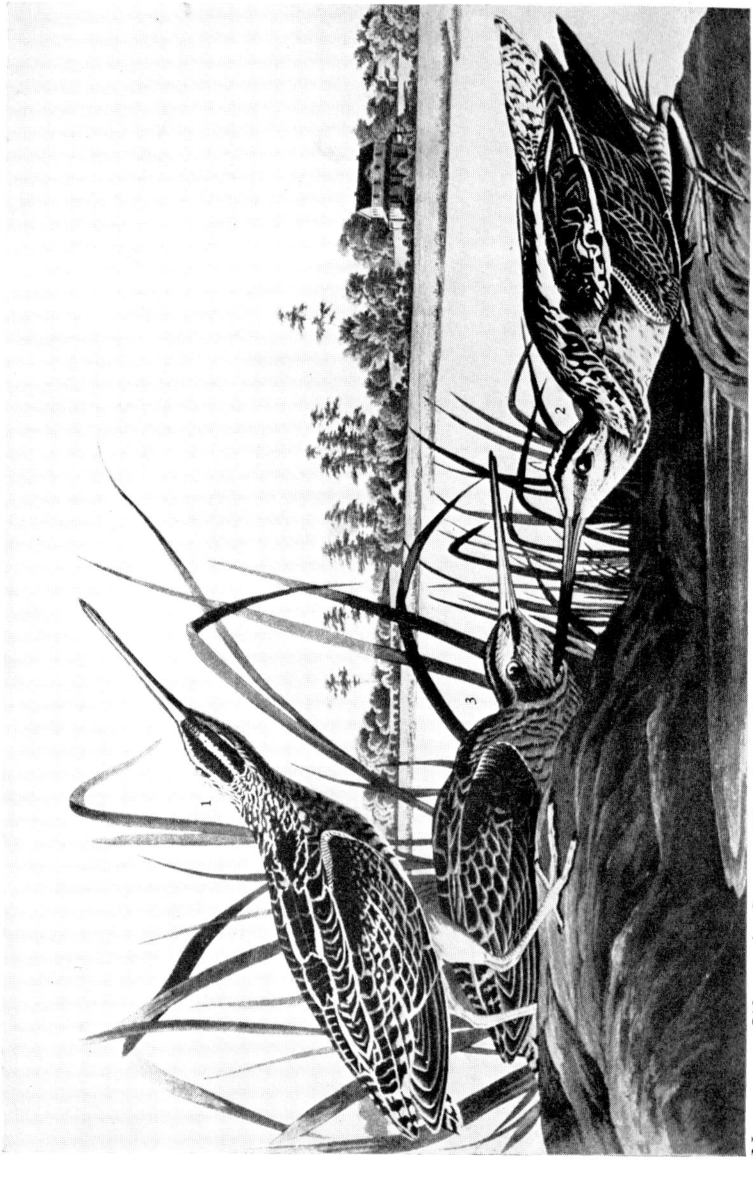

No. 243. WILSON'S SNIPE (*Capella delicata*). FROM J. J. AUDUBON, *The Birds of America*, 1827-1830.

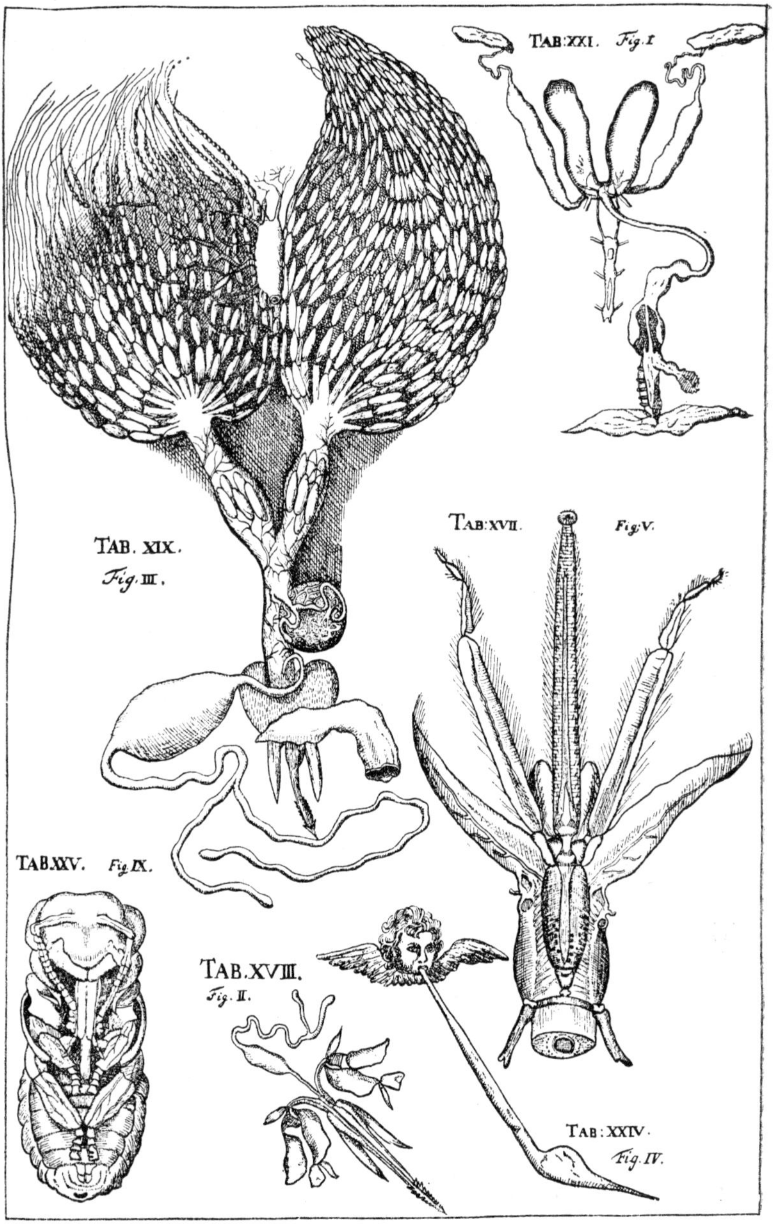

Fig. 25. Drawings from Jan Swammerdam's *Biblia Naturae* concerning the Honeybee. Most important are XIX: III representing the documentation of the establishment of the bee-king as a female; XXI: I that of the drones as males according to their inner genitals; XXIV: IV showing the use of dyed liquids blown into the tracheae as an aid for their preparation.

is more distinct than in the other one. In the bee both touch one another and are somehow connected. One ovary is to the left, the other to the right of the abdomen. They are connected by the tracheae which are running through both halves, so that they can be separated only with difficulty. The external aspect of the ovary is that of a membranaceous, very delicate and tender organ, all the eggs being visible through the membranes. Each of the two parts of the ovary has its special parts, which are called oviducts (read: ovarioles). These are, however, only part of the ovary and the eggs originate of them . . . The eggs take origin and perfection in the ovary which is at once oviduct, Fallopian tube and womb, all this wrapped into thin membraneous ducts. They must be composed of more than one layer of membranes, which, however, the eye unarmed by lenses cannot discern . . .

I was unable to count the oviducts in a female bee, as their number is embarrassing, their soft structure causes them easily to lacerate, especially by their mixing with the tracheae. Likewise I could not count the number of eggs in every oviduct, which are so easily counted in a big moth, being ten big and a few incomplete eggs per oviduct . . . Somewhat later I tried again unsuccessfully to make such a count. But from the great many partial counts which I had made, I assume that over 300 oviducts are in the ovary of the bee. In the mature female there are about 17 eggs per oviduct, which gives us 5100 visible eggs per female. But some eggs are distinctly bigger than others. The lower eggs are the biggest, those towards the waist the smallest ones, and there they are only visible to the specialist with the aid of a lens. The extreme upper ends of the ovaries, where these tiny eggs are situated, is found high upwards in the abdomen (towards the breast); they are folded and bent . . .

Nothing appears more wonderful among all the transformations of nature than that of a caterpillar into a butterfly. Yet, recognising that this usually does not differ from the growth of other animals or the budding and flowering of plants, we realise, that it is not more miracular than the usual metamorphosis of plants, and that the miracle is not in the object, but in our imagination. We simply ignore the properties of the chrysalis, within which the animal is hidden as the flower in its bud . . . It is necessary to realise, that a chrysalis is only a change, or better an increase, enlargement and extension of a larva in its limbs and parts, which has already the shape of the future animal; in other words, it is the growth of a larva into the shape of the adult insect, which continues through the stage of the chrysalis. Thus, the larva does not *change* into a pupa, but *grows* by increase of its organs into a chrysalis, and similarly the chrysalis is not transformed into the winged insect, but the same larva which took after a moult the shape of the chrysalis grows from it into the adult form. This is not different from the change of a chicken into a hen or of a tadpole into a frog. None of them is transformed, but the adult form is reached by successive extension of the body parts of the younger stages (cf. Plate XXXVII).

(*b*) *Plate* XLVI. General comparison of the agreement in the changes or increases of the body-parts and limbs in the eggs, larvae and pupae of insects, of the frog, and of the carnation (abbreviated) :—

Stage.	*Louse.*	*Fly.*	*Frog.*	*Carnation.*
I.	nit	egg	egg	seed
II.	egg-skin	egg-skin	egg-gallerte	rind of seed
III.	neonate larva	neonate maggot	neonate tadpole	small germ
IV.	greater larva	greater maggot	greater tadpole	greater germ
V.	' nymph '	pupa	metamorphosing tadpole	germ with flower bud
VI.	mature louse	adult fly	mature frog	flower ready for fertilisation

63

JOHN RAY (WRAY) (1627-1705), ENGLAND.

The important systematist of plants and animals corresponded widely. In the following letter to Dr. Hans Sloane from February 1684, he explains the aims of his *Historia Plantarum* (1680). It shows us the intensive exchange of dried plants between ' herbalists ' of that period, and of the great difficulties in identifying the long and often ambiguous descriptions of earlier botanists. Anyhow, we learn, that Linnaeus did not rise from an unobscured anarchy in taxonomy. (From E. Lankester, *Correspondence of John Ray*. London, 1848. pp. 138-141).

Mr. Ray to Dr. Hans Sloane. Black Notley, Feb. 11, 1684.

Sir, I thank you for your letter of Jan. 31, which I received by post, with the plants inclosed. The Fungus, upon opening the letter, unluckily slipped out, and was not minded, because not expected, and it being candle-light, and company in the room, trod to pieces of a sudden, before I had taken out the other plants, and read so far as to know it was sent. The other two were a little crumpled, and the *Luciniae* doubled, but without much prejudice I reduced them again to their right situation. They are both plants I had never before the good hap to see. The *Polypodium plumosum* is an elegant plant, and the leaf you sent a perfect one, and well conserved; but I am to seek for the reason of the name, and should be glad to learn its place of growth, and more of its history, from you.

I am not sure that Mr. Newton was the first inventor of that plant I put under his name. I rather suspect Mr. Lawson might be. I mean no more by putting his name to it than that it is published in his work under that name, as I do by the names of other authors, v.g. Abies *Ger. Purk*. However, he was the first who showed it to me, and gave me as much as I have set down of the history of it. Dr. Pluncket's observation of the vesicles on the back side of the leaves deserves to be added to its description.

I am not positive in asserting the plant called *Homiontis* to be only a variety of *Phyllitis* (cf. *Scolopendrium vulgare L.*), and not a distinct species, but only put it down as my suspicion. You that have seen it, and know it better than I do, are better able to judge of that; but the *Hemionitis, vera Dioscoridis* of Lobel I assert to be nothing but a small *Phyllitis* growing in a shady place. Your advice concerning inserting the varieties of sundry species, especially such as are esteemed for their beauty or variety, I approve and shall observe. Howbeit, it is not my intention to supersede the use of any approved botanical authors, but my reasons for attempting this work were—(1) to satisfy the importunity of some friends who solicited me to undertake it. (2) To give some light to young students in the reading and comparing other herbalists, by correcting mistakes, and illustrating what is obscure, and extricating what is perplex and entangled, and in cutting off what is superfluous, or, under different titles, repeated for distinct. (3) To alleviate the charge of such as are not able to purchase many books; to which end I endeavour an enumeration of all the species already described and published. (4) To facilitate the learning of plants, if need be, without a guide or demonstrator, by so methodizing of them, and giving such certain and obvious characteristic notes of the genera, that it shall not be difficult for any man that shall but attend to them, and the description, to find out infallibly any plant that shall be offered to him, especially being assisted by (the) figure of it. And lastly, because no man of our nation hath lately attempted such a work; and those that formerly did, excepting Dr. Turner, were not sufficiently qualified for such an undertaking, and so have acquitted themselves accordingly . . .

The *Polygonum pusillo vermiculato Scopylli folio* of Lobel I do not take to be the *Erica maritima Anglica supina* (cf. *Frankenia laevis L.*), which is well known to me; but I put down such an opinion, because so reputed an herbalist as Mr. Goodyer describes the said *Erica* for that *Polygonum*, at least if I mistake not; and, to say the truth, there is nothing in Lobel's figure, description, or in the place, which contradicts it. Your sample of Lobel's *Polygonum*, if a perfect plant, and well conserved, I should be glad to see. The *Erica* I hope this summer (God willing) to see growing in its natural place. I am in doubt whether the *Polygonifolia per terram sparsa*, etc., belong to this tribe, or rather to the *Asperifolia*. That its spike of flowers is so reflected and turned as theirs are I can assure you, and I think it hath a perfect flower . . .

The 'Hortus Farnesianus' said and supposed to be written by Tobias Aldinus, but indeed Petrus Castellus his work, as appears by his name in capital letters in some preface or epistle to the book, I have not, and should be glad to see.

I render you many thanks for your assistance and communications, and shall, with an honourable mention of you, own what I have or shall receive from you.

I rest, Sir, your very humble servant

John Ray.

64

RUDOLPH JACOB CAMERARIUS (1665-1721), TUEBINGEN.

Camerarius was one of the most progressive botanists of his time. In his *De sexu plantarum epistola* (1694, see Ostwald's Klassiker der exakten Wissenschaften, pp. 2, 12, 24) he gave the experimental proof of the presence of sexes in plants and of their distribution in various individuals in the different species.

Let me begin with a description of the plants, by looking at the flowers. These are the precursors of the seeds and show two peculiarities: petals and stamens. When the latter are fully developed, they have different colours and are a kind of vessel or capsule, each of which is set upon its thread or peduncle, and and they open along furrows. They are at that time filled with a fine homogeneous powder, which is spread about. It is this powder which dyes the nose yellow, when one smells a rose or a lily. If you rub it on the hand it is fine and mealy, and under the microscope it appears in the form of numerous globules, of specific form in each plant, and in some the surface is set with spines. The stamens surround the pistil, which is a prolongation of the seed-case. In many flowers you find the stamen and the pistil, so long as they are still closed, like sticking one to the other; yet with the swelling of the bud they separate and grow distinct when the bud unfolds. The pistil or pistils—according to the species—is always close to the stamina, in such a way that its cloven end must be powdered by the pollen of the stamina early and amply.

The unfolding of the petals and of the stamina is soon followed by their death. Then the lower, remaining part of the pistil swells, whilst its upper part wilts away. Based upon this observation, I opened a great papilionaceous flower before it had unfolded, in order to study the early state of the pod which swells after they have done flowering. Against the light or under the microscope small green vesicles could be recognised in linear arrangement through the cuticle along the suture of the tender pod. By continued observation on various flowers it became obvious, that these vesicles are nothing else than the shells of the future seeds. We thus find the primordia of the fruits in the flowers. Consequently we should expect as many fruits as before flowers were present, were it not that some of them drop off or are torn away by various accidents before maturation.

In some plants the stamina are so far distant from the pistils, that they form a special organ which withers without forming a fruit, whilst at some distance the pistil and the beginnings of the seeds take origin. This is the case, e.g. in the maize. In this cereal the protruding panicle at the end of the stalk is too well known to need a detailed description. After the wilting and drying of this panicle without seed formation farther down those thick cylindrical spadices are taking shape, which with their grains are covered by some leaves and protruding from each grain a long thread, which spread like a tail and which receive the pollen . . .

In certain plants we find another relation of the stamina to the seeds. In the marigold (*Mercurialis*) and in the hop, part of the plants bear flowers, the other seeds. And when we put the mature seeds for germination into the soil we see, that two kinds of plants are produced by them, which are similar and bear the same name, until they prepare for propagation. Then one notices, that the ones bear only flowers, i.e. stamina, and remain without fruit or seed, whilst the other bear fruits, but are definitely wanting in petals and stamina . . . In the second group of plants, in which flowers and fruits are found on the same plant, but separated, I have learned in two cases, how detrimental to the plants is the loss of the stamina. When I took away, in the Ricinus, the round flowerbuds before the unfolding of the stamina and prevented carefully the appearance of new ones, I never obtained from the remaining intact seed buds complete seeds, but the empty seed-coats suspended, wilting and drying. Similar was the case with the maize. When the unfolding tufts were cut early, two ears appeared which were devoid of any seed, and only a great number of empty seed-capsules appeared.

The mulberry tree and the dog's-mercury (*Mercurialis*) exemplify the third group of plants, where fruits and flowers occur on separate plants. A mulberry tree which had none other in its neighbourhood with flowers, yielded berries which, however, did not contain a single seed. Similarly the Mercurialis bearing the seed-part yielded many, but no germinating seeds whatever, when it was isolated from other flowering plants. It seems thus justifiable to ascribe to the stamina the function of the male organs, and then the seed-capsule with its stigma and pistil would correspond to the female organ. Even if we thus have demonstrated the sexual differentiation of the plants, the act of propagation itself remains unclear. In order to solve this difficult problem it would be desirable, to learn from the lynx-eyes of the microscopists what is contained in the pollen of the stamens, how far they (i.e. the pollen grains) enter into the female organ, if they arrive intact at the place, where they unite with the seed-buds and what is exudated from them in this process.

65

JOSEPH PITTON DE TOURNEFORT (1656-1708), PARIS.

Tournefort, a known botanist, was the first to apply in his *Institutiones rei herbariae* (1700. I, p. 94 f.) the concept of the genus by practice into botanical systematics, as is shown here in his treatment of the group of *Malva*.

The well characterised genus *Malva* embraces—enumerated are 50 names—the following species:—
Malva foliis crispis.
Malva rosea folio subrotundo.
Malva arborea veneta dicta, parvo flore.
Malva aceris folio, virginiana.
Malva ulmi folio, semine rostrata.
Malva ulmi folio, semine cum gemino rostro, etc.

66

CAROLUS LINNAEUS (1707-1778), UPPSALA.

With Carl Linné begins the modern taxonomy of animals and plants. The *Systema Naturae* (ed. X. Animals 1758, and the *Species plantarum* for plants 1757) describes every plant and animal by two names: that of the species and that of the genus, replacing thus the long, embarrassing and ambiguous descriptions of species of earlier authors. But Linnaeus was far from being the sober pedant as which he is often regarded. The style of many of his writings is splendid. Our first selection easily explains why he is often regarded as one of the great writers of Swedish romanticism.

(*a*) From the *Marriage of Plants* (*Nuptia Plantarum* 1720, first published 1823. (Translated from the French of K. Hagberg, *Carl Linné, Le Roi des Fleurs.* Paris, 1944. pp. 45-46).

(*b*) The system of plants, arranged according to the sexual characters. Syst. Nat. X ed., Vol. II, p. 837.

(*c*) The classification of the Lemurs. Syst. Nat. X, Vol. I, 1758, p. 29 f.

(*a*) In spring, when the light sun ascends upon our zenith, it awakes in all organisms the life, which throughout the cold winter lies strangled. Then, all creatures feel better, fresher after the heavy, sullen life of the winter; then, all birds begin to sing and chirp, after the long and silent winter; then, all insects come out from their hiding places where they were hidden stone-dead during the winter; then, all the herbs, withered during the winter, start a new life and all trees become green again; yes, even man is as though reborn. Pliny has said wisely: " Sole nihil utilius."

This sun gives an indescribable joy of life to all; we then see the grouse and woodcock make their displays, all animals coming upon heat.

" Et totus fervet Veneris dulcedine mundus,
Omnia vere vigeat et veris tempore floreat."

Yes, love even spreads to the plants, then their males and females, their hermaphrodites even, celebrate their nuptials, and about these I will write here and show which are the sexual organs of the plants, which are males, which females and which hermaphrodites . . .

The leaves of the flowers (the petals) contribute nothing to reproduction, they serve just as nuptial beds which the great Creator has so beautifully arranged, decorated with such noble curtains and perfumed with such sweet odours, in order that the husband may celebrate there the wedding with an enhanced solemnity. Once the bed is thus prepared, it is time for the husband to embrace the dear bride and to offer her his gifts; I mean to say, that now one sees how the testicles open up and distribute the pollen which falls upon the stigma and fecundates the ovary.

(*b*) *The Linnéan system of plants.*

A. *Plants with flowers.*

Aa. Hermaphroditic flowers.

aa. With free stamina.

aaa. With stamina of undetermined length:

Classis	1.	with one stamen	*Monandria*
	2.	with two stamina	*Diandria*
	3.	with three stamina	*Triandria*
	4.	with four stamina	*Tetrandria*
	5.	with five stamina	*Pentandria*
	6.	with six stamina	*Hexandria*
	7.	with seven stamina	*Septandria*
	8.	with eight stamina	*Octandria*
	9.	with nine stamina	*Enneandria*
	10.	with ten stamina	*Decandria*
	11.	with 12-19 stamina	*Dodecandria*
	12.	with 20 or more stamina which sit not upon the bottom of the fruit, but upon the inner side of the calyx	*Icosandria*
	13.	with 20 or more stamina sitting on the bottom of the fruit	*Polyandria*

abb. With stamina of definitely different lengths:

	14.	plants with four stamina, two neighbouring longer and two shorter ones.	*Didynamia*
	15.	plants with six stamina, four longer, and two shorter opposite ones.	*Tetradynamia*

ab. With cohering stamina.

	16.	the stamina are cohering below	*Monadelphia*
	17.	The stamina cohering into two bunches	*Diadelphia*
	18.	Stamina cohering into three or more bushels	*Polyadelphia*
	19.	stamina fused into one cylinder	*Syngenesia*

	20. stamina fused with the pistils	*Gynandria*
Ab. With separate sexes:		
	21. Male and female flowers on the same plant	*Monoecia*
	22. Male and female flowers on different plants	*Dioecia*
	23. Apart of gynandromorphic flowers either male or female flowers or both on one or different plants	*Polygamia*

B. *Plants with neither conspicuous stamina nor pistile which in the other plants are essential parts of the flower.*

24.	*Cryptogamia*

(*c*) *Classis* : *Mammalia*. I. *Primates*.

3. *Genus*: *Lemur*. 4 upper incisors at distances. 6 lower incisors, erect, compressed, parallel and close together. Eye-teeth single, close to their neighbours. Molars some in number, slightly lobed, the anterior ones longer and more pointed.

(1) *tardigradus*. The tailless lemur. Mus. Ad. Fr. 1, p. 3.

A tailless monkey with claws of thumb pointed. Syst. nat. 5, No. 2.

A dog-faced slow mammal. Seba, Museum I, p. 55, pl. 35, 1, 2 ; pl. 47, 1.

The most elegant animal of Robinson. Ray, Quadrup. 161.

Lives in Ceylon.

Squirrel sized, reddish with a brown dorsal stripe and a white throat. With a white stripe between the eyes, and with covered face. Ears pitched (*urceolatae*), within folded (*bifoliatae*). The inner hands and the soles of the feet are bare. Claws roundish, only those of the legs pointed (*subulati*). Almost tailless. Two nipples on the breast and two close to it on the abdomen.

A slowly-walking animal, of excellent hearing, monogamous.

(2) *catta* Lemur with annulate tail.

The squirrel-monkey of Madagascar or the Maucauco. Edwards, Aves 199, pl. 199.

Lives in Madagascar.

Cat-sized, greyish. Eye-region black. Head acuminate, ears covered. The long tail with alternating black and grey rings. Claws pointed, more rounded on the thumbs of the feet. Climbs like a monkey, but slowly. Of similar behaviour as the *Genetta auctorum*.

(3) *volans* A tailed lemur which flies by a membrane surrounding it.

A flying cat-monkey. Petiver, Gazophylacticum 14, pl. 9, fig. 8; Acta Angl. 277, No. 1065.

The wonderful bat. Bontius, Java 68, pl. 69.

A flying *terneata*. Cat. Seba, Museum I, p. 93, pl. 58, 2, 3.

Lives in Asia.

Flies like a squirrel or a flying mouse with the help of a membrane extending from the head to the hands, from the hands along the sides to the feet and from there to the end of the tail. Claws sharp. Two nipples on the breast. Eats the fruits of trees. By its nipples it appears to be close to the lemurs or the monkeys, but I have not seen it, and specimens must be examined further on.

67

STEPHEN HALES (1677-1761), ENGLAND.

The priest Stephen Hales was one of the founders of modern plant and animal physiology. His outstanding observations and experiments are concentrated in his book *Statick Essays*, containing the Vegetable Staticks and the Haemastaticks (2 vols. 1733). Of these we bring:

(*a*) On the tearing of plants (I, p. 108—ed. 1738).

(*b*) Measuring the size of bloodpressure in horses (II, p. 1 f.).

(*a*) *On the tearing of plants. First Experiment:* On the 30th of March at 3 h.p.m. I cut a grapevine which stood on a western side 7 inches above the soil. The remaining stump (c) had no branches, was 4 to 5 years old and 16 mm. thick. Upon the free end of this stump I put a glass tube (bf) which was 7 feet long and 6 mm. wide. The joining part at (b) was made impenetrable by a thick layer of a mixture of wax and turpentine held together by some threads. A second tube (fg) was put upon the first one and upon this a third one (ga), so that this tube composed of three tubes was 25 feet long. As the plant did not immediately start to ' tear,' I poured about two feet of water into the tube, which the stem rapidly sucked in, so that at 8 h.p.m. only three inches of it still stood above the stump. From the early morning to 10 h.p.m. of the 31st of March the sap had risen to a height of $8\frac{1}{4}$ inches. The next day at 6 h.a.m., when it was still freezing, the sap had risen a further $3\frac{1}{4}$ inches. Thus, the liquid within

Fig. 26. Rise of liquid by root pressure of a grape-vine eight metres high. (From Stephen Hales *Vegetable Staticks*. 1727. Pl. IV: 17).

the tube rose day by day until it reached a height of 21 feet. It would probably have risen higher, if the joint (b) would not have leaked. When this was repaired, the sap rose sometimes one inch in three minutes. In the season when the grapevine teared most, the sap did rise by day and by night, but more so by day, and most during the hot hours. This experiment demonstrates the great power which has its seat in the root and presses the sap upwards, when the plant is tearing. The next experiment was intended to learn, if this power is still present in the grapevine after the termination of the season of tearing. . . .

In the first experiment when I had arranged a tube upon a very short stump of a grapevine which had no twigs we observed that the sap rose uninterruptedly through the whole day, most speedily during the hot hours. In the second experiment the sap dropped continuously in the same measure as the noon heat rose. When we remember the strong transpiration of the trees, we must conclude, that the drop of the sap is caused by the transpiration of the branches, which is stronger at noon than at any other hour of the day, decreasing towards evening, and probably ceasing entirely at night, when dew is falling. In late April the grapevine has gained a much greater surface because of the unfolding of many leaves. Therewith the transpiration increases and the excess of sap which has found its outlet by tearing ceases until the next spring.

This is the case with all trees which are tearing or bleeding. This phenomenon always ceases as soon as the younger leaves have a sufficient surface for strong transpiration. We also observe that the bark of oak and of many other trees is easily loosened in consequence of a slipperiness caused by the superabundance of sap. But as soon as the leaves form a sufficiently great surface to give off this sap by transpiration, the bark is not anymore so easily loosened, but sits fasts upon the wood.

(*b*) *From the 'Haemastatick.' Experiment I. Measuring the blood pressure in horses.* In December I caused a mare to be tied down alive on her back . . . Having laid open the left crural artery about three inches from her belly, I inserted into it a brass pipe, and by means of another brass pipe which was adapted to it, I fixed a glass tube which was nine feet in length. Then untying the ligature on the artery, the blood rose in the tube 8 feet 3 inches perpendicular above the level of the left ventricle of the heart. But it did not attain its full height at once; it rushed up about half way in an instant, and afterwards gradually at each pulse, 12,—, 6, 4, 2 and sometimes one inch. When it was at its full height, it would rise and fall at and after each pulse 2, 3 or 4 inches; and sometimes it would fall 12 or 14 inches, and have there for a time the same vibrations up and down at and after each pulse, as it had, when it was at its full height . . .

14. As this experiment shows how much the force of the blood in the arteries is abated by different degrees of bleeding; so it may be of use to direct what quantity to let out at a time in bleeding; for whatever the real quantity of the circulating blood be, it is certain that the estimate of what can be with safety let out at once, must be taken from the proportion which that bears to the whole quantity of blood, which will flow out of the vein or artery of the animal till it dies.

15. We see also from this experiment, the reasonableness of the practice of bleeding at several distant times, where it is requisite to take away a great quantity of blood, and not to do it all at once, which would too much weaken the force of the blood. For since it was found by several instances in this experiment, that when the force of the blood was much depressed by bleedings, it would be considerably raised again by the actions of the muscles out of whose very fine and long capillary vessels it moves but slowly, as also by the motion of all parts of the mare; so the case is doubtless the same when the vigour of the blood is in any degree rebated in the large vessels, by bloodletting, that vigour will in some measure be in a little time restored again, not only by the action of the several parts of his body, whereby the blood would have time to flow in from all parts, to supply the most evacuated vessels, whereby there would be a just proportionate evacuation of all parts; but also because the vessels themselves would thereby have time to contract themselves in some proportion to the degree of their evacuation.

68

ABRAHAM TREMBLEY (1700-1784), GENEVE.

Naturalist belonging to the circle of C. Bonnet. He is famous by his long continued observations on the life-history, the reproduction and the regeneration of the green freshwater polype.

From 1744 *Mémoires pour servir à l'histoire d'un genre de polypes d'eau douce.* English translation by G. Adams (1746. p. 159 f.).

This reproduction is performed sooner or later, as the weather is more or less warm. In the height of summer the arms will sometimes begin to shoot in 24 hours, and in two days have been in a state to eat, but in cold weather it will be 15 or 20 days before the head is formed.

If a polyp, having young ones, be cut transversely, the young ones do grow after the section. It often happens, that the second parts which have had no young ones at the time of the section, have had young shoots before itself could eat, and before it had arms.

In whatsoever place a polyp was cut, whether at the middle or near either end, the experiment equally succeeded, and each portion became a complete polyp, which walked, eat and multiplied. A polyp being cut close under the arms, though small as it was, it became a complete polyp, which at the beginning was all arms. If a polyp be cut transversally into three or four pieces, the posterior end of the first produces a tail, the anterior of the last a head, and the intermediate pieces acquire both head and tail.

To cut a polyp lengthwise, it must be made to contract as much as possible, because the more it is contracted the larger the body is: Therefore put the polyp upon a slip of white paper in a small drop of water, and when by touching it is very much contracted, drain away the water . . .; then with a sharp pair of scissors cut through both, paper and polyp; the divided parts will adhere to the paper like a jelly, . . . until thrown in water . . .

The polyps are not big enough to permit to cut them at one time into many more (than four) pieces. But I have instead cut them successively into many pieces. I cut one polyp into four parts, fed these pieces well and when they had grown to a certain size, I cut them again into two or three pieces, into as many as their size permitted. Finally, I bred these pieces and cut them again. In this way I cut one polyp into fifty pieces; then I stopped the experiment, as I felt that I had sufficiently demonstrated my point. All these fifty pieces had developed into complete polyps, which were quite normal in all their functions. Some of them I kept for over two years and they multiplied well . . .

I also fed for over two years some polyps with which I never had experimented before, together with the offspring of cut polyps. When the other circumstances were equal, I never found any difference in their reproduction (fertility) . . .

69

CHARLES BONNET (1720-1793), GENEVE.

Bonnet was one of the centres of biology during a period of great revolutions. As he became early almost blind, his observations are restricted to his youth. We bring here the fascinating story of his observations on the parthenogenesis of aphids (from *Traité d'Insectologie.* 1745. 2 vols. Oeuvres d'Histoire Naturelle et de Philosophie, 1779, Vol. I, p. 16 ff).

Is there, then, no copulation among plant lice? So as to have something more than conjecture on this matter, M. de Réaumur proposed an experiment which he tried four or five times without success: this was to take a plant louse on emergence from its mother's body and rear it in such a way that it could not have commerce with any insect of its species. "If a plant louse which would thus have been raised by itself," said M. de Réaumur "produced plant lice, it must either have done so without copulating or else have copulated within its mother's body."

Stimulated by the invitation of M. de Réaumur, I undertook in 1740 to try the experiment on the plant louse of the spindle tree.

First Observation. First experiment on an aphid of the spindle tree to decide whether plant lice reproduce without copulating. Several methods of raising a plant louse in isolation present themselves. Here is that which I decided upon. In a flower pot filled with ordinary earth, I pushed down as far as its neck a vial full of water. Into this vial I introduced the cut end of a small branch of the spindle tree on which I left only 5 to 6 leaves after examing them on both sides with the utmost care. I then placed on one of these leaves a plant louse whose mother lacked wings and had just given it birth while I looked on. I finally covered the little branch with a glass vase whose rim was applied exactly against the surface of the earth in the flower pot; from which I was surer of my prisoner's conduct than Acrisius was of Danae's, although she was locked up by his order in a bronze tower.

It was on May 20th, at 5 p.m. that my plant louse was committed from birth to the confinement which I have just described. I took care from then on to keep an exact diary of its life. In this record, I noted its least movement; not a step it took was a matter of indifference to me. Not only did I watch it every day, starting usually at 4 or 6 in the morning and scarcely ever stopping before 9 or 10 at night, but I likewise observed it several times in the same hour and always with an eyeglass to render the observation more exact and to inform myself of the most secret action of our little solitary prisoner. But if this application cost me some trouble and kept me somewhat confined, I received in return something for which I could congratulate myself that I had gone through with it.

My louse changed its skin four times; on the evening of the 23rd; at 2 p.m. on the 26th; at 7 a.m. on the 29th; and at 7 p.m. on the 31st.

It was immediately after ridding itself of its skin that my louse was working to turn over. With its two hind legs, as with two arms, it embraced its cast skin, trying to raise it so as to disengage the barbs attaching it to the leaf or twig on which it moults. It repeated its efforts in a different direction. Little by little it managed to disengage one of its legs and finally all the others. As soon as the cast skin was no longer attached, the louse pushed it up into the air and abandoned it. All this was a somewhat tough task for a louse whose legs had not had time to strengthen themselves.

Perhaps I may be accused of childishness if I record the misgivings which my louse caused me at this last moult. Although it was always shut up in such a way as to leave no fear that some other insect might slip into its solitude, I found it so swollen and so shiny that it seemed to be in the same condition as lice which are nourishing worms inside themselves. What added to my fear and increased my chagrin was that it appeared to be motionless. Unfortunately I had only the light of a candle by which to observe it. Having finally realised that it was only changing its skin I was somewhat reassured; but I was still not without misgivings. It had been lying on its side, and soon was on its back so that its belly was entirely visible. I saw it move its legs, which up to that time it had held to its chest, as nymphs do, it shook them several times as if it desired to use them to change its position; but weak as they were, having just cast their old skin, they did not seem strong enough to acquit themselves of the task. In its position, and on an almost vertical leaf, the louse was held only by its cast, to the end of which its body was still fastened. It was thus exposed to the possibility of a fatal fall. This crisis disquieted me so that I regained my peace of mind only as, little by little, it righted itself.

But it is time to come to the most interesting moment in the life of our little hermit. Happily delivered from four maladies which might have carried it off, it finally arrived at the stage to which I had hoped by my pains to bring it. It had become a full-grown plant louse. On June the 1st, around 7 o'clock in the evening, I saw, to my great happiness, that it was about to give birth; after which I felt it mandatory to refer to it as a she-louse. Beginning on that day, and until the 21st inclusive, she had 95 young, all born alive and most of them arriving while I looked at . . .

Finally, to finish the story of our louse, I have only to say that, being obliged to absent myself all the 25th until 5 o'clock the next morning, I was chagrined on my return, not to find her where I had left her, nor anywhere in the neighbourhood which I searched in vain. Inasmuch as, since she had begun to give birth, I had not considered it necessary to keep her completely confined, she no doubt took advantage of this fact to depart and to finish her days in other parts. One will easily judge that I was not insensible to this loss. I had seen her born, I had followed her carefully for a month, and I would have enjoyed continuing to observe her with the same care until death.

These plant lice give birth to living young during the good season, but lay eggs at the end of autumn, signs of copulation. The reader will ask impatiently: Why does copulation occur in insects which do not need one another (sex), and which are fit to reproduce in isolation . . . In the good season these plant lice give birth to living young, i.e. they lay living young. In the middle of autumn they lay true eggs. This I discovered in autumn, 1740, and was confirmed afterwards by famous observers. I showed in my book that the females are able to change all their behaviour when 'laying' living young or eggs. I described these eggs, how careful they are laid and what occurs before the time of laying them, during oviposition and afterwards. First I thought these eggs to be immature embryos. But then I showed them to be true eggs, and I enumerated the reasons for this conclusion.

70

JULIEN OFFREY DE LAMETTRIE (LA METTRIE) (1709-1751), PARIS.

De Lamettrie extended the theory of Descartes and regarded man also as a machine (*L'Homme Machine.* 1747), becoming thus one of the founders of first modern representatives of materialism in biology. (English translation by G. C. Bussey. Chicago, 1912, ed. 1778 Leyde, p. 71, 108).

The soul is therefore but an empty word, of which no one has any idea, and which an enlighted man should use only to signify the parts in us that thinks. Given the least principle of motion, animated bodies will have all that is necessary for moving, feeling, thinking, repenting, or in a word for conducting themselves in the physical realm, and in the moral realm which depends upon it.

Yet we take nothing for granted; those who perhaps think that all the difficulties have not yet been removed shall now read of experiments that will completely satisfy them.

(1) The flesh of all animals palpitates after death. This palpitation continues longer, the more cold-blooded the animal is and the less it perspires. Tortoises, lizards, serpents, etc., are evidence of this.

(2) Muscles separated from the body contract when they are stimulated.

(3) The intestines keep up their peristaltic or vermicular motion for a long time.

(4) According to Cowper, a simple injection of hot water reanimates the heart and the muscles.

(5) A frog's heart moves for an hour or more after it has been removed from the body, especially when exposed to the sun or better still when placed on a hot table or chair. If this movement seems totally lost, one has only to stimulate the heart, and that hollow muscle beats again. Harvey made this same observation on toads.

(6) Bacon of Verulam, in his treatise *Sylva Sylvarum*, cites a case of a man convicted of treason, who was opened alive, and whose heart, thrown into hot water leaped several times, each time less high, to the perpendicular height of two feet.

(7) Take a tiny chicken still in the egg, cut out the heart and you will observe the same phenomena as before, under almost the same conditions. The warmth of the breath alone reanimates an animal about to perish in the air pump. The same experiments which we owe to Boyle and to Stenon, are made on pigeons, dogs, and rabbits. Pieces of their hearts beat as their whole hearts would. The same movements can be seen in paws that have been cut off from moles.

8. The caterpillar, the worm, the spider, the fly, the eel—all exhibit the same phenomena; and in hot water, because of the fire it contains, the movement of the detached parts increases.

(9) A drunken soldier cut off with one stroke of his sabre an Indian rooster's head. The animal remained standing, then walked, and ran; happening to run against a wall, it turned round, beat its wings still running, and finally fell down. As it lay on the ground, all the muscles of this rooster kept on moving. That is what I saw myself, and almost the same phenomena can easily be observed in kittens or puppies with their head cut off.

(10) Polyps do more than move after they have been cut in pieces. In a week they regenerate to form as many animals as there are pieces. I am sorry that these facts speak against the naturalist's system of generation; or rather I am very glad of it, for let this discovery teach us never to reach a general conclusion even on the ground of all known (and most decisive) experiments.

Here we have many more facts than are needed to prove, in an incontestable way, that each tiny fibre or part of an organized body moves by a principle which belongs to it. Its activity, unlike voluntary motions, does not depend in any way on the nerves, since the movements in question occur in parts of the body which have no connection with the circulation . . .

We will not say that after death every machine or every animal disappears entirely or gets another form, as we have no knowledge thereabout. Hence, whoever says that an immortal machine is pure imagination or something reasonable, judges like the caterpillar seeing the moults of their likes and crying bitterly over the fate of their likes which have come to perdition. The soul of these caterpillars (and every animal has a soul of its own) is too limited in order to understand all the natural processes. Even the most clever ones of them cannot imagine themselves that they will become a butterfly. The same is true for us. In fact, what do we know about our future fate, as all we know is only past history? Let us be contented with our ignorance which we cannot overcome . . .

Let us then conclude boldly that man is a machine, and that in the whole universe there is but a single substance differently modified.

71

RÉNE ANTOINE FERCHAULT DE RÉAUMUR (1683-1757), LA ROCHELLE (FRANCE).

This discoverer of a stable thermometer was in his youth a broad physiological experimentator. Later he concentrated upon his marvellous *Mémoires pour servir à l'histoire des insects* (6 vols. 1737-1748), the fundament until this day for the knowledge of the life histories of the insects. An unpublished volume, *The Natural History of the Ants*, was edited in 1926 (pp. 157, 163) by Wheeler as volume VIIa. From this part we give the discovery of trophobiosis of ants and plant lice, and of the sexual character of the winged ants.

(*a*) Ants climb trees in such numbers that it would be inexpedient to try to prevent them . . . They in no wise seek to injure the vegetation, but they know that there are on the trees certain insects that work for them. Whenever we follow the ants to a spot where they stop, we discover plant-lice or scale-insects either on the leaves or on the branches. I have stated elsewhere that I have had no better guides than the ants in discovering species of both these insects. The plant-lice emit from their posterior end and through two tubes near it a sugary liquid for which the ants are very greedy. A similar liquid exudes, perhaps, from divers parts of the body of the scale insects, and the ants like to lap it up. In brief, both the plant-lice and the scale-insects produce an effusion of liquid by which the ants profit. This is demonstrated by our study on the plant-lice (Mém. Vol. III) and on the scale-insects (Mém. Vol IV) . . . To one examining the ants that in very dense files ascend and descend the trunk of a tree, they seem all to belong to the same species and to all appearances to the same colony. It would seem that the plant-lice and the scale-insects that have been discovered on a tree are the property of the ant-hill by which they were discovered. At least they remain in the peaceful possession of the colony that first finds them. Alien ants happening in few individuals are too feeble to encroach on the army already in possession, and this army always resists an attempt on the part of strangers to participate in the harvest of the sugary liquids emitted by the plant-lice and scale-insects. Still there are occasions when ants of different species contend in combat for the possession of a tree . . .

(*b*) Are there then in an ant-colony two kinds of females, winged and wingless? No there is only one kind, both of them are the same individuals seen at different seasons . . . When we follow the ants through their various stages we see that those that are born without wings pass their lives without having them, whereas those that are to be winged have wings from the moment of their pupal hatching, like flies or butterflies . . . It is not only certain that the small winged ants are the males, but it is equally certain that their mating, which has remained concealed for many centuries, is more easily witnessed than that of any other insect . . . Being on the road to Poitou and finding myself on the levée of the Loire, very near Tours, on one of the first days of September, 1731, I descended from my coach, enticed to stroll about by the beauty of the spot and the mild temperature of the air, which was the more agreeable because the earlier hours of the day had been warm The sun was within about an hour of setting. During my stroll I noticed a lot of small mounds of sandy and earthy particles rising above the openings that led the ants to their subterranean abode. Many of them were at that time out of doors; they were red, or rather reddish, of medium size. I stopped to examine several of these earthen hillocks and noticed on each among the wingless ants a number of winged ones of two very different sizes. Some of them had abdomens no larger than those of the wingless ants, and to judge from unaided vision one of the larger winged individuals must have weighed more than two or three times as much as one of the smaller. Over the beautiful levée, where I was enjoying my walk, there appeared in the air in places not very far apart small clouds of large fliers which flew about in circling paths. They might have been taken for gnats or crane-

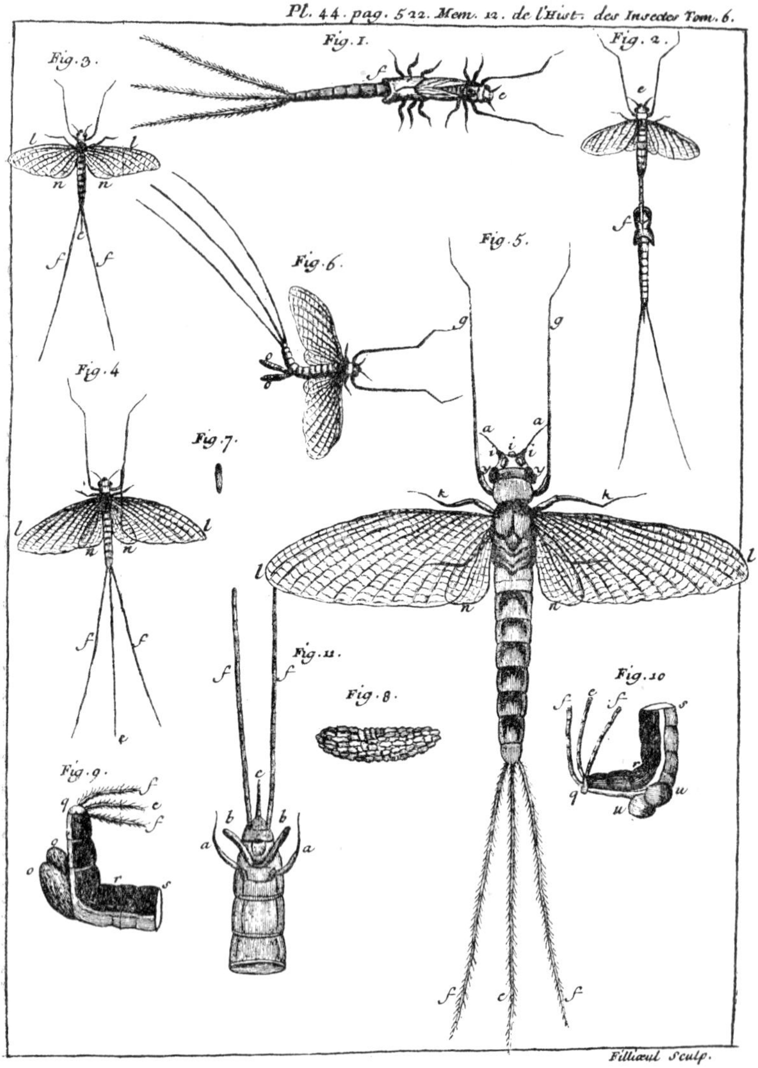

Fig. 27. Mayflies. From the *Mémoires pour servir a l'étude des Insectes* of Réaumur. Paris, (1734-1742). 6 vol. edition.

flies or may-flies. Often the small cloud hung in the air at a height within reach of the hand . . . I observed—and the observation was as important as it was easy to make—that I almost invariably captured them in pairs. Not only did I almost always find in my hand one large and one small ant, but most frequently I took them copulating and held them for some time before they separated. The small ant was resting on the large just as among common flies the male while mating rests on the female. The posterior end of the small ant was curved downward so as to apply itself to that of the female and it adhered so firmly that force was necessary to separate the pair. The abdomen of this small male was scarcely half as long as that of the large female . . . I compressed the abdomen of the large ants and caused clusters of eggs to exude . . . It is therefore in mid-air that the nuptials must be celebrated of those ants that pass the greater portion of their lives underground and the remainder of their lives crawling on its surface or at most on walls, plants or trees . . . I have always seen the ants return one by one to their ant-hill. Thus it is in the air that mating begins and continues. Then it is the task of the female to support the male, contrary to the damselflies, among which the male carries the female . . . It has thus been established that the wings are necessary to the ants, both males and females, in order that they may mate, and that these organs are given to them only for this purpose. At least it is certain that the females do not long retain their wings after they have fecundated. The males also shed theirs, but apparently they retain them much longer . . . I confined a single very large winged female in a glass beaker, and towards noon of the following day all four wings had fallen off . . . Of four males which I imprisoned in a glass tube with females that no longer bore wings, two lost theirs after two or three days . . .

72

GEORGE LOUIS LECLERC, COMTE DE BUFFON (1707-1788), PARIS.

The great work of Buffon are the 42 volumes of his *Histoire Naturelle générale et particulière* (1779-1840), a description of the world whose splendid style is shown by the chapter of the nightingale. Yet many general problems are treated in a tendency to create a general biology.

(*a*) The Nightingale. (English translation. London, 1812. Vol. 15, p. 296).

(*b*) Note on Classification. (Vol. I: p. 13).

(*c*) On the Degeneration (read: variation) of Animals. (English translation by W. Smellie. London, 1791, Vol. IV, p. 7 f.).

(*a*) To every person whose ear is not totally insensible to melody, the name of Nightingale must record the charms of those soft evenings in spring, when the air is still and serene, and all nature seems to listen to the songster of the grove. Other birds . . . excel in the several parts which they perform; but the Nightingale combines the whole, and joins sweetness of tone with variety and extent of execution. His notes assume each diversity of character, and receive every change of modulation; not a part is repeated without variation; and the attention is kept perpetually awake, and charmed by the endless flexibility of strains. The leader of the vernal chorus begins the prelude with a low and timid voice, and he prepares for the hymn to nature by essaying his powers, and attuning his organs: by degrees the sound opens and swells; it bursts with loud and vivid flashes; it flows with smooth volubility; it faints and murmurs; it shapes with rapid articulations: the soft breathings of love and joy are poured from his inmost soul, and every heart beats unison, and melts with delicious langour. But this continual richness may satiate the ear. The strains are at times relieved by pauses, which bestow dignity and elevation. The mild silence of evenings heightens the general effect, and not a rival interrupts the solemn scene.

The Nightingale excels all birds in the softness and mellowness, and also in the duration, of its warble, which sometimes lasts without interruption twenty minutes . . . The Nightingales differ much in the quality of their song; in some it is so inferior that they are reckoned as not worth keeping . . . A Nightingale may happen to hear the sweet music of some other birds, and, in the glow of emulation, improve his own; he communicates the melody to his young; and thus it is transmitted, with various alterations, through the succeeeding races. After the month of

June, the Nightingale's warble is gone; a raucous croaking remains . . .

(*b*) Nothing is well defined what is not well described. In order to describe well you have to see the object, to see it again and to examine it without any prejudice, without any preconceived system, and if you do otherwise, the description is wanting the stamp of truth.

Would it not be more simple, natural and true to call an ass an ass, a cat a cat, instead of pretending that an ass is a horse (read: *Equus asinus*) or a cat a lynx (read: *Felis catus*)?

(*c*) *Of the Degeneration of Animals* . . . Thus the temperature of the climate, the quality of the food, and the evils produced by slavery (read: domestication), are the three causes of the changes and degeneration of animals. The effects of each merit a separate examination; and their relations, when viewed in detail, will exhibit a picture of nature in her present condition, and of what she was before her degradation.

Let us compare our pitiful sheep with the mouflon from which they derived their origin. The mouflon is a big animal. It is fleet as a stag, armed with horns and thick hoofs, covered with coarse hair, and dreads neither the inclemency of the sky, nor the voracity of the wolf. It not only escapes from its enemies by the swiftness of his course, but it resists them by the strength of its body, and the solidity of the arms with which its head and feet are fortified. How different from our sheep, which subsist with difficulty in flocks, which are unable to defend themselves by their numbers, which cannot endure the cold of our winters without shelter, and which would perish if man withdrew his protection. In the warmest climates of Asia and Africa, the mouflon, the common parent of all races of sheep, appears to be less degenerated than in any other region. Though reduced to domestic state, it has preserved its stature and its hair; but the size of its horns is diminished. Of all domestic sheep, those of the Senegal and India are the largest, and their nature has suffered least degeneration. The sheep of Barbary, Egypt, Arabia, Persia, Calmuckia, etc., have undergone greater changes. In relation to man, they are improved in some articles, and vitiated in others. But with regard to nature, improvement and degeneration are the same thing; for they both imply an alteration of original constitution. Their coarse hair is changed into fine wool. Their tail, loaded with a mass of fat, has acquired a magnitude so incommodious, that the animals trail it with pain. While swollen with superfluous matter, and adorned with a beautiful fleece, their strength, agility, magnitude, and arms are diminished: these long tailed sheep are half the size only of the mouflon. They can neither fly from danger, nor resist the enemy. To preserve and multiply the species, they require the constant care and support of man.

The degeneration of the original species is still greater in our climates. Of all the qualities of the mouflon, our ewes and rams have retained nothing but a small portion of vivacity, which yields to the crook of the shepherd. Timidity, weakness, resignation, and stupidity, are the only melancholy remains of their degraded nature. To restore their original size and strength, our Flanders sheep should be united with the mouflon, and prevented from propagating with inferior races; and, if we would devote the species to the more useful purposes of affording us good mutton and wool, we should imitate some neighbouring nations in propagating the Barbary race of sheep, which, after being transported into Spain and Britain, have succeeded very well. Strength and magnitude are male attributes; plumpness and beauty of skin are female qualities. To obtain fine wool, therefore, our rams should have Barbary ewes; and, to augment the size, our ewes should be served with the male mouflon . . .

The wild animals, not being under the immediate dominion of man, are not subject to such great changes as the domestic kinds. Their nature seems to vary with different climates; but it is nowhere degraded: if they were capable of choosing their climate and their food, the changes they undergo would be still less. But, as they have at all times been hunted and banished by man, or even by the strongest and most ferocious quadrupeds, most of them have been obliged to abandon their native country, and to occupy lands less friendly to their constitution. Those whose nature had ductility enough to accomodate themselves to this new situation, have diffused over vast territories; while others have had no other recourse than to confine themselves in the deserts adjacent to their own country. There is no animal which, like man, has spread over the whole surface of the earth . . .

As climate and food have little influence on wild animals, and the empire of man still less, the chief varieties amongst them proceed from another cause. They depend on the number of individuals of those which produce, as well as of those that are produced. In those species in which the male attaches himself to one female, as in that of the roebuck, the young demonstrate the fidelity of their parents by their entire resemblance to them. In those, the females of which often change the male, as in that of the stag, the varieties are numerous; and as in all nature there is not one individual perfectly similar to another, the varieties among animals are proportioned to the number and frequency of the produce. In species, the females of which bring forth five or six young three or four times a year, the number of varieties must be much greater than in those which produce a single young once a year . . .

73

ALBRECHT VON HALLER (1708-1777), GOETTINGEN AND BERN.

From this important physiologist we quote a chapter of his *Dissertation sur les parties irritables et sensibles des Animaux.* Lausanne, 1755, p. 39 f. (Latin edition. Goettingen, 1752). Haller understands contractility where he writes irritability.

Irritability (read: contractility) is so different from sensibility, that the most irritable parts are not sensitive, and the most sensitive are not irritable. I shall also demonstrate, that irritability does not depend upon the nerves, but upon the primordial structure of the irritable parts.

The nerves, those organs of every sensation, have no irritability at all. This is astonishing, but true. When I irritate a nerve, the muscle which it innervates shows immediately convulsions. This is always the case, as I have often shown for the diaphragm and for the abdominal muscles of a rat, or for the leg muscles of a frog. . . . But when one irritates the nerves within these muscles, the nerves never show contraction, as I have ascertained often in dogs and frogs . . . I have (also) used a mathematical instrument divided into very small sections. This I laid along the nerve, which remained perfectly immobile during strong stimulation. This experiment, by the way, does not agree with the oscillations attributed often to nerves . . .

On cutting the crural nerve of a dog, its legs became insensitive; one may subject it to any kind of maltreatment and the animal will not show any sign of suffering. But when this cut nerve is stimulated, the muscles of the leg show convulsions . . .

I have also ligatured (lié) in small animals the nerve-trunks which go to the extremities. I thus paralysed the legs and made them insensitive. When I afterwards stimulated the muscles, they contracted as before, in spite of their not being subject any longer to the governance of the soul.

74

CASPAR FRIEDERICH WOLFF (1738-1794), HALLE.

Wolff was the founder of the modern school of epigenesis of the embryos of animals and plants (*Theoria generationis.* 1759).

(*a*) Epigenesis of the chicken embryo and the principles involved.
(*b*) From the early development of the chicken.
(*a*) and (*b*) see Ostwald's Klassiker der exakten Wissenschaften No. 84/85. Leipzig 1896.
(*c*) Review of the *Theoria Generationis* by A. von Haller. (*Göttinger Anzeigen*, No. 143. Halle, 1760) (3) reprinted in ed. 1774 p. XXXVIII.

(*a*) (The youngest chicken observed after 28 hours of incubation, is) a mass consisting only of cells (= spherules) of a certain shape and position. The cells are, however, little connected with each other and simply heaped together; they are transparent, mobile, and almost liquid. The mass shows neither heart nor vessels, nor trace of red blood (§166).

We could not say in general, that what is not accessible to our senses is therefore non-existent. Applied to our observations, however, this principle is more a sophism than a truth. The particles of which all animal organs are composed, in their primordial condition are spherules which can always be discerned by a microscope of medium magnification. Could we really maintain that one is unable to see a body because of its smallness when the parts which compose it are easily recognisable? Nobody has so far discovered with a strongly magnifying lens parts which are invisible with less powerful lenses. Either one cannot see them at all or they appear sufficiently large. Therefore, that those parts were hidden because of their infinite smallness and that they then only gradually come forth is a fable. The way in which nature produces organic parts can be very well recognised in the history of the development of the limbs and of the kidneys . . . (§240).

The first primordia of the extremities are little elevations raised above the other cell-substances. The cellular substance, however, which surrounds the vertebral column, and the adjacent substance, furnish the raw material for the elevation which will be structurally organised later on. There can be no doubt about this for one who himself has observed the successive transformation of this substance (§224).

We understand, hence, why the formation of a part and its organisation are not accomplished in the individual by one and the same act, so that a formed part would be *eo ipso* organised. Rather, a part is first formed, and only then organised . . . Furthermore, the organisation of the separate part, which is a process different from its first formation is continually perfected by the rise of parts still to be organised . . . (§240).

(*b*) The intestine of the chicken embryo is first a simple membrane . . . This longitudinal stripe of the double plate, first plane, begins to swell into a cylindric shape and then resembles the primitive intestine. We are thus certain, that this instestine is a new formation and that it could not have existed as such before and just now only have been unwrapped . . . No doubt remains with regard to the truth of epigenesis.

The principles of Wolff's theory are contained in the following three principles which are applied to plant- and animal-development alike.

By the ' essential force ' (*vis essentialis*) liquids are collected from the surrounding earth, forced into the plant and spread into it, stored in some parts, whilst other are excreted . . . The chicken embryo at the beginning of its development takes food from the egg substance. This absorption occurs by a force, which is not the contraction of the heart, the arteries or the pressure caused by them in neighbouring veins nor their compression by muscular activity, and in general it does not enter by definite channels, and thus it is identical with the essential force of the plants (I, 11 ; II, 3). (§242).

Besides this essential force it is the stiffening force of the nutritive liquids which together are a sufficient principle to explain the development of plants and animals (II, 60).

The developing organism is not a machine, but composed of not organised substance. This developing substance should be distinguished from the machine into which it is wrapped. The machine (read: structure) is its product (§253).

(*c*) For a long time we have not read such an important book as that of C. F. Wolff of Berlin, who has defended epigenesis in a thesis in Halle, entitled *Theoria Generationis* (24. 11. 1759) . . . He begins with the herbs, the most simple creatures. In the beginning nothing is in the leaf except many vesicles (*i.e.* cells), and in the young root we recognise the same structure or a transparent one without seeing any vessels. This is very important for Mr. Wolff, as he mainly tries to prove, that the vessels are in this (early) stage of the plant not simply too small or too transparent, but are entirely wanting. To this purpose he quotes some experiments: One may change in a young leaf with a needle the shape of these vesicles, may push one entire vesicle from one place to another, may press two vesicles into one and separate them thereafter or may even empty out until it collapses. One may even produce new vessels by moving the drops, giving them another between two vesicles or may unite two (vessels) into one. Wolff concludes from these experiments, that the delicate structure of (young) plants has no vessels at all, but only vesicles. The vessels are at the beginning only spaces, without walls, and only in the full-grown herb do they assume a solid covering, so-called skins. The growth of the leaves is performed by new vesicles which push themselves between the older ones, and the vessels originate as part of the solidifying sap changes into a wall between the vessels and into the space around the vessels. From this thickened sap which concentrates within the vessels and the vesicles both parts become more and more perfect. Yet at the beginning the matter of the plants is nothing but a mere mixture, from which slowly leaves and vessels differentiate. Both are the consequence and not the cause of the movement of the sap. From both these groups the entire plant is composed. Wolff follows the growth of the white cabbage and finds its origin in a curved pointed part which grows out of the seed. This point is transformed slowly into leaves, which also grow perfect by and by, and new points for further growth appear. All this is a sap, which goes out of the plant at one end and becomes slowly thicker and firmer. The young root grows from sap which enters from the outside through its bark, entering its venous woody structure, and there produces ' returning ' vessels which are not arteries. From the rest of Wolff's remarks upon the plants we note only that he regards fertilization as a sexual process, but sees its main importance in the fact, that the male pollen is extremely nutritious. In this entire process there is no need to assume any other basic force except movement the *vis essentialis*, and the thickening of the sap.

With regard to the development of animals one has to accept one principle only, namely that nothing exists except what can be seen. Everything in animals is composed of vesicles which are visible. Hence, we cannot accept, that invisible parts are present in addition . . .

75

CAPTAIN JAMES COOK (1728-1779)

DESCRIBES AUSTRALIA'S ANIMAL LIFE.

During his three great explorations of Australia and the Pacific World, Captain Cook and his companions observed the plant and animal life of the new territories.

(*a*) On the Kangaroo. From J. Hawkesworth, *Account of Voyages for Discoveries in the Southern Hemisphere*, by J. Cook. Vol. III, p. 157. London, 1773.

(*b*) Of Animals and Ants. From J. D. Hooker *Journal of the R.H. Sir Joseph Banks*. London, 1896. 1770, p. 208 f, 301 f. Animals of New Holland.

(*a*) Early on the 24th June, 1770. As I was walking this morning at a little distance from the ship, I saw myself one of the animals which had so often been described: it was of a light mouse colour, and in size and shape very much resembling a greyhound; it had a long tail also, which it carried like a greyhound; and I should have taken it for a wild dog, if instead of running, it had not leapt like a horse or deer: its legs were said to be very slender, and the print of its foot to be like that of a goat; but where I saw it, the grass was so high that the legs were concealed, and the ground was too hard to receive the touch;

Mr. Banks also had an imperfect view of this animal, and was of the opinion that its species were hitherto unknown.

(*b*) Quadrupeds we saw but few, and were able to catch but few of those we did. The largest was called by the natives Kangooroo; it is different from any European, and indeed, any animal I have heard or read of, except the jerboa of Egypt, which is not larger than a rat, while this is as large as a middling lamb. The largest we shot weighed 84 lbs. It may, however, be easily known from all other animals by the singular property of running, or rather hopping, upon only its hinder legs, carrying its fore-feet close to its breast. In this manner it hops so fast that in the rocky bad ground where it is commonly found, it easily beat my greyhound, who, though he was fairly started at several, killed only one, and that quite a young one. Another animal was called by the natives *je-quoll;* it is about the size of, and something like, a pole-cat, of a light brown, spotted with white on the back, and white under the belly. The third of the opossum kind, and much resembled that called by de Buffon *Phalanger*. On these two last I took only one individual of each. Bats here were many: one small one was much, if not identically, the same as that described by de Buffon under the name *Fer de Cheval*. Another sort was as large as, or larger than a partridge; but of this species we were not fortunate enought to take one. We supposed it, however, to be the Rousette or Rougettee of the same author. Besides these, wolves were, I believe, seen by several of our people, and some other animals described . . . (p. 208).

Of sea-fowl there are several species (gulls, shags, gannets, boobies, etc.). In the rivers were ducks, curlews . . . The land birds were crows, very like our English ones, most beautiful parrots and parroquets, white and black cockatoos, pigeons, beautiful doves, bustards, and many others which did not at all resemble those of Europe. Most of these were extremely shy, so that it was with difficulty that we shot any of them . . .

P. 303. Of insects there were but few sorts, and among them only the ants were troublesome to us . . . One species is green as a leaf and lives upon trees, where it built a nest in size between a man's head and his fist, by bending the leaves together, and gluing them with a whitish papery substance which held them firmly together. In doing this their management was most curious: they bend down four leaves broader than a man's hand,

and placed them in such a direction as they choose. This requires a much larger force than these animals seem capable of; many thousands indeed are employed in the joint work. I have seen as many as could stand by one another, holding down such a leaf, each drawing down with all its might, while others within were employed to fasten the glue. How they had bent it down, I had no opportunity of seeing, but that it was held down by main strength, I easily proved by disturbing a part of them, on which the leaf, bursting from the rest, returned to its natural situation, and I had an opportunity of trying with my finger the strength that these little animals must have used to get it down . . .

Another sort there were, quite black, whose manner of living was most extraordinary. They inhabited the inside of the branches of one sort of tree, the pith of which they hollowed out almost to the very end of the branches, nevertheless the tree flourished as well to all appearance as if no such accident had happened to it. When first we found the tree, we of course gathered the branches, were surprised to find our hands instantly covered with legions of these small animals, who stung most intolerably . . . Rumphius mentions a similar instance to this in his *Herbarium Amboinense* (II: 257); his tree, however, does not at all resemble ours.

A third sort nested inside the root of a plant which grew upon the bark of trees in the same manner as mistletoe (*Myrmecodia* = *Myrmecia beccarii* Hf.). The root was the size of a large turnip, and often much larger; when cut, the inside showed innumerable winding passages in which these ants lived. The plant itself throve to all appearance not a bit the worse for its numerous inhabitants. Several hundreds have I seen, and never one but what was inhabited; though some were so young as not to be much larger than a hazel-nut. The ants themselves were very small, not about half as large as our red ants in England; they sting indeed, but so little that it was scarcely felt. The chief inconvenience in handling the roots came from the infinite number . . . The fourth kind were perfectly harmless, though they resembled almost exactly the white ants of the East Indies . . . Their architecture was, however, far superior to that of any other species. They had two kinds of houses, one suspended (p. 305) on the branches of trees, the other standing upright on the ground. The first sort were generally three or four times as large as a man's head; they were built of a brittle substance,

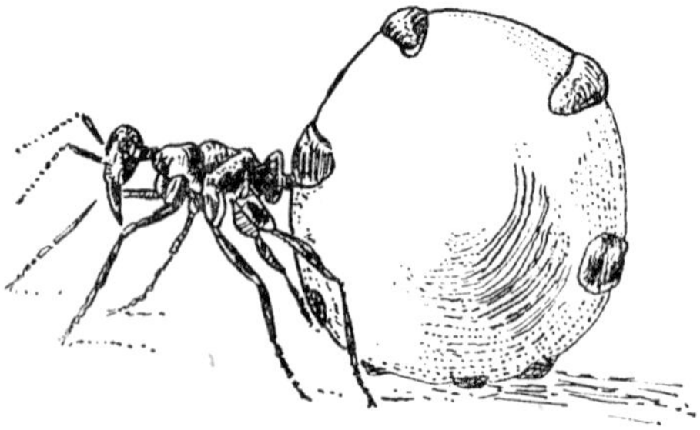

Fig. 28. Scenes from the Honey Ant Totem of the Warramunga Tribe in Central Australia. A. A totem ceremony, in which the performer represents a woman ancestor searching for and gathering ants on which she feeds. B. The typical Australian Honeypot Ant *Melophorus inflatus* Lubb. (From Spencer and Gillen, McKeown.)

seemingly made of small parts of vegetables kneeded together with some glutinous matter, probably afforded by themselves. On breaking this outer crust innumerable cells appeared, full of inhabitants, winding in all directions, communicating with each other, as well as with divers doors which led from the nest. From each of these an arched passage led to different parts of the tree, and generally one large one to the ground . . . The second kind of a house was very often built near the foot of a tree, on the bark of which their covered ways, though but seldom in the first kind of house, were always to be found. It was formed like an irregularly sided one, and was sometimes more than six feet high, and nearly as much in diameter. The smaller ones were generally flat-sided, and resembled very much the old remains of Druidical worship. The outer crust of these was five cms. thick at least, of hard, well tempered clay, under which were their cells; to these no doors were to be seen. All their passages were undergound, where probably they were carried on till they met the root of some tree, up which they ascended and so up the trunks and branches by the covered way before mentioned. This I should suppose to be the houses to which they retire in the winter season, as they are undoubtedly able to defend them from any rain that can fall, while the others, though generally built under the shelter of some overhanging branch, must, from the thinness of the covering, be but a slight defence against a heavy rain . . .

The sea, however, made some amends for the barrenness of the land. Fish, though not so plentiful as they generally are in the higher latitudes, were far from scarce . . .

76

LAZAR[R]O SPALLANZANI (1729-1799), BOLOGNA.

Spallanzani is the founder of modern experimental biology. The extent of his observations and experiments is almost unbelievable. Hence, the great number of selections.

(*a*) On the Reproduction of Amphibians. (Dissert. Nat. Hist. Animals and Vegetables. English edition, 1784. pp. 6 ff).

(*b*) On the Regeneration of the Head of a Snail. (Programme ou précis d'un ouvrage sur les réproductions animales. Trad. Franç, De la Schiame. Génève, 1768. pp. 57-67).

(*c*) On the Resuscitation of dried rotatorians, Tardigrada, and nematodes. (From Opuscules de Physique Animale et Vegétale. Trad. Franç, J. Senebier. Génève, 1777. Vol. II. p. 308 ff.).
(*d*) Life is killed by Heat. (Ibidem. Vol. I. p. 12).
(*e*) On animal Respiration. (Mémoires sur la respiration. Posthumously ed. by J. Senebier. Génève, 1803. p. 184 ff).

(*a*) The amores of the green frog begin in April and May being influenced by the temperature of the atmosphere; during it the male maintains an incessant croaking. In autumn and winter the immature eggs lie in the ovary which is divided into two lobes: some eggs being very small, scarcely visible to the naked eyes, others seven to eight times larger, both globular . . . p. 7. If the eggs are again examined in spring, we find them in the ovary, but the larger ones considerably enlarged and they will be found to be mature when the male is coupled with the female. The copulation is as observed by Swammerdam and Roesel . . .

P. 8. Vallisnieri asserts elsewhere, that the female does not discharge her eggs when she is kept constantly separate from the male. I say constantly; for if they be pulled asunder when the eggs are descended into the cavity of the uterus, they are discharged, though the female is kept separate, but they are not prolific.

If the situation of the eggs, at the time of copulation, is examined, during the first days they will be found in the sac of the ovary; and during the succeeding, partly in the ovary and partly in the oviducts, and at last all in the uterus, except the small ones, which remain attached to the ovary. The eggs, when in the ovary, are smaller than when in the oviducts and uterus. In these situations they are enveloped with that viscid transparent mucilage, which is improperly called frog's seed.

Of the various experiments I have made, in order to ascertain whether eggs taken from the ovary, the oviducts and the uterus, when the male is embracing the female, would be prolific. P. 9. I must own that not one has succeeded. As this is a point of extreme importance, I repeated my experiments to satiety; and in my journals I find that I have opened 156 females, while they were embraced by the males, of not one of which did the eggs ever bring forth young, though I immediately placed them in water; whereas those that were expelled spontaneously by the females, all were prolific. I have taken further pains: the discharge of the eggs lasts about one hour; during this process I killed a female, and put the eggs that remained in the body in the water into which those discharged by the female fell; but the latter produced tadpoles, while the former became an offensive putrid mass. From the facts I concluded, that the fecundation of the eggs does not take place within, but without the body; whence it appears how far Linnaeus was mistaken, when he pronounced, in his usual decisive tone: " Nullum in rerum natura, in ullo vivente corpore fieri fecundationem vel ovi impregnationem extra corpus matris (cf. Artedi, Ichthyology II: 32).

Hence we likewise see the falsehood of the strange opinion of Professor Menzius, that while the male embraces the female so closely, the seed is emitted from the fleshy prominence of the toe, and passing through many windings unknown to us, penetrates into the thorax and there impregnates the eggs . . . p. 14. The reader will probably be surprised at this description, whence it appears, that the tadpole does not come out of the eggs, but that the egg is transmuted into the tadpole . . . These phenomena were new and unexpected . . . I am obliged to call these globules tadpoles or fetuses instead of eggs

P. 113. My celebrated friend Mr. Bonnet, since I communicated to him in 1767, my discovery of the pre-existence of the germ in the frogs, has never ceased urging me to attempt the artificial fecundation of this animal. I began with the terrestrial toad with red eyes and dermal tubercles which begins earliest in the spring to propagate its species . . . Just before parturition, of which I was apprised by the excessive swelling of the belly, I parted the male from the female, and set the latter by herself in a vessel full of water. In a few hours the cords of the eggs began to appear; as soon as about the length of a foot was expelled, I cut them off, and left one in the vessel, while I took out the other, in order to wet it with semen, which I procured from the male that had just been separated from the female. It is easy for any one who has the slightest skill in comparative anatomy, to find the seminal vesicles . . . At the time of coupling they are always full. I laid open the vesicles, and securing the liquor which had the transparency of water, into a watch glass, I spread it on the piece of cord with a pencil; but the quantity was only sufficient to go over 2/3 (inches); after the operation I placed this piece in a vessel of the same water as that in which the unimpregnated portion lay. It was the 16th of March, the weather cold and unfavourable for the evolution of the tadpoles. I was forced to wait the longer in suspense for the event, about which I was not a little anxious . . . After five days many of the tadpoles in the 2/3 (inches) of the cord over which the pencil had passed, began to assume an elongated figure, whilst the others remained round . . . P. 115. All the tadpoles in the portion of the cord wetted with semen were not evolved. Of 176, 63 were spoiled, probably because they were not touched by the fecundatory fluid.

(*b*) Chap. VI. *On the reproduction* (*read: reparation or regeneration*) *of the head and other parts of terrestrial snails and of the feelers of slugs* (*limaces*). The head of a snail is much more complex than one would think, with its big brain in two lobes, constricted in the middle, with two nerves going out below, which correspond to our spinal cord, and ten forward into the head . . . , two enter the feelers at the end of each of which is placed one eye with two membranes and three humours (aqueous, cristalline and vitreous). Some muscles govern the more or less irregular movements of the head and of the feelers . . . The snail has not only a mouth, but also lips, a tongue, a palate, teeth, etc. These teeth are so intimately connected that often they appear as one only. First, this snail is able to regenerate its feelers . . . But nature does not always follow *one* way in the regeneration of the snail's feelers . . . , yet it appears always, no matter where on the feeler the operation has been made . . . When, however, you cut off the entire head of the snail, you see another appear and in a most peculiar manner. When, e.g., one cuts the head and the tail off an earthworm, the regenerating parts form an organic unity and whole, i.e. a head and tail similar in details to the parts cut off, to which nothing is lacking but a fuller development of all the segments. But from the trunk of the snail something organically whole does *not* come forth, having all the parts of the head which was cut off. These different parts are at the beginning often separated one from the other: or they appear one *after* the other at different times and only after a rather long time they unite, become consolidated and form a whole, fairly resembling the earlier model . . . The regenerate is sometimes only a small ball with elements of the two lips, the two small feelers, of the mouth and of the teeth. This small ball is situated on the centre of the trunk and does not contain all the parts such as the two big feelers or the anterior part of the snail's foot which is in contact with the head in normal snails . . . Whilst time effaces the scars, the line of section on the neck remains visible in some individuals for over two years. And even after such a long interval the regeneration of the head is not yet completed in all: one is lacking one feeler, another more feelers; sometimes the new heads are smaller than their natural size, or they are monstrous and filled with tumours . . . This shows quite easily what are the bizarrities observed in the regeneration of the head of snails which are cut off at the same time. But are these really *ludi naturae* or are they variations based upon constant and unchangeable laws? . . .

That the regenerated heads fed normally I regarded as a certain indication of a complete regeneration of all parts composing the head. Yet, in order to be sure I dissected some snails with fully regenerated heads. I learned that the new head is not only in possession of those similar and dissimilar parts which I have already mentioned, but also of others which together with the former make the head perfect. Every part of the new regenerates joins so well the old parts (tissues), that one would not think that the snail was ever mutilated, if the greyish line cutting transversally over its neck were not a certain indicator of the section. All these things always occur if the section has been made before or behind the brain. In the latter case the rest of the spinal cord produces a new brain from which grows out later the ten nerves mentioned above. It was natural to think that the snail should regenerate also other, more simple parts of its body. The neck and the foot regenerate perfectly . . . The small number of experiments with slugs shows that they also regenerate their feelers as easily as the snail, whilst the other parts of the head is much more difficultly regenerated in slugs than in snails.

(*c*) Leeuwenhoek and Baker had discovered the revival of dried-out Rotatoria. S. made systematic experiments, and discovered the same for the previously unknown Tardigrada and for nematodes.

P. 308. As long as the drop of water which I observed was full, the three Rotatoria moved easily, they swam everywhere with their snout between the sand grains, which they moved as if they were searching for food; they, however, did not leave the fluid when they approached its borders. Having arrived there they turned round promptly. When the drop was conspicuously evaporating the Rotatoria moved more slowly, and still slower when the drop had entirely evaporated, when they could not change their place anymore. They remained on the same spot, turned continuously around and allongated themselves. These movements occurred especially in the tail and in the head, which

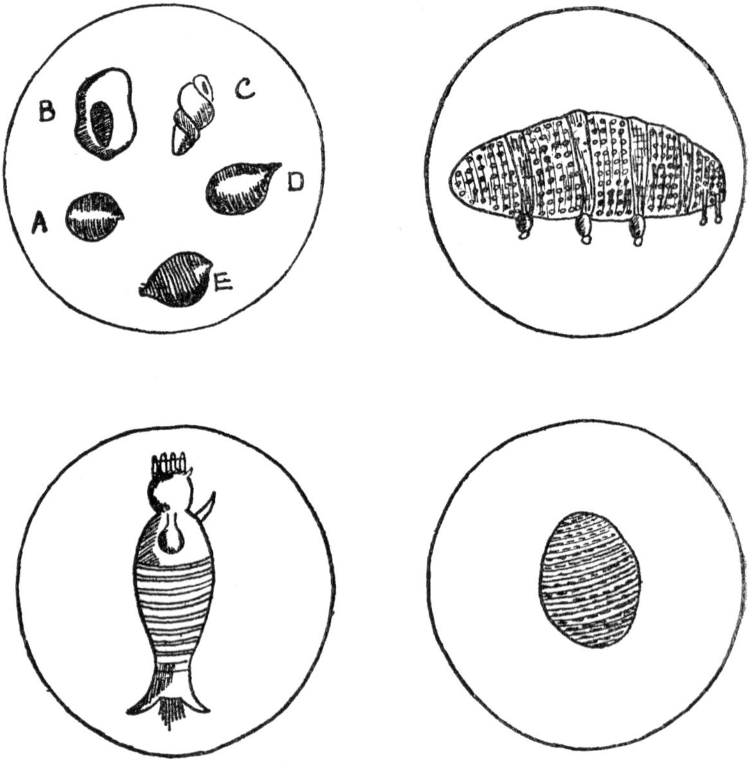

Fig. 29. From Lazaro Spallanzani's experiments on the resuscitation of animals from dried moss and sand (1777). Above left—(IV, 4) A and C are dried out Rotatorians which begin (D-E) to resuscitate after having been moistened. Above right—(IV, 8): A living Tardigrad, a group which was described here for the first time. Below left—(IV, 3): Living Rotatoria. Below right—V, 9): The same animal dried out.

came out of the body of these animals and was then retracted. In this retracted position they remained after the evaporation was completed. The Rotatoria then changed their aspect, not only by complete loss of movement and of being alive, but their size was considerably reduced. They had become three small corpuscles, so deformed, that it was impossible to recognise what they had been (IV A—C).

They remained for an hour in this condition of apparent death. Then I dropped another drop of the same water on the place of the evaporated one. The reader may well imagine with what attention I followed their resurrection, which succeeded fully. After a few minutes the Rotatoria were slightly swollen, one of their sides becoming pointed (IV D). The pointed part began to move by growing longer and shorter. Soon also the opposite end became pointed and moved like the other one. I soon realized that the two pointed parts were the head and the tail of the animal, which came little by little out of the body, into which they were retracted when the first drop had evaporated. The transverse rings, the longitudinal lines, the inner and outer organs, all reappeared and the three Rotatoria assumed their earlier shape and size, and this in very little time. They soon began to move amongst the grains of sand as quickly as before.

After having discovered in the sand of drains, which I put into water, some more Rotatoria, I repeated these experiments and found that they always became alive, no matter how long they were dry. I have now the remarkable case of sand which I had kept dry for four years. Yet after putting it into water, the Rotatoria revived promptly . . . I have eleven times dried the same sand and wettened it as many times. Always I observed the Rotatoria to die after the drying out of the water and to begin their life again, as soon as I wetted it again.

(*d*) (Resumé of his experiments against Needham J. T., that in infusions after boiling no infusion-animals appear, when the vials are hermetically closed): I found Needham's conclusions not convincing, not only because I suspected that the vials had not been exposed sufficiently to heat to kill the seeds within the vial, but mainly because these seeds could have easily entered the vials (afterwards) and have given birth to the infusion animalcules. He had the vials closed only by very porous cork-stoppers. I repeated therefore the experiments more exactly, by closing the vials hermetically, by keeping them for one hour in boiling water.

When I opened and examined these vials after a convenient time, I did not find the least sign of life of these animalcules, after the close observation of the infusions of 19 different vials.

(*e*) Conclusions from hundreds of experiments on the respiration of *Helix nemoralis L.* :

(1) The snail has organs of respiration.
(2) These snails destroy the oxygen of usual air, they cannot live without it, and they destroy it not completely.
(3) They also destroy the azote gas (nitrogen), but much less.
(4) These worms cannot make use of the *eudiometre*, as a perfect eudiometre should show the complete destruction of the oxygen and conserve completely the nitrogen.
(5) The higher the temperature, the speedier is the destruction of the oxygen by them, and the quicker they die.
(6) When the temperature drops to —1°, the destruction of oxygen is interrupted, but also the pulsation of the heart and the circulation of their body-fluids.
(7) Probably this suspension of the movements of the heart and of the fluids lasts in these snails throughout the winter.
(8) Dead as well as living snails destroy the oxygen. Hence follows, that their lungs are not the only cause of the change in the oxygen, but that other parts in their body produce also this effect.
(9) The shells themselves of the snails destroy the oxygen gas and continue to destroy it, even long after their inhabitant snails have been taken out of the shells.
(10) The shells lose the property of destroying the oxygen gas only when they are decomposed.
(11) The carbonic acid gas is produced because of the destruction of the oxygen gas.
(12) In the vessels wherein the snails are kept a certain humidity prevails.

77

JOSEPH PRIESTLEY (1733-1804), BIRMINGHAM.

Priestley discovered the oxygen and its vital importance for animals. His many experiments are described in: *Experiments and Observations on Different Kinds of Air*. (3 vols. 1774, 1775, 1777). A surprising kind of air, the dephlogisticated air, is produced from the gas generated by red mercury oxide.

The reason of my great expectations from this mode of experimenting is simply that, by exhibiting substances in the form of air (read: gas), we have an opportunity of examining them in a less composed state and are advanced one step nearer to their primitive elements.

His experiments lead Priestley to discover the rejuvenation of the air made putrid by animal respiration and asphyxiation through the presence of green plants. (See: 2nd edition. Vol. I. 1775. p. 70 ff.).

Section IV. Of Air infected with Animal Respiration, or Putrefaction. p. 86 ff. This observation led me to conclude, that plants, instead of affecting the air in the same manner with animal respiration, reverse the effects of breathing, and stand to keep the atmosphere sweet and wholesome when it is become (p. 87) noxious, in consequence of animals either living and breathing, or dying and putrefying in it.

In order to ascertain this, I took a quantity of air, made thoroughly noxious, by mice breathing and dying in it, and divided it into two parts; one of which I put into a phial immersed in water; and to the other (which was contained in a glass jar, standing in water) I put a sprig of mint. This was about the beginning of August, 1771, and after eight or nine days, I found that a mouse lived perfectly well in that part of the air, in which the sprig of mint had grown, but died the moment it was put into the other part of the same original quantity of air, and which I had kept in the very same exposure, but without any plant growing in it.

This experiment I have several times repeated; sometimes using air in which animals had breathed and died, and at other times using air, tainted with vegetable or animal putrefaction; and generally with the same success.

Once, I let a mouse live and die in a quantity of air which had been noxious, but which had been restored by this process, and it lived nearly as long as I conjectured it might have done in an equal quantity of fresh air; but this is so exceedingly various, that it is not easy to form (p. 88) any judgement from it; and in this case the symptom of 'difficult respiration' seemed to begin earlier than it would have done in common air.

Since the plants that I made use of manifestly grow and thrive in putrid air; since putrid matter is well known to afford proper nourishment for the roots of plants; and since it is likewise certain that they receive nourishment by their leaves as well as by their roots, it seems to be exceedingly probable, that the putrid effluvium is in some measure extracted from the air by means of the leaves of plants, and therefore that they render the remainder more fit for respiration. . . .

On the 8th of March, 1775, I put a mouse into a glass vessel, containing two ounce measures of the air from mercurius

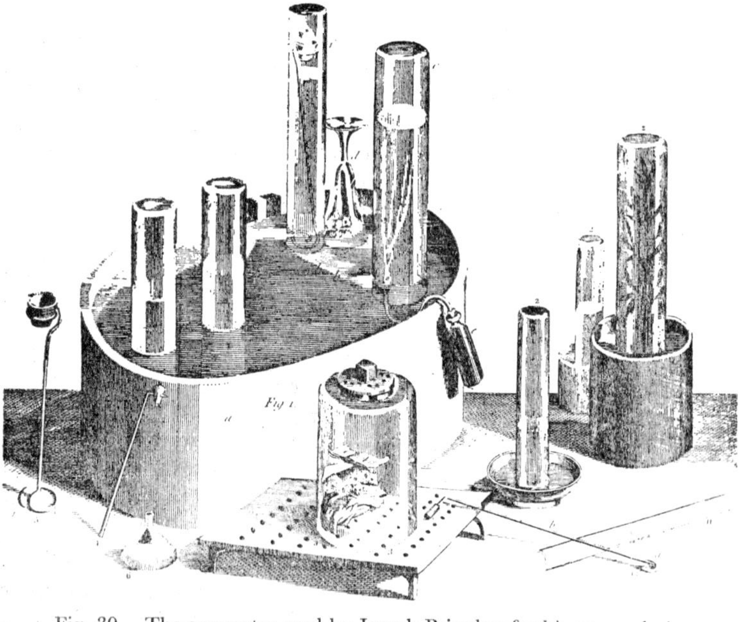

Fig. 30. The apparatus used by Joseph Priestley for his gas analysis. Vol. I, 1774, oppos. frontispiece.

calcinatus. Had it been common air, a full-grown mouse, as this was, would have lived in it about of a quarter of an hour. In this air, however, my mouse lived a full half hour and though it was taken out seemingly dead, it appeared to have been only exceedingly chilled; for, upon being held to fire, it presently revived. By this I was confirmed in my conclusion that the air extracted from the mercurius calcinatus was at least as good as common air; but I did not certainly conclude that it was any better, because I knew it was not impossible but that another mouse might have lived in it half an hour. So little accuracy is there in this method of ascertaining the goodness of air . . . The day after I applied the test of a small part of that very air which the mouse had breathed so long. Had it been common air, it must have been very nearly as noxious as possible, so as not to be affected by nitrous air; but to my surprise I found that though it had been breathed so long it was still better than common air. After mixing it with nitrous air in the usual proportion of two to one, the nitrous air had made it speedily 2/9 less than before, whereas I had never found that, in the longest time, any common air was reduced more than 1/5 of its bulk by any proportion of nitrous air, not more than 1/4 by any phlogistic process whatever. Thinking of this extraordinary fact upon my pillow, the next morning I put another measure of nitrous air to the same mixture, and to my utter astonishment found that it was farther diminished to almost 1/2 of its original quantity; but a third measure could not diminish it any further, yet left it one measure less than it was even after the mouse had been taken out.

Being now fully satisfied that this air, even after the mouse had breathed it half an hour, was much better than common air; and having a quantity of it still left, 1½ ounces, I put the mouse into it. When I observed that it seemed to feel no shock upon being put into it, but that it remained perfectly at its ease another full half hour, when I took it out quite lively and vigorous. Measuring the air next day, I found it to be reduced from 1½ to 2/3 of an ounce measure. And after this it was nearly as good in the nitrous air test as common air. It was evident from the mouse having been taken out quite vigorous, that the air could not have been rendered very noxious.

For my further satisfaction I procured another mouse, and putting it into less than 2 ounce measures of air extracted from mercurius calcinatus and air from red precipitate, it lived three-quarters of an hour. But not having had the precaution to set the vessel in a warm place, I suspect the mouse died of cold. However, as it had lived three times as long as it could probably have lived in the same quantity of common air, and I did not expect much accuracy from this kind of test, I did not think it necessary to make any more experiments with mice.

(From further measurements and experiments) I conclude that it was 4 to 5 times as good as common air. Since, I have procured better air than this, between 5 to 6 times as good as the best common air.

78

JOHANN FRIEDRICH BLUMENBACH (1752-1840), GOETTINGEN.

With his *De Generis Humani Varietate Nativa*, 1775, Blumenbach laid the foundation to modern anthropology. English translation by T. Bendyshe, 1865, p. 264, 275.

Innumerable varieties of mankind run into one another by insensible degrees. We have now completed a universal survey of the genuine varieties of mankind. And as, on the one hand, we have not found a single one which does not even among other warm-blooded animals, especially the domestic ones, very plainly take place under our eyes, and deduce its origin from manifest causes of degeneration; so, on the other hand, no variety exists, whether of colour, countenance, or stature, etc., so singular as not to be connected with others of the same kind by such an imperceptible transition, that it is very clear they all are related, or only differ from each other in degree.

Five principal varieties of mankind may be reckoned. Even among these arbitrary kinds of divisions, one is said to be better and preferable to another. After long and attentive consideration, all mankind seems to me as if it may best be divided into the following five varieties: *Caucasian*, *Mongolian*, *Ethiopian*, *American*, and *Malay*. I have alloted the first place to the Caucasian, as I esteem it the primeval one. This diverges in both directions into two, most remote and very different from each other, on the one side into the Ethiopian, and on the other into the Mongolian. The remaining two occupy the intermediate positions between the primeval one and these two varieties: the American between the Caucasian and the Mongolian, the Malay between the same Caucasian and Ethiopian . . .

Conclusion. Thus, too, there is this insensible transition by which all varieties run together, and which according to the causes and ways of degeneration in man and other domestic animals, brings us to the conclusion, which seems to flow spontaneously from physiological principles applied by the aid of critical zoology to the natural history of mankind: *that no doubt can any longer remain but that we are with great probability right in referring all and singular as many varieties of man as are at present known to one and the same species.*

79

ANTOINE LAVOISIER (1743-1794), PARIS.

Lavoisier explains the respiration of animals as oxydation, in his *Expériences sur la respiration des animaux.* Hist. Acad. Sci., 1777. Paris, 1780, p. 185 ff.

Of all phenomena of the animal economy, none is more striking, more worthy the attention of philosophers and physiologists than those which accompany respiration. This function is essential to life and cannot be suspended for any time without exposing the animal to the danger of immediate death . . . After the experiments of Hales and Cigna, those of Dr. Priestley have shown by a number of very ingenious, delicate and novel experiments, that the respiration of animals has the property of phlogisticating air, in a similar manner to what is effected by the calcination of metals and many other chemical processes, and that the air ceases not to be respirable till the instant when it becomes surcharged, or at least saturated with phlogiston.

Certain doubts led me to experiments on a different plane and I found myself led irresistibly, by the consequences of my experiments, to very different conclusions: air which has served for the calcination of metals is, as we have already seen, nothing but the mephitic residuum of atmospheric air, the highly respirable part of which has combined with the mercury, during the calcination; and the air which has served the purposes of respiration, when deprived of the fixed air, is exactly the same; and, in fact, having combined with the latter residuum about one half of its bulk of dephlogisticated air, extracted from the calx of mercury, I re-established it in its former state and rendered it equally fit for respiration, combustion, etc., as common air, by the same method as that I pursued with air vitiated by the calcination of mercury.

The result of these experiments is that, to restore air that has been vitiated by respiration to the state of common respirable air, two effects must be produced: first to deprive it of the fixed air (CO_2) it contains, by means of quicklime or caustic alcali; secondly, to restore to it a quantity of highly respirable or dephlogisticated air, equal to that which it has lost. Respiration, therefore, acts inversely to these two effects, and I feel myself led to two consequences equally probable, between which my present experience does not permit me to choose . . . The first opinion is supported by an experiment which I communicated to the Academy in 1775, where I showed that dephlogisticated air may be wholly converted into fixed air by an addition of powdered charcoal; and later I proved that this conversion may be effected by several other methods. It is possible, therefore, that respiration may possess the same property, and that dephlogisticated air, when taken into the lungs, is thrown out again as fixed air . . . Does it not then follow, from all these facts, that this pure species of air has the property of combining with the blood and that this combination constitutes its red colour? But, which ever of these two opinions we embrace, whether that the respirable portion of the air combines with the blood, or that it is changed into fixed air in passing through the lungs; or lastly, as I am inclined to believe, that both these effects take place in the act of respiration, we may, from facts alone, consider as proved:

I. That respiration acts only on the portion of pure or dephlogisticated air contained in the atmosphere; that the residuum or mephitic part is a merely passive medium which enters into the lungs and departs from them nearly in the same state, without change or alteration.

II. That the calcination of metals, in a given quantity of atmospheric air, is effected, as often declared, only in proportion as the dephlogisticated air, which it contains, has been drained and combined with the metal.

III. That, in like manner, if an animal be confined in a given quantity of air, it will perish as soon as it has absorbed, or converted into fixed air, the major part of the respirable portion of air, and the remainder is reduced to a mephitic state.

IV. That the species of mephitic air, which remains after the calcination of metals, is in no wise different from that remaining after the respiration of animals, according to all my experiments;

Provided always that the latter residuum has been freed from its fixed air; that these two residua may be substituted for each other in every experiment, and that they may each be restored to the state of atmospheric air by a quantity of dephlogisticated air equal to that of which they have been deprived. A new proof of this last fact is that, if the proportion of this highly respirable air, contained in a given quantity of the atmospheric, be increased or diminished in such proportion will be the quantity of metal which we shall be capable of calcining in it, and, to a certain point, the time which animals will be capable of living in it.

80

JAN INGENHOUSZ (INGEN-HOUSZ) (1730-1799), UTRECHT, LEYDEN AND VIENNA.

Ingenhousz discovered in his " Experiments upon Vegetables," London, 1779, p. 14, 28, the assimilation of plants.

Section II. On the manner in which the dephlogisticated air (= oxygen) is obtained from the leaves of plants.

As the leaves of plants yield dephlogisticated air only in the clear daylight, or in the sunshine, and begin their operation only after they have been in a certain manner prepared, by the influence of the same light, for beginning it; they are to be put in a very transparent glass vessel, or jar, filled with fresh pump water (which seems the most adapted to promote this operation of the leaves, or at least not to obstruct it); which, being inverted in a tub full of the same water, is to be immediately exposed to the open air, or rather to the sunshine: thus the leaves continuing to live, continue also to perform the office they performed out of the water, as far as the water does not obstruct it. The water prevents only new atmospheric air being absorbed by the leaves, but does not prevent that air, which already existed in the leaves, from coming out. This air, prepared in the leaves by the influence of the light of the sun, appears soon upon the surface of the leaves in different forms, most generally in the form of round bubbles, which, increasing gradually in size, and detaching themselves from the leaves, rise up and settle at the inverted bottom of the jar: they are succeeded by new bubbles, till the leaves, not being in the way of supplying themselves with new atmospheric air, become exhausted. This air, gathered in this manner, is really dephlogisticated air, of a more or less good quality, according to the nature of the plant from which the leaves are taken, and the clearness of the daylight to which they were exposed. It is not very rare to see these bubbles so quickly succeeding one another, that they rise from the same spot almost in a continual stream: I saw this more than once, principally in the *Nymphaea alba*.

Section VI. The production of the dephlogisticated air from the leaves is not owing to the warmth of the sun, but chiefly, if not only, to the light.

If the sun caused this air to ooze out of the leaves by rarifying the air in heating the water, it would follow that, if a leaf, warmed in the middle of the sunshine upon the tree, was immediately placed in water drawn directly from the pump, and thus being very cold, the air bubbles would not appear till, at least, some degree of warmth was communicated to the water; but quite the contrary happens. The leaves taken from trees or plants in the midst of a warm day, and plunged immediately into cold water, are remarkably quick in forming air bubbles, and yielding the best dephlogisticated air.

If it was the warmth of the sun, and not its light, that produced this operation, it would follow, that, by warming the water near the fire about as much as it would have been in the sun, this very air would be produced; but this is far from being the case.

I placed some leaves in pump water, inverted the jar, and kept it near the fire as was required to receive a moderate warmth, near as much as a similar jar, filled with leaves of the same plant, and placed in the open air, at the same time received from the sun. The result was, that the air obtained by the fire was very bad, and that obtained in the sun was dephlogisticated air.

A jar full of walnut tree leaves was placed under the shade of other plants, and near a wall, so that no rays of the sun could reach it. It stood there the whole day, so that the water in the jar had received thereabout the same degree of warmth as the surrounding air (the thermometer being then at 76°); the air obtained was worse than common air, whereas the air obtained from other jars kept in the sunshine during such a little time that the water had by no means received a degree of warmth approaching that of the atmosphere, was fine dephlogisticated air.

No dephlogisticated air is obtained in a warm room, if the sun does not shine upon the jar containing the leaves.

81

GILBERT WHITE (1720-1793), SELBORNE.

White, clergyman in a small parish, became one of the initiators of modern watching and loving nature. His *Natural History of Selborne* appeared in 1789. We render here the Letter XL of this good, and still today widely read book, which has appeared in many editions (e.g. ed. 1937, p. 107-111).

Dear Sir (Thomas Pennant). Selborne, Sept. 2, 1774.

Before your letter arrived, and of my own accord, I had been remarking and comparing the tails of the male and female swallow, and this ere any young broods appeared; so that there was no danger of confounding the dams with their pulli: and besides, as they were then always in pairs, and busied in the employ of nidification, there could be no room for mistaking the sexes, nor the individuals of different chimneys the one for the other. From all my observations, it constantly appeared that each sex has the long feathers in its tail and give it that forked shape; with this difference, that they are longer in the tail of the male than in that of the female.

Nightingales, when their young first come abroad, and are helpless, make a plaintive and a jarring noise; and also are snapping or cracking, pursuing people along the hedges as they walk: these last sounds seem intended for menace and defiance.

The grasshopper-lark chirps all night in the height of summer.

Swans turn white the second year, and breed the third.

Weasels prey on moles, as appears by their being sometimes caught in mole-traps.

Sparrow-hawks sometimes breed in old crows' nests, and the kestrel in churches and ruins.

There are supposed to be two sorts of eels in the island of Ely. The threads sometimes discovered in eels are perhaps their young; the generation of eels is very dark and mysterious.

Hen-harriers breed on the ground, and seem never to settle on trees.

When redstarts shake their tails they move them horizontally, as dogs do when they fawn: the tail of a wagtail, when in motion, bobs up and down like that of a jaded horse.

Hedge sparrows have a remarkable flirt with their wings in breeding time: as soon as frosty mornings come they make a very piping plaintive noise.

Many birds which become silent about midsummer reassume their notes again in September; as the thrush, blackbird, woodlark, willow-wren, etc.; hence August is by much the most mute month, the spring, summer, and autumn through. Are birds induced to sing again because the temperament of autumn resembles that of the spring?

Linnaeus ranges plants geographically; palms inhabit the tropics, grasses the temperate zones, and mosses and lichens the polar circles; no doubt animals may be classed in the same manner with propriety.

House-sparrows build under eaves in the spring; as the weather becomes hotter they get out for coolness, and nest in plum-trees and apple-trees. These birds have been known sometimes to build in rooks' nests, and sometimes in the forks of boughs under rooks' nests.

As my neighbour was housing a rick he observed that his dogs devoured all the little red mice that they could catch, but rejected the common mice; and that his cats ate the common mice, refusing the red.

Red-breasts sing all through the spring, summer, and autumn. The reason that they are called autumn songsters is, because in the two first seasons their voices are drowned and lost in the general chorus; in the latter their song becomes distinguishable. Many songsters of the autumn seem to be the young cock red-breasts of that year: notwithstanding the prejudices in their favour, they do much mischief in gardens to the summer-fruits.

The titmouse, which early in February begins to make two quaint notes, like the whetting of a saw, is the marsh titmouse: the great titmouse sings with three cheerful joyous notes, and begins about the same time.

Wrens sing all the winter through, frost excepted.

House-martins came remarkably late this year both in Hampshire and Devonshire: is this circumstance for or against either hiding or migration?

Most birds drink sipping at intervals; but pigeons take a long continued draught, like quadrupeds.

Notwithstanding what I have said in a former letter, no grey crows were ever known to breed on Dartmoor; it was my mistake.

The appearance and flying of the *scarabaeus solstitialis*, or fern-chafer, commence with the month of July, and cease about the end of it. These scarabs are the constant food of *caprimulgi*, or fern owls, through that period. They abound on the chalky downs and in some sandy districts, but not in the clays.

In the garden of the Black-bear inn in the town of Reading is a stream or canal running under the stables and out into the fields on the other side of the road: in this water are many carps, which lie rolling about in sight, being fed by travellers, who amuse themselves by tossing them bread: but as soon as the weather grows at all severe these fishes are no longer seen, because they retire under the stables, where they remain till the return of spring. Do they lie in a torpid state? If they do not, how are they supported?

The note of the white-throat, which is continually repeated, and often attended with odd gesticulations on the wing, is harsh and displeasing. These birds seem of a pugnacious disposition; for they sing with an erected crest and attitudes of rivalry and defiance; are shy and wild in breeding-time, avoiding neighbourhoods, and haunting lonely lanes and commons; nay even the very tops of the Sussex-downs, where there are bushes and covert; but in July and August they bring their broods into gardens and orchards, and make great havoc among the summer-fruits.

The black-cap has in common a full, sweet, deep, loud, and wild pipe; yet that strain is of short continuance, and his motions are desultory; but when that bird sits calmly and engages in song in earnest, he pours forth very sweet, but inward melody, and expresses great variety of soft and gentle modulations, superior perhaps to any of our warblers, the nightingale excepted.

Black-caps mostly haunt orchards and gardens; while they warble their throats are wonderfully distended.

The song of the redstart is superior, though somewhat like that of the white-throat: some birds have a few more notes than others. Sitting very placidly on the top of a tall tree in a village, the cock sings from morning to night: he affects neighbourhoods, and avoids solitude, and loves to build in orchards and about houses; with us he perches on the vane of a tall maypole.

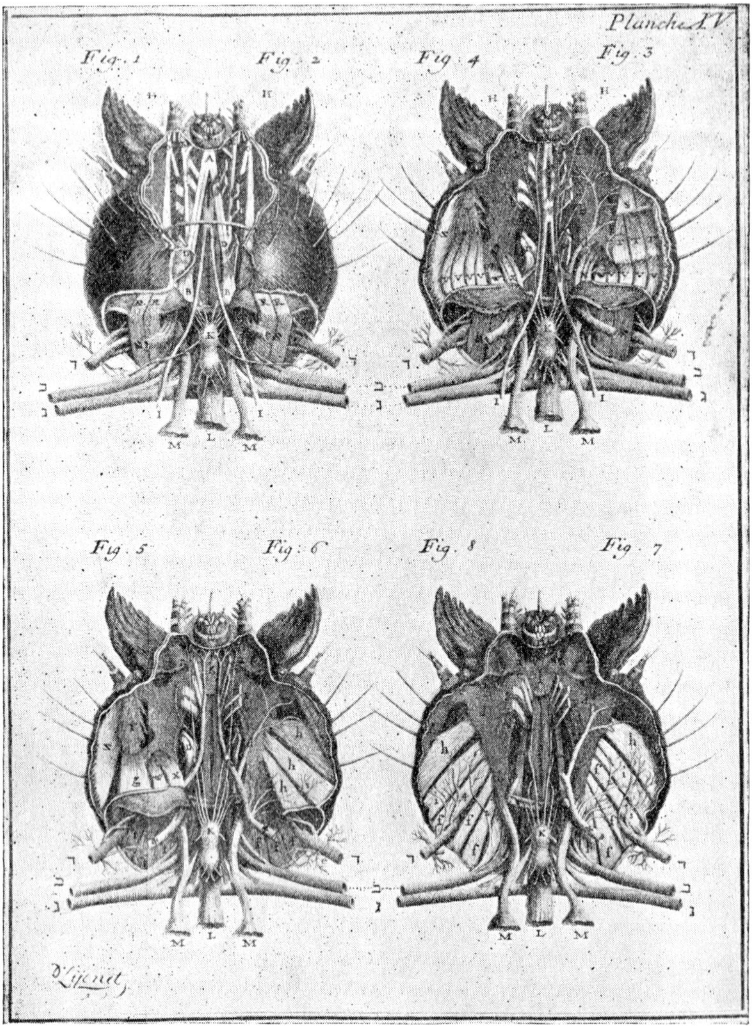

Fig. 31. From the famous work of Lyonnet describing the detailed anatomy of the caterpillar of the moth *Cossus cossus*. From *Traite anatomique de la Chenille qui ronge le Bois de Saule*, (1760), Pl. XV.

The fly-catcher is of all our summer birds the most mute and the most familiar; it also appears the last of any. It builds in a vine, or a sweetbriar, against the wall of a house, or in the hole of a wall, or on the end of a beam or plate, and often close to the post of a door where people are going in and out all day long. This bird does not make the least pretension to song, but uses a little inward wailing note when it thinks it's young in danger from cats or other annoyances: it breeds but once, and retires early.

Selborne parish alone can and has exhibited at times more than half the birds that are ever seen in all Sweden; the former has produced more than 120 species, the latter only 221. Let me add also that it has shewn near half the species that were ever known in Great Britain (252).

On a retrospect, I observe that my long letter carries with it a quaint and magisterial air, and is very sententious; but, when I recollect that you requested stricture and anecdote, I hope you will pardon the didactic manner for the sake of the information it may happen to contain.

82

JOHANN WOLFGANG VON GOETHE (1749-1832), WEIMAR.

This great genius, mainly known as poet, was actively interested in many problems of natural history, such as theory of colours, breeding of insects, comparative osteology, etc. His main interest was to find the 'primary type' of plants and animals. Apart from two series of papers, a poem is dedicated to each of these ideas. My friend, Dr. S. Kahn, reader of American Literature at the Hebrew University, Jerusalem, kindly put at our disposal the following translation of the poem named " The Metamorphosis of Plants" (ab. 1792), which offered many linguistic difficulties for a translation

The Metamorphosis of the Plants.

Thou art bewildered, Beloved, by the thousandfold medley
  Of these flowers that all about the garden throng;
Thou hearest many names, and their barbaric sounds
  Ever crowd one another, into thine ear and out.
All the Forms are the same, and no one is like the other;

And thus their manifold chorus points to a secret law,
To a sacred riddle. O, could I but at once, sweet Friend,
Happily hand to you the riddle-solving Word!
Observe but the growth of the plant: how, little by little,
It is gradually led to take form as blossoms and fruit.
Out of the seed it develops, as soon as the earth's quiet
Fructifying womb discharges it graciously into life,
And entrusts it then to the magic of light, the holy,
Ever-bestirred—a most delicate structure of sprouting leaves.
Simply in the seed the power slept; a pattern,
Incipient, self-enclosed, lay bent beneath its sheath,
Leaf and root and germ, half-formed and without colour;
Thus, the dry kernel preserves its peaceful life,
Swells and strives upward, yielding to gentle rain,
And suddenly raises itself, out of surrounding night.
But simple remains the Form of its first appearance;
And so it is too among plants that the child defines itself.
Immediately follows a shoot, rising, node upon node,
Ever renewed and higher, the original structure preserving.
Yet not always the same; for, as thou seest, the leaf,
Which follows after, maturing, begets variety,
Extended, jagged, divided into points and parts,
That, entwined before, had slept in the organ below.
And thus, at first, it attains the highest, fixed development,
Which, in many a genus, moves thee to such astonishment.
Berribed, serrated, on the sleek and succulent surface,
The fullness of the inner force seems endless and free.
Yet Nature here, with powerful hands, restrains
Development, guiding it gently towards the more perfect.
More frugally she conveys the sap now, constricts the vessels,
And the ribs of the stalks rise to greater perfection.
But suddenly, leafless, the delicate stem upheaves,
And a wonderful creation attracts the observer.
Arranged in a circle, counted and yet without number,
Now each tiny leaf beside its brother stands.
About the axle crowding, the hidden calyx, decisive,
Develops its colourful crown to its proudest stature.
Now does Nature vaunt her most resplendent garb,
Showing, all in a row, member on member ranked.
Ever freshly astonished, thou seest blooms on their stems
Stirring o'er slender scaffolds of alternating leaves.

This glory but proclaims a new creation,
Aye, the coloured leaf perceives the hand divine.
And quickly it contracts; the most delicate forms
At once are striving forward, determined to be united.
Intimately now, the lovely couples, together,
Arrange their numbers in order about the sacred altar.
Hymen hovers near, and powerful, glorious perfumes
Scatter their sweet scents about, enlivening all.
Now also unnumbered buds, quickly, separately swell,
Pleasantly sheathed in the womb of swelling fruits.
And here Nature completes the ring of eternal forces;
Yet at once a new force follows the one before it,
So that the chain may extend down through all the ages,
And, like each separate link, the Whole be filled with life.
Turn thy glance now, O Beloved, to the motley crowd
Which no longer swarms before thy soul to confuse thee.
Now each of the plants proclaims to thee eternal laws,
Each of the flowers speaks to thee more and more clearly.
But when thou here decipherest the goddess's hieroglyphics,
Thou learnest to see them in all, in other metamorphoses also.
Let the crawling caterpillar hesitate, the butterfly hurry busily,
Let man himself mould and change his established Form!
O! remember then also how, out of the germ of acquaintance,
Little by little in us delightful habit arose,
Friendship, deep within us, in all its power, was unveiled,
And how at last Amour produced its flowers and fruits.
Think, how manifold the Forms, first this, then another,
Silently unfolding, Nature lent to our feelings!
Rejoice thou then, too, in this day! For holy Love
Strives up towards the noblest fruit of convictions shared,
That a couple may see things alike and, bound together
In harmonious contemplation, may find the higher world.

83

CHRISTIAN CONRAD SPRENGEL (1750-1816), SPANDAU.

Sprengel, teacher in a small town, discovered the ' secret of nature,' namely the fecundation of flowers by insects, in his *Das entdeckte Geheimnis der Natur im Bau und in der Befruchtung der Blumen* (1793, p. 1 ff).

When I examined carefully, in the summer of 1787, the flowers of *Geranium sylvaticum*, I found that the lowermost part of their corolla leaves was beset on the inner side and on both margins with fine, soft hairs. Convinced that the wise creator of nature has not produced even one small hair without a definite purpose, I wondered what might be the purpose of these hairs. And I considered, that on the supposition that the five drops of nectar which are secreted by the same number of glands, are apportioned to certain insects as food, it is not improbable, that care was taken

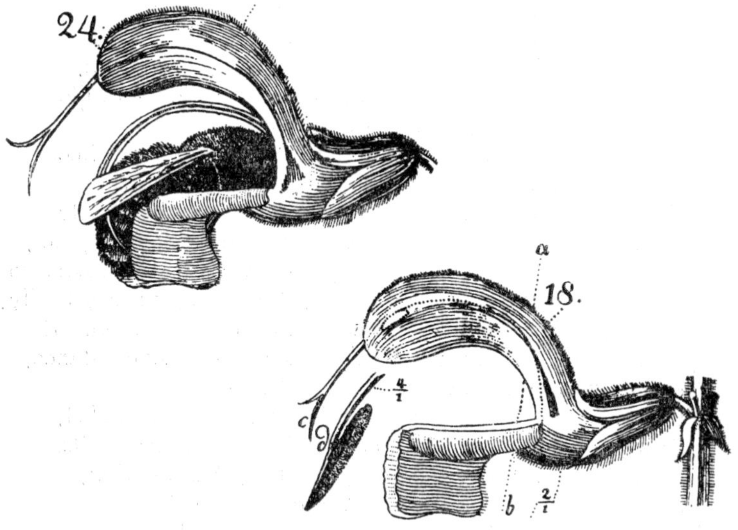

Fig. 32. Pollination of *Salvia pratensis* as represented by Sprengel (1793).

to protect the nectar from being spoiled by the rain and that to that purpose these hairs were formed there.

The accompanying figures may illustrate this theory. They concern *Geranium palustre*, which is very similar to the *G. sylvaticum*. Each drop of nectar lies upon its gland immediately beneath the hairs on the margins of the two nearest corolla-leaves. As the flower is erect and fairly big, drops of rain must enter it when it rains. But no rain-drop can reach the nectar-drops and mix with them, as it is arrested by the hairs above the nectar, just as a drop of sweat running over our forehead is arrested by the eyebrows and eyelashes, before it reaches our eyes. Yet an insect is not prevented by these hairs from reaching the drops of nectar. I examined other flowers and found, that some of them had certain peculiarities in their structure which seemed to serve the same purpose. The longer I continued these studies, the more I discovered, that those flowers which contain nectar are so arranged, that the insects can easily reach it, but that rain cannot spoil it. I concluded, that the sap of these flowers is primarily secreted for insects, and that the nectar is protected against rain, in order that the insects may delight in it, pure and unspoiled.

In the following summer I studied the forget-me-not, and found that its flower has nectar, and that this is completely protected against the rain. At the same time I paid attention to the yellow ring which surrounds the opening of the tube of the corolla and which contrasts so beautifully with the sky-blue colour of the corolla-margin. Could this also be in relation to insects? Could nature have coloured this ring so beautifully, that it may show the insects the path to the nectary? I examined other flowers with this in mind, and most of them confirmed the assumption. I learned, that those flowers which have multicoloured corolla-leaves, have these spots, figures, lines or dots of special colour always there, where the entrance to the nectary is placed. I now concluded from the part to the whole. If the corolla be specially coloured in particular regions for the sake of insects, it is in general coloured for the sake of insects; and if the peculiar colour of a part of the corolla serves to guide an insect, which rests upon the flower, into the proper path to the nectar, the colour of the corolla serves to make it conspicuous from far off to the insects which fly about in the air in search of food.

When I then studied some species of iris, I soon discovered that their flowers can only be fertilised by insects. I looked for other flowers which might be built in such a way that they could be fertilised by insects only. Thus, this feeding is an end for the insects, but for the flowers the one and only means of their fertilization. The whole structure of such flowers may be explained when the following points are kept in mind:

1. These flowers must be fertilised by this or that species of insects or by some of them.
2. This must be effected in such a way, that the insects in their search for the nectar of the flowers alight on the flowers, creep into them and thereby necessarily brush off the pollen from the

stamens and bring it to the stigma. The latter, to receive it, is either covered with fine hairs or by a sticky liquid.

In this way I discovered on *Epilobium angustifolium* something which otherwise I would never have seen. This hermaphrodite flower is fertilised by humble-bees and bees, not every flower by its own pollen, but the older flowers by the pollen brought by these insects from the younger flowers. This discovery threw light upon many of my earlier discoveries.

When, then, I studied the common spurge, I found in it an arrangement which is just the opposite of that just mentioned. This flower is fertilised by insects in such a way that they bring the pollen of the older flowers to the stamina of the younger ones.

These are the main discoveries on which my theory of flowers is based.

84

EDWARD JENNER (1749-1823), LONDON.

Jenner liberated by the inoculation against the small-pox, Europe from one of its most serious scourges. See his *Inquiry into the causes and effects of the variolae vaccinae, or cow-pox.* 1798. p. 32, 66.

Thus the cow-pox make progress from an inflammation and swelling in the heel of the horse from there by careless dressings to the nipples of the cows, and from the cow to the dairymaids.

In support of so extraordinary a fact, I shall lay before my reader some cases:

Case XVII (entire): The more accurately to observe the progress of the infection I selected a healthy boy, about eight years old, for the purpose of inoculation for the cow-pox. The matter was taken from a sore on the hand of a dairymaid, who was infected by her master's cows, and it was inserted, on the 14th of May, 1796, into the arm of the boy by means of superficial incisions, barely penetrating the cutis, each about half an inch long.

On the seventh day he complained of uneasiness in the axilla, and on the ninth he became a little chilly, lost his appetite, and had a slight headache. During the whole of this day he was perceptibly indisposed, and spent the night with some degree of restlessness, but on the day following he was perfectly well . . . In order to ascertain whether the boy, after feeling so slight an affection of the system from the cow-pox virus, was secure from the contagion of the small-pox, he was inoculated the 1st of July following with variolous matter, immediately taken from a pustule, but no disease followed. Several months afterwards he was again inoculated with variolous matter, but no sensible effect was produced . . .

Should it be asked whether this investigation is a matter of mere curiosity, or whether it tends to any beneficial purpose, I should answer that, notwithstanding the happy effects of inoculation, with all the improvements which the practice has received since its first introduction into this country, it not very unfrequently produces deformity of the skin, and sometimes, under the best management, proves fatal.

These circumstances must naturally create in every instance some degree of painful solicitude for its consequences. But as I have never known fatal effects arise from the cow-pox, even when impressed in the most unfavourable manner, producing extensive inflammations and suppurations on the hands; and as it clearly appears that this disease leaves the constitution in a state of perfect security from the infection of the small-pox, may we not infer that a mode of inoculation may be introduced preferable to that at present adopted?

85

THOMAS MALTHUS (1766-1834), ENGLAND (HAILEYBURY).

Observations of the social poverty in the slums of the first industrial towns of England induced the clergyman Malthus to analyse the reasons for this poverty. In his *Essay on the Principles of Population* (1798), he tried to explain that the different rate of growth of the means of subsistence and of population growth be the source for the increasing poverty. We print here parts of chapters 1 and 2 of the Essay.

It is observed by Dr. Franklin (1751) that there is no bound to the prolific nature of plants or animals but what is made by their crowding and interfering with each other's means of subsistence. Were the face of the earth, he says, vacant of other plants, it might be gradually sowed and overspread with one kind only, as for instance with fennel; and were it empty of other inhabitants, it might in a few ages be replenished from one nation only, as for instance with Englishmen.

This is incontrovertibly true. Through the animal and vegetable kingdoms Nature has scattered the seeds of life abroad with the most profuse and liberal hand; but has been comparatively sparing in the room and the nourishment necessary to rear them. The germs of existence contained in this earth, if they could freely develop themselves, would fill millions of worlds in the course of a few thousand years. Necessity, that imperious, all pervading law of nature, restrains them within the prescribed bounds. The race of plants and the race of animals shrink under this great restrictive law; and man cannot by any efforts of reason escape from it. In plants and irrational animals, the view of the subject is simple. They are all impelled by a powerful instinct to the increase of their species; and this instinct is interrupted by no doubts about providing for their offspring. Wherever, therefore there is liberty, the power of increase is exerted; the superabundant effects are repressed afterwards by want of room and nourishment.

The effects of this check on man are more complicated. Impelled to the increase of his species by an equally powerful instinct, reason interrupts his career, and asks him whether he may not bring beings into the world for whom he cannot provide the means of support. If he attend to this natural suggestion, the restriction too frequently produces vice. If he hear it not, the human race will be constantly endeavouring to increase beyond the means of subsistence. But as, by that law of our nature which makes food necessary to the life of man, population can never actually increase beyond the lowest nourishment capable of supporting it, a strong check on population, from the difficulty of acquiring food, must be constantly in operation. This difficulty must fall somewhere, and must necessarily be severely felt in some or other of the various forms of misery, or the fear of misery, by a large proportion of mankind.

That population has this constant tendency to increase beyond the means of subsistence, and that it is kept to the necessary level by these causes, will sufficiently appear from a review of the different states of society in which man has existed. But, before we proceed to this review, the subject will, perhaps, be seen in a clearer light if we endeavour to ascertain what would be the natural increase of population if left to exert itself with perfect freedom; and what might be expected to be the rate of increase in the productions of the earth under the most favourable circumstances of human industry. It will be allowed that no country has hitherto been known where the manners were so pure and simple, and the means of subsistence so abundant, that no check whatever has existed to early marriages from the difficulty of providing for a family, and that no waste of the human species has been occasioned by vicious customs, by unhealthy occupations, or too severe labour. Consequently in no state that we have yet known has the power of population been left to exert itself with perfect freedom . . .

According to a table of Euler, calculated on a mortality of 1 in 36, if the births be to the deaths in the proportion of 3 to 1, the period of doubling will be only 12 years and 4/5. And this proportion is not only a possible supposition, but has actually occurred for short periods in more countries than one . . . It may safely be pronounced, therefore, that population, when unchecked, goes on doubling itself every 25 years, or increases in a geometrical ratio.

The rate according to which the productions of the earth may be supposed to increase, it will not be so easy to determine. We may be perfectly certain that the ratio of their increase in a limited territory must be of a totally different nature from the ratio of the population increase. A thousand millions are just as easily doubled every 25 years by the power of population as a thousand. But the food to support the increase from the greater number will by no means be obtained with the same facility. Man is necessarily confined in room. When acre has been added to acre till all the fertile land is occupied, the yearly increase of food must depend upon the amelioration of the land already in possession. This is a fund, which, from the nature of all soils, instead of increasing, must be gradually diminishing. But population, could it be supplied with food, would go on with unexhausted vigour; and the increase of one period would furnish the flower of a greater increase the next, and this without limit . . . Let us suppose that the yearly additions which might be made to the former average produce, instead of decreasing, which they certainly would do, were to remain the same; and that the produce of this island might be increased every twenty-five years by a quantity equal to what it at present produces. The most enthusiastic speculator cannot suppose a greater increase than this. In a few centuries it would make every acre of land in the island like a garden . . . It may be fairly pronounced, therefore, that, considering the present average state of the earth, the means of subsistence, under circumstances the most favourable to human industry, could not possibly be made to increase faster than in an arithmetical ratio.

The necessary effects of these two different rates of increase, when brought together, will be very striking . . . Taking the whole earth, emigration would, of course, be excluded; and supposing the present population equal to a thousand millions, the human species would increase as the numbers, 1, 2, 4, 8, 16, 32, 64, 128, 256, and subsistence as 1, 2, 3, 4, 5, 6, 7, 8, 9. In two centuries the population would be to the means of subsistence as 256 to 9, and in 2000 years the difference would be almost incalculable. In this supposition no limits are placed to the produce of the earth. Yet the power of population being in every period so much superior, the increase of the human species can only be kept down to the level of the means of subsistence by the constant operation of the strong law of necessity, acting as a check upon the greater power . . .

In savage life, where there is no regular price of labour, it is little to be doubted that similar oscillations take place. When population has increased nearly to the utmost limits of the food, all the preventative and the positive checks will naturally operate with increased force. Vicious habits with respect to sex will be more general, the exposing of children more frequent, and both the probability and fatality of wars and epidemics will be considerably greater; and these causes will probably continue their operation till the population is sunk below the level of the food; and then the return to comparative plenty will again produce and increase, and, after a certain period, its further progress will again be checked by the same causes.

The following propositions are intended to be proved:

1. Population is necessarily limited by the means of subsistence.
2. Population invariably increases where the means of subsistence increase, unless prevented by some very powerful and obvious checks.

3. These checks, and the checks which repress the superior power of population and keep its effects on a level with the means of subsistence, are all resolvable into moral restraint, vice and misery.

The first of these propositions scarcely needs illustration. The second and third will be established by a review of the immediate checks to population in the past and present state of society, given in the following chapters.

86

NICHOLAS THEODORE DE SAUSSURE (1767-1845), GENEVE.

The plant physiologist De Saussure opened new ways by qualitative studies of plant nutrition in his *Recherches chimiques sur la végétation*, 1806, p. 4. The following paragraph demonstrates the increase of dry-matter and of carbon in beans growing in distilled water, at the expense of the air only.

These observations prove that plants decompose carbonic acid gas in closed vessels if it is mixed with atmospheric air in a higher ratio than occurs in the natural condition of the latter. We must now inquire, whether this decomposition is also effected in free air, which scarcely contains 1 vol. % of carbonic acid gas. It has recently been maintained that plants which grow in pure water and free air increase their volume only from the water, and contain after their development is completed a smaller total quantity of carbonic acid than that which was present at first. I made several experiments which all yielded results entirely opposed to this opinion . . .

Experiment 2: I put four beans of 6.368 g. between pebbles in glass jars, and wetted them with distilled water. After growing for three months under the open sky the bean-plants weighed immediately after the flowering 87.149 g. green weight. After drying the weight dropped to 10.721 g., which means that the beans had almost doubled their dry weight during their growth. When carbonised in closed vessels these plants yielded 2.703 g. carbon, against the weight of 1.209 g. carbon of four beans similar to those with which the experiment was made. The carbon of the beans growing in water and free air had thus doubled in quantity. There cannot be any doubt, that this was effected only by the decomposition of the carbonic-acid gas in the atmosphere. As we have seen, plants which vegetate in vessels filled with pure atmospheric air do not increase their carbon.

87

THOMAS ANDREW KNIGHT (1759-1835), ELTON.

Knight was an original research worker in plant physiology. We render here the author's own résumé of his experiments which laid the foundations for plant tropisms, in this case for geotropism. From the Proceedings of Roy. Soc., London, 1806. pp. 218-220.

In whatever position a seed is placed to germinate, its radicle always makes an effort to descend towards the centre of the earth, whilst the elongated germen takes a precisely opposite direction: and it has been proved by Du Hamel, that if a seed, during its germination, be frequently inverted, the points, both of the radicle and germen, will return to their first direction. These opposite effects have, by some naturalists, been attributed to gravitation. If they really proceeded from that cause, those effects would take place only whilst the seed remained at rest, in the same position with respect to the attraction of the earth, and that the operation of gravitation would be suspended by a constant and rapid change of position in the germinating seed, and might be counteracted by the agency of centrifugal forces. The foundation of this opinion was tested in the following experiments:

Knight moved by running water in his garden, vertically, a wheel of 11 inches diameter. Round the circumference of this wheel, several seeds of the garden-bean, previously soaked in water, were bound in such a manner that their radicles were made to point in every direction. The wheel made rather more than 150 revolutions in a minute. In a few days the seeds began to germinate, and these radicles, in whatever direction were protruded, turned their points outwards from the circumference of the wheel, and in their subsequent growth receded still further from it. The germens, on the contrary, took the opposite direction, and in a few days their points met at the centre of the wheel. Three of these plants were suffered to remain on the wheel; their stems soon extended beyond its centre, but their points returned, and met again at the centre.

The experiment was repeated in a different manner, by adding to the former apparatus another wheel, also of 11 inches diameter, which moved horizontally, and to which he could give different degrees of velocity. Round the circumference of this horizontal wheel, seeds of the garden-bean were bound, as in the former experiment, and the wheel was made to perform 250 revolutions in a minute. The effect produced by this motion soon became obvious; for the radicles now pointed downwards about 10 degrees below the horizontal line of the wheel's motion, whilst the germens pointed the same number of degrees above it; but when the motion of the wheel was diminished to 80 revolutions in a minute, the radicles pointed about 45 degrees below the horizontal line, and the germen as much above it; the one always receding from the axis of the wheel, the other approaching to it.

These experiments prove that the radicles of the germinating seed are made to descend, and the germens to ascend, by some external cause, and not by any power inherent in vegetable life; and there is little reason to doubt that gravitation is the principal, if not the only, agent employed in this case by nature. The radicle is increased in length only by parts successively added to its point; whereas the germen, on the contrary, is elongated by a general extension of its parts previously organized; and its vessels and fibres appear to extend themselves in proportion to the quantity of nutriment they receive. When the germen deviates from the perpendicular direction, the sap accumulates on its under side, and consequently, as the vessels and fibres on that side elongate more rapidly than those of the upper side, the point of the germen must always turn upwards. This increased elongation of the vessels and fibres of the under side produces also the most extensive effect in the subsequent growth of the trunk and branches of trees. The immediate effect of gravitation is to occasion the depression of the branches; but, by the above-mentioned increased longitudinal extension of the under side, their depression is prevented, and they are even enabled to raise themselves above the natural level

It has been objected by Du Hamel, that gravitation can have little influence on the germen when it points perpendicularly downwards. Experiments with the seeds of the horse-chestnut and of the bean showed, that the radicle of the bean, when made to point perpendicularly upwards, formed a considerable curvation in the course of a few hours. The germen was more sluggish, but, in spite of any efforts made by us to prevent it, constantly changed its direction in less than 24 hours.

It may also be objected, that few of the branches of trees rise perpendicularly upwards, and that their roots always spread horizontally. Respecting the first of these objections, it may be observed that luxuriant shoots, which abound in sap, constantly turn upwards, and endeavour to acquire a perpendicular direction; but that the feeble and more slender shoots grow in almost every direction, probably from their fibres being more dry, and their vessels less amply supplied with sap, so that they are less affected by gravitation. Concerning the second objection it may be answered, that the compression of the radicle, as it penetrates the soil, obstructs the motion of the sap, and occasions the generations of numerous lateral roots; and as their substance is less succulent than that of the radicle first emitted, they are less obedient to gravitation, and consequently extend horizontally in every direction.

88

CHARLES BELL (1774-1842), EDINBURGH AND LONDON.

Bell is well known in the history of the exploration of the anatomy and physiology of the central nervous system. Here we translate a letter (see: E. Ebstein, Aerztebriefe. Berlin, 1920. pp. 76-78) to his brother G. J. Bell on the motory and the sensory roots of the cordal spine.

26th Nov., 1807. I have done a more interesting Nova Anatomia Cerebri than it is possible to conceive. I lectured it yesterday. I prosecuted it last night till one o'clock. And I am sure that it will be well received.

Dec. 5th, 1807. My New Anatomy of the Brain occupies my head almost entirely. I hinted to you that I was 'burning,' or, on the eve of a grand discovery. I consider the organs of the outward senses as forming a distinct class of nerves from the others. I take five tubercles within the brain as the internal senses. I trace the nerves of the nose, eye, ear, and tongue to these. Here I see established connection—there the great mass of the brain receives processes from the central tubercles. Again, the great masses of the cerebrum send down processes or crura, which give off all the common nerves of voluntary motion, etc. I establish thus a kind of circulation, as it were. In this inquiry I describe many new connections—the whole opens up a new and simple light, and the whole accords with the phenomena, with the pathology, and is supported by interesting views. My object is not to publish this, but to lecture it, to lecture it to my friends, to lecture it to Sir Joseph Banks' cotery of old women, to make the town ring with it, as it is really the only new thing that appeared in anatomy since the days of Hunter; and, if I make it out, as interesting as the circulation, or the doctrine of absorption. But I must still have time: now is the end of a week and I will be at it again.

2nd March, 1810. I write to tell you that I am going to establish my Anatomy of the Brain on facts, the most important that have been discovered in the history of the science. You recollect that I have entertained the idea that the parts of the brain were distinct in function; and that the cerebrum was in a particular manner the organ of mind; and this from other circumstances than what I am now to detail to you.

It occurred to me that, as there were four grand divisions of the brain so were four divisions of the spinal marrow; first, a lateral division, then a division into the back and forepart. Next it occurred to me that all the spinal nerves had within the sheath of the spinal marrow two roots, one from the back part, another from before. Whenever this occurred to me I thought that I had obtained a method of inquiring into the functions of the parts of the brain.

Experiment 1: I opened the spine, and pricked and injured the posterior filaments of the nerves; no motion of the muscles followed. I then touched the anterior division, immediately the parts were convulsed.

Experiment 2: I now destroyed the posterior part of the spinal marrow by the point of a needle, no convulsive movement followed. I injured the anterior part and the animal was convulsed.

It is almost superfluous to say that the part of the spinal marrow having sensibility is what comes from the cerebrum; the posterior and insensible part belongs to the cerebellum.

Taking these facts as they stand, is it not most curious that there should be thus established a distinction in the parts of a nerve, and that a nerve should be insensible? But then, as the foundation of a great system, if I can but sustain them by repeated experiments, I am made; and a real gratification ensured for a large portion of my existence.

89

ALEXANDER WILSON (1766-1813), PHILADELPHIA.

90

JOHN JAMES AUDUBON (1785-1851), NEW YORK.

Both these men initiated the exploration of the birds of North America. From Wilson's *American Ornithology*, Vol. I, 1808, we select the observation of the Solitary Sandpiper (*Tringa Solitaria*). From Audubon's famous *Birds of America* (1827-1838) we render the plate representing Wilson's Snipe (*Capella delicata*), No. 243.

This new species inhabits the watery solitudes of our highest mountains during the summer, from Kentucky to New York; but is nowhere numerous, seldom more than one or two being seen together. It takes short, low flights; runs nimbly about the mossy margins of the mountain springs, brooks and pools, occasionally stopping, looking at you, and perpetually nodding the head. It is so unsuspicious, or so little acquainted with man, as to permit one to approach within a few yards of it, without appearing to take any notice, or to be the least alarmed. At the approach of cold weather, it descends to the muddy shores of our large rivers, where it is occasionally met with, singly, on its way to the south. I have made many long and close searches for the nest of this bird, without success. They regularly breed on Pocano mountain . . . in Pennsylvania, arriving there early in May, and departing in September. It is usually silent, unless when suddenly flushed, when it utters a sharp whistle.

91

JEAN BAPTISTE DE MONET LAMARCK (1744-1829), PARIS.

The genial taxonomer of animals and plants, the first who arranged the systematics of the Evertebrates in a modern manner, is best known for his *Philosophie Zoologique*, (1809), the first full exposition of the idea of the transformation of species. Yet this concept had to wait for Darwin, 55 years, until it conquered the scientific world. We give here parts of chapter VII (p. 218 ff.), in which environment and habits are described as causes of the transformation of species.

It is indeed long since the influence of the various states of our organisation on our character, inclinations, activities and even ideas has been recognised; but I do not think that anyone has yet drawn attention to the influence of our activities and habits even on our organisation. Now since these activities and habits depend entirely on the environment in which we are habitually placed, I shall endeavour to show how great is the influence exerted by that environment on the general shape, state of the parts and even organisation of living bodies. It is, then, with this very positive fact that we have to do in the present chapter . . .

The influence of the environment as a matter of fact is in all times and places operative on living bodies; but what makes this influence difficult to perceive is that its effects become perceptible or recognisable (especially in animals) after a long period of time . . .

Races of animals living in any locality must retain their habits as long as conditions remain constant: hence the apparent constancy of the races that we call species—a constancy which has raised in us the belief that these races are as old as nature. But in the various habitable parts of the earth's surface, the character and situation of places and climates constitute both for animals and plants environmental influences of extreme variability. The animals living in these various localities must therefore differ among themselves, not only by reason of the state of complexity of organisation attained in each race, but also by reason of the habits which each race is forced to acquire; thus when the observing naturalist travels over large portions of the earth's surface and sees conspicuous changes occurring in the environment, he invariably finds the characters of species undergo a corresponding change. Now the true principle to be noted in all this is as follows:

1. Every fairly considerable and permanent alteration in the environment of any race of animals works a real alteration in the needs of that race.
2. Every change in the needs of animals necessitates new activities on their part for the satisfaction of those needs, and hence new habits.
3. Every new need, necessitating new activities for its satisfaction, requires the animal, either to make more frequent use of some of its parts which it previously used less, and thus greatly to develop and enlarge them; or else to make use of entirely new parts, to which the needs have imperceptibly given birth by efforts of its inner feeling; this I shall shortly prove by means of known facts.

Thus to obtain a knowledge of the true causes of that great diversity of shapes and habits found in the various known animals, we must reflect that the infinitely diversified but slowly changing environment in which the animals of each race have successively been placed, has involved each of them in new needs and corresponding alterations in their habits. This is a truth which, once recognised, cannot be disputed. Now we shall easily discern how the new needs may have been satisfied, and the new habits acquired, if we pay attention to the two following laws of nature, which are always verified by observation:

First Law : In every animal which has not passed the limit of its development, a more frequent and continuous use of any organ gradually strengthens, developes and enlarges that organ, and gives it a power proportional to the length of time it has been so used; while the permanent disuse of any organ imperceptibly weakens and deteriorates it, and progressively diminishes its functional capacity, until it finally disappears.

Second Law: All the acquisitions or losses wrought by nature on individuals, through the influence of the environment in which their race has long been placed, and hence through the influence of the predominant use or permanent disuse of any organ; all these are preserved by reproduction to the new individuals which arise, provided that the acquired modifications are common to both sexes, or at least to the individuals which produce the young.

Here we have two permanent truths, which can only be doubted by those who have never observed or followed the operations of nature, or by those who have allowed themselves to be drawn into the error which I shall now proceed to combat. Naturalists have remarked that the structure of animals is always in perfect adaptation to their functions, and have inferred that the shape and condition of their parts have determined the use of them. Now this is a mistake: for it may be easily proved by observation that it is on the contrary the needs and uses of the parts which have caused the development of these same parts, which have even given birth to them when they did not exist, and which consequently have given rise to the condition that we find in each animal . . .

The permanent disuse of an organ, arising from a change of habits, causes a gradual shrinkage and ultimately the disappearance and even extinction of that organ . . .

Olivier's *Spalax*, which lives underground like the mole, and is apparently exposed to daylight even less than the mole, has altogether lost the use of sight; so that it shows nothing more than vestiges of this organ. Even these vestiges are entirely hidden under the skin and other parts, which cover them up and do not leave the slightest access to light. The *Proteus*, an aquatic reptile allied to the salamanders, and living in deep dark caves under the water, has, like the *Spalax*, only vestiges of the organ of sight, vestiges which are covered up and hidden in the same way . . .

The frequent use of any organ, when confirmed by habit, increases the functions of that organ, leads to its development and endows it with a size and power that it does not possess in animals which exercise it less. We have seen that the disuse of any organ modifies, reduces and finally extinguishes it. I shall now prove that the constant use of any organ, accompanied by efforts to get the most out of it, strengthens and enlarges that organ, or creates new ones to carry on functions that have become necessary.

The bird which is drawn to the water by its need of finding there the prey on which it lives, separates the digits of its feet in trying to strike the water and move about on the surface. The skin which unites these digits at their base acquires the habit of being stretched by these continually repeated separations of the digits; thus in course of time there are formed large webs which unite the digits of ducks, geese, etc., as we actually find them. In the same way efforts to swim, that is to push against the water so as to move about in it, have stretched the membranes between the digits of frogs, sea-tortoises, the otter, the beaver, etc. On the other hand, a bird which is accustomed to perch on trees and which springs from individuals all of whom had acquired this habit, necessarily has longer digits on its feet and differently shaped from those of the aquatic animals that I have just named. Its claws in time become lengthened, sharpened and curved into hooks, to clasp the branches on which the animal so often rests. We find in the same way that the bird of the water-side which does not like swimming and yet is in need of going to the water's edge to secure its prey, is continually liable to sink in the mud. Now this bird tries to act in such a way that its body should not be immersed in the liquid, and hence makes its best efforts to stretch and lengthen its legs. The long-established habit acquired by this bird and all its race of continually stretching and lengthening its legs, results in the individuals of this race becoming raised as though on stilts, and gradually obtaining long, bare legs, denuded of feathers up to the thighs and often higher still. We note again that this same bird wants to fish without wetting its body, and is thus obliged to make continual efforts to lengthen its neck. Now these habitual efforts in this individual and its race must have resulted in course of time in a remarkable lengthening, as indeed we actually find in the long necks of all water-side birds. If some swimming birds like the swan and goose have short legs and yet a very long neck, the reason is that these birds while moving about on the water acquire the habit of plunging their head as deeply as they can into it in order to get the aquatic larvae and various animals on which they feed; whereas they make no effort to lengthen their legs.

If an animal, for the satisfaction of its needs, makes repeated efforts to lengthen its tongue, it will acquire a considerable length (anteater, green woodpecker); if it requires to seize anything with this same organ, its tongue will then divide and become forked. Proofs of my statement are found in the humming-birds which use their tongues for grasping things, and in lizards and snakes, which use theirs to palpate and identify objects in front of them . . .

Conclusion adopted hitherto: Nature (or her Author) in creating animals, foresaw all the possible kinds of environment in which they would have to live, and endowed each species with a fixed organisation and with a definite and invariable shape, which compel each species to live in the places and climates where we actually find them, and there to maintain the habits which we know in them.

My individual conclusion: Nature has produced all the species of animals in succession, beginning with the most imperfect or simplest, and ending her work with the most perfect, so as to create a gradually increasing complexity in their organisation; these animals have spread at large throughout all the habitable regions of the globe, and every species has derived from its environment the habits that we find in it and the structural modifications which observation shows us . . .

In order to show that this second conclusion is baseless, it must first be proved that no point on the surface of the earth ever undergoes variation as to its nature, exposure, high or low situation, climate, etc.; it must then be proven that no part of animals undergoes even after long periods of time any modification due to a change of environment or to the necessity which forces them into a different kind of life and activity from what has been customary to them. Now if a single case is sufficient to prove that an animal which has long been in domestication differs from the wild species whence it sprang, and if in any such domesticated species, great differences of conformation are found between the individuals exposed to such a habit and those which are forced into different habits, it will then be certain that the first conclusion is not consistent with the laws of nature, while the second, on the contrary, is entirely in accordance with them. Everything then combines to prove my statement, namely: that it is not the shape either of the body or its parts which gives rise to the habits of animals and their mode of life; but that it is, on the contrary, the habits, mode of life and all the other influences of the environment which have in course of time built up the shape of the body and of the parts of animals. With new shapes, new faculties have been acquired, and little by little nature has succeeded in fashioning animals as we actually see them. Can there be any more important conclusion in the range of natural history, or any to which more attention should be paid than that which I have just set forth?

92

GEORGES LEOPOLD CHRÉTIEN FREDERIC DAGOBERT CUVIER (1769-1832), PARIS.

This famous zoologist, comparative anatomist and palaeontologist is best known by his *Discours sur les révolutions de la surface du globe* (1812), of which exist various English translations, e.g. Mitchill, 1818, New York. p. 98 f.

(*a*) On the reconstruction of fossil animals based upon the law of correlation.

(*b*) The geological periods as exposed by the strata of different animal fossils are explained as periods ended by cataclysms.

(*a*) *Reconstruction of fossil animals.* Fortunately, comparative anatomy possesses a principle, which, properly developed, is capable of clearing up all difficulties: I refer to the principle of the correlation of forms in the organism, by means of which every species may, by rigorous scrutiny, be recognised by any fragment of any of its parts. Every organism forms a whole, a unique and perfect system, whose parts are mutually correspondent and concur in the same definite action by a reciprocal reaction. None of these parts can change without the whole changing; and consequently each of them, separately considered, points out and marks all the others.

If, for instance, the intestines of an animal are so organized as only to digest fresh meat, it follows that its jaws must be constructed to devour a prey, its claws to seize and tear it; its teeth to cut and divide it; the whole structure of its locomotory organs such as to pursue and to catch it; its sensory organs to perceive it at a distance; and nature must have put into its brain the necessary instinct to know how to conceal itself and how to ambush its victims. Such will be the general requirements from a carnivore; every animal of this diet will invariably unite these qualifications, for its species could not survive without all of them. But apart from these general requirements there are particular ones, relating to the size, species and haunts of their prey. And each of these peculiar conditions results from modifications of the morphological details, which they derive from the general conditions. Thus, not only the class, but also the order, the genus, and even the species are detected in the formation of every part of the body.

Indeed, in order that the jaws be enabled to seize the prey, the condyle of its articulation must be suitably shaped; it must show a certain relation between the position of the resisting power and that of the strength employed with the point of support (*fulcrum*), a certain volume of the temporal muscle, requiring an equivalent extent in the groove hollow which receives it, and a certain convexity of its zygomatic arch under which it passes; this zygomatic arcade must also be sufficiently strong to give strength to the masseter muscle. In order that the animal be able to carry off its prey, the muscles which support the head must be fairly strong, whence results a definite shape of those vertebrae where these muscles are attached, and of the occiput where they are inserted. In order that the teeth can cut the flesh, they must be sharp, and this the more or less as their more or less exclusive task is to cut the flesh. Their roots should be more solid the more solid bones they have to break. All these circumstances will in a like manner influence the development of all those parts which serve to move the jaw.

In order that the claws be able to seize the prey, they must have a certain mobility of the talons, a certain strength in the nails, whence will result determinate peculiar formations in all the claws, and the necessary distribution of muscles and tendons; the forearm must have a certain facility of turning, whence again will result a definite shape of the bones, which compose it. But the bone of the forearm, which articulates in the shoulderbone, cannot change its structure without a corresponding change in the shape of the latter. The shoulderblade must have a certain degree of strength in these animals which use their forelegs to seize a prey, and they will thence obtain a peculiar structure. The movements of all these parts will require certain properties from all their muscles, and the impressions of all the muscles thus proportioned will more fully determine the structure of the bones.

It is easy to see that similar conclusions impose themselves for the posterior extremities, which contribute to the rapidity of locomotion; as to the formation of the rump and of the vertebrae, which influence the ease and flexibility of the motions; as to the shape of the bones of the nose, of the socket of the eye, of the ear, whose mutual relation to the perfection of the senses of smelling, seeing and hearing are obvious. In one word: the shape of the tooth bespeaks the structure of the articulation of the jaw, that of the shoulderblade that of the claws, just as the equation of a mathematical curve embraces all its properties. And when we take every property separately, as the basis of a particular equation, we should find again both the ordinary equation as well as all the other properties. And the same holds good for the claw, the shoulderblade, the articulation of the jaw, the thigh bone and for every other single bone, which will determine a certain tooth, or the tooth requires them reciprocally. By starting by any one bone, he who possesses a knowledge of the *law of organic economy* will be able to construct the whole animal.

This principle is sufficiently self-evident, in this general formulation, and does not require a further demonstration. But when the principle is to be applied, there are many cases in which our theoretical knowledge of the morphological correlations of the structure is insufficient, if it would not be supported by observation . . .

As these coincidences are constant, they must have a satisfactory cause; but as we do not know it, we ought to supply the defect of the theory by observation. The latter serves us to establish suppositious laws, which become almost as certain as the laws of reasoning, when they are based upon often repeated observations. Thus, to-day, when we see only the tracks of a cleft foot we can conclude that the animal which left it is a ruminant. And this conclusion is as sure as any other in physics or morality. This foot-mark alone is sufficient to give the observer the shape of the teeth, the form of the jaws, the structure of the vertebrae, and the form of all the bones of the legs, thighs, shoulders, and of the pelvis of the animal which passed the road . . .

But in thus adopting the method of observation as an additional means, when theory forsakes us, we arrive at astonishing results. The smallest prominence of a bone, the tiniest apophysis, have a determined character relative to the class, order, genus and species to which they belong; so that whenever we have only the well preserved extremity of a bone, we may by scrutinizing it, and by applying analogical skill and close comparison, determine all these things as certainly as if we had the whole animal before our eyes. I have often in this way experimented on portions of known animals, before I put my entire confidence in its application to fossil animals; but it has always given so successful results, that I have no longer any doubt on the certainty of the results which it affords . . .

(*b*) Thus we find a sequence of changes in the organic world, caused or running parallel with a change of the environment. When the sea regressed the last time from our continents, its inhabitants differed little from those creatures which it nourishes to-day. We said ' the last time ' as any careful study of the organic remainders instructs us that even in midst of the oldest marine formations we find layers which are filled with animals and plants of continents and of freshwater. Thus, repeated catastrophes changing the stratification of the layers, have not made emerge one after the other the various parts of our continents from the waves, but repeatedly dry parts of the earth were submerged again, either by sinking down or being flooded by a rising sea. It is notable, that these repeated irruptions and regressions did not occur slowly. On the contrary, most of the catastrophes which induced them occurred suddenly, what is most easily proven for the last catastrophe. This has left behind in the high north the cadavers of giant mammals, which were enclosed by the ice and thus remained conserved until to-day with skin and hairs. If the freezing and their death would not have occurred at the same time, they would have been decomposed by putrefaction. And on the other hand, this eternal frost could not have ruled in the places before, as the animals could not have existed under such conditions. It must have been the same moment which killed these mammals and covered the land which they had inhabited with ice. And this must have taken place suddenly and not slowly. What can be demonstrated so conspicuously for this last catastrophe, is only slightly less conspicuous for the earlier ones. The ruptures, flexions and overturnings of the oldest layers do not permit any doubt, that sudden and strong causes have brought these layers into their present conditions.

Terrible events have thus often interfered into the life of our planet. Countless creatures became the sacrifices of these catastrophes. The one, inhabitants of the continents were devoured by the waves, the others, children of the sea were put upon dry soil together with the suddenly risen bottom of the sea. Such species were exterminated for ever and left only scattered remainders which even the naturalist can recognise only with difficulty.

93

LOUIS ADALBERT VON CHAMISSO (1781-1838), BERLIN.

The poet and soldier Chamisso participated as naturalist in the big expedition of the Russian ship *Rurik.* He discovered at that opportunity (in October, 1815) the alternation of generations of *Salpa,* which he first published in the *Tagebuch der Weltumsegelung des Rurik* (1815-1817), which were published in 1821, and which are added to all editions of the works of the poet. A Latin dissertation *De Salpa* appeared already in Berlin in 1819.

The species of *Salpa* appear in two forms, the progeny bearing no resemblance to their mother during their entire lifespan, and on their part producing offspring which will resemble her. Thus, any given salpa resembles neither her mother nor her daughters, but on the other hand is similar to her grandmother, to her granddaughters, and to her sisters.

Salpa, like all the soft headless animals, are either bisexual or female only (parthenogenetic). Each bears live offspring (she bears them living), but the one form bears only one solitary animal, which on her part bears many offspring; while the other form produces a process (*stolon prolifer*) of creatures, ineluctibly bound together, each of which, in her turn, creates only one foetus. These bear, therefore, individual progeny.

These two regularly alternating forms of a well defined species may be termed an ' alternation of generations.' There are, then, the single parturition, of the solitary generation, (that is to say, the solitary salpa), and the proliferous parturition, or the proliferous generation (the proliferating salpa).

Many animals of this genus lay a chain of eggs, from each of which issues an animal directly resembling the mother.

In contrast to this stands the single parturition of the salpas in place of egg-laying—the proliferating animals, and the solitary salpa, which resembles the original mother, eventually cut off from these, that is to say, from the proliferating salpas (as from an egg), one from each animal.

Here, particularly, the salpa engrossed Escholtz and myself, for here we discovered, in these soft, transparent creatures, the phenomenon which to us appears of such importance—that this same species appears in two essentially differing forms. The solitary free-swimming salpa bears live offspring of another form, strung together almost like polyps, every single one of which, in her turn, brings forth a solitary, free-swimming animal, in which reappears the form of the original generation. This phenomenon has a seeming resemblance to that of the generation of the butterfly by the caterpillar, and the subsequent regeneration of a caterpillar by the butterfly.

94

MARIE FRANCOIS XAVIER BICHAT (1771-1802), PARIS.

The founder of modern histology in his *Anatomie Générale* (1801). English translation by G. Hayward. Boston, 1822. See English translation by Coffyn and Calvert, London, 1824, Vol. I, p. LI ff.

Observations upon the Organisation of Animals. All animals are an assemblage of different organs, which, executing each a function, concur in their own manner, to the preservation of the whole. It is several separate machines in a general one, that constitutes the individual. Now these separate machines are themselves formed by many textures of a very different nature, and which really compose the elements of these organs. Chemistry has its simple bodies, which form, by the combinations of which they are susceptible, the compound bodies; such as caloric, light, hydrogen, oxygen, carbon, azote, phosphorus, etc. In the same way anatomy has its simple textures, which, by their combinations four with four, six with six, eight with eight, etc., make the organs. These textures are: 1st—the cellulae; 2nd—the nervous of animal life; 3rd—the nervous of organic life; 4th—the arterial; 5th—the venous; 6th—the texture of the exhalants; 7th—that of the absorbents and their glands; 8th—the osseous; 9th—the medullary; 10th—the cartilagineous; 11th—the fibrous; 12th—the fibro-cartilaginous; 13th—the muscular of animal life; 14th—the muscular of organic life; 15th—the mucous; 16th—the serous; 17th—the synovial; 18th—the glandular; 19th—the dermoid; 20th—the epidermoid; 21st—the pilous.

These are the true elements of our bodies. Their nature is constantly the same, wherever they are met with. As in chemistry, the simple bodies do not alter, notwithstanding the different compound ones they form. The organized elements of man form the particular object of this work.

The idea of thus considering abstractedly the different simple textures of our bodies, is not the work of the imagination; it rests upon the most substantial foundation, and I think it will have a powerful influence upon physiology as well as practical medicine. Under whatever point of view we examine them, it will be found that they do not resemble each other; it is nature and not science that has drawn the lines of distinction between them.

First : Their forms are everywhere different; here they are flat, there round. We see the simple textures arranged as membranes, canals, fibrous fasciae, etc. No one has the same external character with another, considered as to their attributes of thickness or size. These differences of form, however, can only be accidental, and the same texture is sometimes seen under many different appearances; for example, the nervous appears as a membrane in the retina, and as cords in the nerves. This has nothing to do with their nature; it is, then, from the organization and the properties, that the principal differences should be drawn.

Secondly: There is no analogy in the organization of the simple textures. We shall see that this organization results from parts that are common to all, and from those that are peculiar to each; but the common parts are all differently arranged in each texture. Some unite in abundance the cellular texture, the blood vessels and the nerves; in others, one or two of these three commoner parts are scarcely evident or entirely wanting. Here there are only the exhalants and absorbents of nutrition; there the vessels are more numerous for other purposes. A capillary network, wonderfully multiplied, exists in certain textures, in others this network can hardly be demonstrated. As to the peculiar part, which essentially distinguishes the texture, the differences are striking. Colour, thickness, hardness, density, resistance, etc., nothing is similar. Mere inspection is sufficient to show a number of characteristic attributes of each, clearly different from the others. Here is a fibrous arrangement, there a granulated one: here it is lamellated, there circular. Notwithstanding these differences, authors are not agreed as to the limits of the different textures. I have had recourse, in order to leave no doubt upon this point, to the action of different re-agents. I have examined every texture, submitted them to the action of caloric, air, water, the acids, alkalies, the neutral salts, etc., drying, putrefaction, maceration, boiling, etc.; the products of many of these actions have altered in a different manner each kind of texture. Now it will be seen that the results have been almost all different, that in these various changes, each acts in a particular way, each gives results of its own, no one resembling another. There has been considerable inquiry to ascertain whether the arterial coats are fleshy, whether the veins are of an analogous nature, etc. By comparing the results of my experiments upon the different textures, the question is easily resolved. It would seem at first view that all these experiments upon the intimate texture of systems, answer but little purpose; I think, however, that they have effected a useful object, in fixing with precision the limits of each organized texture; for the nature of these textures being unknown, their difference can be ascertained only by the different results they furnish.

Thirdly: In giving to each system a different organic arrangement, nature has also endowed them with different properties. You will see in the subsequent part of this work, that what we call texture (read: tissue) presents degrees infinitely varying, from the muscles, the skin, the cellular membrane, etc., which enjoy it in the highest degree, to the cartilages, the tendons, the bones, etc., which are almost destitute of it. Shall I speak of the vital properties? See the animal sensibility predominant in the nerves, contractility of the same kind particularly marked in the voluntary muscles, sensible organic contractility, forming the peculiar property of the involuntary, insensible contractility and sensibility of the same nature, which is not separated from it more than from the preceding, characterizing especially the glands, the skin, the serous surfaces, etc., etc. See each of these simple textures combining, in different degrees, more or less of these properties, and consequently living with more or less energy.

Independently of this general difference, each texture has a particular kind of force, of sensibility, etc. Upon this principle rests the whole theory of secretion, of exhalation, of absorption, and of nutrition. The blood is a common reservoir, from which each texture chooses, that which is adapted to its sensibility, to appropriate and keep it, or afterwards reject it . . . It is evident that the greatest part of the organs being composed of very different simple textures, the idea of a peculiar life can only apply to these simple textures, and not to the organs themselves . . . When we study a function, it is necessary carefully to consider in a general manner, the compound organ that performs it; but when you wish to know the properties and life of this organ, it is absolutely necessary to decompose it. In the same way, if you would have only general notions of anatomy, you can study each organ as a whole; but it is essential to separate the textures, if you have a desire to analyse with accuracy its intimate structure.

95

JOHANNES PETER MÜLLER (1801-1858), BERLIN.

This inspiring comparative physiologist has developed the specifity of the sensation of every sensory organ. From: *Zur vergleichenden Physiologie des Gesichtssinnes des Menschen und der Tiere.* II : 2. 1826. See also J. Müller, Elements of Physiology. English translation W. Baly. London, 1842. Vol. II, p. 1059-1087.

Of the Nerves of Special Sense. The nerves have always been regarded as conductors, through the medium of which we are made conscious of external impressions. Thus the nerves of the senses have been looked upon as mere passive conductors, through which the impressions made by the properties of bodies were supposed to be transmitted unchanged to the sensorium. More recently physiologists have begun to analyse these opinions. If the nerves are mere passive conductors of the impressions of light, sonorous vibrations, and odours, how does it happen that the nerve which perceives odours is sensible to this kind of impressions only, and to no others, while by another nerve odours are not perceived; that the nerve which is sensible to the matter of light, or the luminous oscillations, is insensible to the vibrations of sonorous bodies; that the auditory nerve is not sensible to light, nor the nerve of taste to odours; while, to the common sensitive nerve, the vibrations of bodies give the sensation, not of sound, but merely of tremours? These considerations have induced physiologists to ascribe to the individual nerves of the senses a special sensibility to certain impressions, by which they are supposed to be rendered conductors of certain qualities of bodies, and not of others.

This last theory, of which ten or twenty years since no one doubted the correctness, on being subjected to a comparison with facts, was found unsatisfactory. For the same stimulus, for example, electricity may act simultaneously on all the organs of sense,—all are sensible to its action; but the nerve of each sense is affected in a different way,—becomes the seat of a different sensation: in one, the sensation of light is produced ; in another, that of sound; in a third, taste; while, in a fourth, pain and the sensation of a shock are felt. Mechanical irritation excites in one nerve a luminous spectrum; in another, a humming sound; in a third, pain. An increase of the stimulus of the blood causes in one organ spontaneous sensations of light; in another, sound; in a third, itching, pain, etc. A consideration of such facts could not but lead to the inference that the special susceptibility of nerves for certain impressions is not a satisfactory theory, and that the nerves of the senses are not mere passive conductors, but that each peculiar nerve of sense has special powers or qualities which the exciting causes merely render manifest.

Sensation, therefore, consists in the communication to the sensorium, not of the quality or state of the external body, but of the condition of the nerves themselves, excited by the external cause . . .

Sensation consists in the sensorium receiving through the medium of the nerves, and as the result of the action of an external cause, a knowledge of certain qualities or conditions, not of the external bodies, but of the nerves of sense themselves; and these qualities of the nerves of sense are in all different, the nerve of each sense having its own peculiar quality or energy . . .

The sensation of sound, therefore, is the peculiar ' energy ' or ' quality ' of the auditory nerve ; the sensation of light and colours that of the optic nerve; and so of the other nerves of sense. An exact analysis of what takes place in the production of a sensation would of itself have led to this conclusion. The sensations of heat and cold, for example, make us acquainted with the existence of the imponderable matter of caloric, or of peculiar vibrations in the vicinity of our nerves of feeling. But the nature of this caloric cannot be elucidated by sensation, which is in reality merely a particular state of our nerves; it must be learned by the study of the physical properties of this agent, namely, of the laws of its radiation, its development from the latent state; its property of combining with and producing expansion of other bodies, etc. All this again, however, does not explain the peculiarity of the sensation of warmth as a condition of the nerves. The simple fact devoid of the theory is this, that warmth, as a sensation, is produced whenever the matter of caloric acts upon the nerves of feeling; and that cold, as a sensation, results from this matter of caloric being extracted from a nerve of feeling.

So, also, the sensation of sound is produced when a certain number of impulses or vibrations are imparted, within a certain time, to the auditory nerve: but sound, as we perceive it, is a very different thing from a succession of vibrations. The vibrations of a tuning-fork, which to the ear give the impression of sound, produce in a nerve of feeling or touch the sensation of tickling; something besides the vibrations must consequently be necessary for the production of the sensation of sound, and that something is possessed by the auditory nerve alone . . .

From the foregoing considerations we have learnt most clearly that the nerves of the senses are not mere conductors of the properties of bodies to our sensorium, and that we are made acquainted with external objects merely by virtue of certain properties of our nerves, and of their faculty of being affected in a greater or less degree by external bodies. Even the sensation of touch in our hands makes us acquainted, not absolutely with the state of the surfaces of the body touched, but with changes produced in the parts of our body affected by the act of touch. By imagination and reason a mere sensation is interpreted as something quite different.

The accuracy of our discrimination by means of the senses depends on the different manner in which the conditions of our nerves are affected by different bodies; but the preceding considerations show us the impossibility that our senses can ever reveal to us the true nature and essence of the material world. In our intercourse with external nature it is always our own sensations that we become acquainted with, and from them we form conceptions of the properties of external objects, which may be relatively correct; but we can never submit the nature of the objects themselves to that immediate perception to which the states of the different parts of our own body are subjected in the sensorium.

96

KARL ERNST VON BAER (1792-1876), KOENIGSBERG.

Baer is the discoverer of the mammalian egg (1827) and the main founder of comparative embryology. In his famous memoir *Ueber die Entwicklungsgeschichte der Tiere* (1828) he demonstrates the homology in the embryological development of all vertebrates, which in their later development show progressive differentiation. In his autobiography (*Nachrichten ueber Leben und Schriften* (*mitgetheilt von ihm selbst*), St. Petersburg, 1866) we find some notes of great historical interest.

(*a*) Discovery of the mammalian egg (1866. pp. 306-316).
(*b*) Evolution and epigenesis (1866. pp. 318-320).
(*c*) The homology in the embryology of early vertebrate embryos. (1828. pp. 392-399).

(*a*) *Discovery of the mammalian egg.* I was most attracted by the development of mammals, that of their embryos and of their eggs. I could obtain, however, only very few early embryos. These showed such a great agreement with the corresponding stages of the chicken, that no doubt seemed possible about the general agreement of their development. It was well established that the eggs of the various mammalian families differ sharply in their egg-membranes and in their general shape. But recent research, especially of Dutrochet and Cuvier, has started to reduce these differences to one basic type which shows a great agreement with the egg-membranes of the older chick-embryos. This agreement grew as earlier stages of development were compared. I tried to establish these similarities first in the dog. In these studies I I approached closer and closer to the original shape. I was able to conclude that the younger the embryo of the dog was, the more similar it was to that of the young chicken, in the shape of head and trunk, and of the intestinal channel which was closed only at the anterior and posterior ends, but was connected for most of its length with the yolk-sac through a slit. In a still earlier stage the entire embryo was spread flat upon the yolk. The egg had only tiny, almost unrecognizable villi. Under the microscope it appeared very similar to a very small bird's egg, of course without its hard shell. Tracing back the eggs to still earlier stages

I found in the oviducts very small, half translucent, and hence not easily visible, vesicles which showed under the microscope a roundish spot; and still smaller opaque bodies, round and granular. Thus I was led almost by force to the discovery of the egg as it lies before fertilization within the ovary, whilst I would never have found the courage to attempt to arrive directly at such a result.

In order to show the importance of this discovery and why I had not the courage to start with this research straight away, I will say something about much older studies of the same problem . . .

The learned Haller (died 1777) made, in 1753, in Goettingen—together with the student Kuhlemann—systematic autopsies on sheep at definite times after copulation. In 40 sheep, Haller realized that the Graafian vesicle or eggs were important for later egg-development, as these vesicles were always closed before the rutting season, but they were, when he examined them, either open or filled with a yellow mass, the so-called yellow body. Yet Haller did find the egg in the uterus though only on the 17th or 19th day after copulation, when it was already fairly big. He concluded, that a liquid is poured out and enters the uterus, where it is transformed into a slime from which after some long time the egg coagulates . . .

Few physiologists resisted the authority of Haller, either continuing to assume that the Graafian vesicles were the true eggs which migrate through the oviduct, or coming to no conclusion.

Then, in 1797, the Englishman Cruikshank asserted that he had found in rabbits the true eggs in the oviducts on the third day after copulation. These were very much smaller than the Graafian vesicles of the same animal. Yet his work found little acceptance, else the true situation would soon have been elucidated. Only in 1824 did Prévost and Dumas rediscover these small eggs in dogs and rabbits and made other precise observations. But they refused to accept the small egg-size indicated by Cruikshank. The smallest eggs observed by them in dogs had 1 to 3 mm. diameter. The small eggs of Cruikshank could never have been visible, as they are too small and transparent. It seems that they even saw the true eggs in the ovary, but did not recognize them as such, as they were not transparent.

This was the situation when I began to study the problem. In 1826 I had already found some small eggs of 1 to 3 mm. diameter in the *cornua uteri* and in the oviduct, such as Prévost and Dumas had described . . . In spring 1827 I found some much smaller ones in the oviducts, which eggs were, however, not transparent and therefore easily visible. I did not doubt that these were eggs and that the yolk is originally not transparent in mammalian eggs. I discussed this with Burdach in late April and told him that I would like to obtain a bitch which had been in heat (but not copulated) for a few days. According to the observations of Prévost and Dumas the Graafian vesicle would then be still closed. Burdach had by chance just such a bitch in his house and she was killed. When I opened her, I found some burst Graafian vesicles, but none near to bursting. This depressed me, until I found a yellow spot in one of the vesicles, and then in most of them, but only one such spot in each. What could that mean? I opened one vesicle and carefully put this spot with a knife in a watch-glass filled with water. This I brought under the microscope. At my first look into it, I rebounded as though struck by lightning when I discovered a very small, well developed yellow globule of yolk. It took me some time to recover before I regained the courage to look again into the microscope, as I feared to have seen a phantom. It is strange, that a sight which I had expected and which I had longed to see, caused alarm when it were actually seen. However, there was something unexpected: I had not thought that the content of the mammalian egg would be so similar to the yolk (egg) of birds. But as I had used a simple microscope only, the enlargement was moderate and the yellow colour remained visible, which in stronger enlargements and light from below would have appeared as black. I was afraid, because I had seen small, regular globules, well circumscribed, enclosed in a thick membrane, which differed from the bird's egg only by the elevated external membrane. That was confirmed by me and Burdach for some of the other yellow spots.

Thus, the true egg of the dog had been found! It does not float freely in the thick liquid of the Graafian vesicle, but is pressed onto its wall, where it is held by a broad corona of bigger cells, continuous with the layer of cells of the inner wall of the vesicle . . .

Of course I have looked for this egg also in other mammals and in man. There it was more whitish, rarely yellowish and usually not visible without opening the vesicle and without the aid of a microscope, though it was sometimes visible in the pig. In sheep it is almost whitish and cannot be seen, if not especially looked for. Hence, Haller did not find it . . .

The mammalian egg is thus essentially a yolk globule like that of the birds, only smaller. In dogs it measured 1/10 to 1/5 mm. . .

I can thus claim the discovery of the true mammalian egg, including that of man. I concede, however, that this discovery was due less to a very industrious research or to a great perspicacity, than to the sharpness of my eyes in younger years and to the general conception which I had arrived at when studying the development of the chicken.

(*b*) *Evolution* (= *Praeformation*) *and epigenesis.* The general result of these studies differs considerably from the conclusions of Wolff, but this difference is not as big as it may appear. Wolff was fighting especially the then ruling theory of preformation, according to which the embryo is ready formed *ab initio* in the egg, being merely too small to be visible in its details. Against this Wolff proposed the principle of epigenesis, namely that of a real new formation of all parts as well as of the entire embryo. That was going too far. It is true that neither head, limbs, nor various other organs are present at the beginning (of egg-development), but they are formed not as real new formations. but by transformation of something already existing. I would prefer the term evolution (read: unfolding) to epigenesis. It is not the body which is preformed, but the trend of development, which the embryo has to pass through in its development. In this way, it is just the *invisible*, the general trend of development, which is predetermined, and which leads to the same shape into which already the parents have developed. We may say, hence, that the life-process is continual, through the entire line of offspring. It slumbers only from time to time—namely during egg-formation—and creates new individuals. . . .

(*c*) The further back we trace development, so much the more agreement do we find among the most widely different animals, which leads us to the question: Are not all animals essentially similar at the beginning of their development, have they not all a common primary form? In all true ova probably exists a distinct germinal disc, whilst the germ-granules (found in asexual reproduction) are apparently wanting in them. They appear originally to be solid . . . and eventually became hollow, as seemed to me to be in the germ-granules of the Cercariae and Bucephali, yet we perceive that the first act of their vital activity is to acquire a cavity, whereby they become thick-walled, hollow vesicles. The germ in the egg is also to be regarded as a vesicle, which in the bird's egg only gradually surrounds the yolk, but from the very first is completed as an investment by the vitellary membrane; in the frog's egg it has the vesicular form before the type of the vertebrates appears, and in the mammalian from the very first it seems to surround the small mass of the yolk. Since, however, the germ is the rudimentary animal itself, it may be said, not without reason, that the simple vesicle is the common fundamental form from which all animals are developed, not only ideally, but actually and historically. The germ-granule passes into this primitive form of the independent animal immediately by its own power; the egg, however, only after its feminine nature has been destroyed by fecundation. After this influence, the differentiation of germ and yolk, or of body and nutritive substance, arises. The excavation of the germ-granule is nothing else. In the egg, however, there is at first a solid nutritive matter (the yolk), and a fluid in the central cavity; yet the solid nutritive matter soon becomes fluid. To find a correspondence between two animal species, we must go back in development the farther the more different the two are. Hence, we deduce:

1. The law of individual development. The more general characters of a large group of animals appear earlier in their embryos than the more specialised characters. Thus also the vesicle is the most primitive form.

2. From the more general forms the less general are developed, and so on, until finally the most special arises, as has been shown before for birds and articulates.

3. Every embryo of a given animal form, instead of passing through the other forms, rather becomes separated from them.

4. Fundamentally, therefore, the embryo of a higher form never resembles any other form, but only its embryo.

It is only because the least developed forms of animals are but little removed from the embryonic condition, that they retain a certain similarity to the embryos of higher forms of animals. This resemblance is in our opinion not the determining condition of the course of development of the higher animals, but only a consequence of the organisation of the lower forms.

. . . Since the embryo becomes gradually perfected by progressive histological and morphological differentiation, it must in this respect have the more resemblance to less perfect animals, the younger it is. Furthermore, the different forms of animals are sometimes more, sometimes less remote from the principal type. The type itself never exists pure, but only with certain modifications, and the highest developed forms are the farthest distance from these types. If predominant central organs arise, especially a central part of the nervous system, the type becomes considerably modified. The worms and myriapods with their evenly annulated body are nearer the type than the butterfly. When it be true, that during development the principal type appears first and later its modifications, the young butterfly must be more similar to the perfect *Scolopendra* or worm, than the young myriopod or worm to the perfect butterfly. Now, . . . we may say that the butterfly is at first a worm. And in the vertebrates, fishes are less distant from the fundamental type than mammalia and especially as man with his great brain. The mammalian embryo should hence be more similar to the fish than the embryo of the fish to the mammalian. Now the fish is nothing else than an imperfectly developed vertebrate (a baseless assumption!), the mammalian must be regarded as a more highly developed fish, and then it is quite logical to say that the embryo of a vertebrate is at first a fish . . . But the fish is not merely an imperfect vertebrate animal; it has besides its proper fish-characters . . . . The development of an individual of a certain animal is thus determined by two conditions: (1) by a progressive development of the animal by increasing histological and morphological differentiation; (2) by the metamorphosis of a more general form into a more special one . . .

If the general result of our deliberations is well based and true, then there is *one* fundamental thought which runs through all forms and grades of animal development, and regulates all their peculiar relations. It is the same thought which collected the masses scattered through space into spheres, and united them into systems of suns; it is that which called forth into living forms the dust weathered from the surface of the metallic planet. But this thought is nothing less than Life itself, and the words and syllables in which it is expressed are the multitudinous forms of the Living.

97

FRIEDRICH WOEHLER (1800-1887), GOETTINGEN.

The chemist Woehler is usually believed to have made the first synthesis of an organic substance, namely urea. Yet in this synthesis organic substances were used, and Woehler never made publicly the claim to have definitely proven that organic substances can be produced from inorganic ones without action of a *vis vitalis*. We reproduce here the extract *Artificial formation of Urea* published in the Quart. J. Sci. Lit. and Art, London, 1828, pt.I, pp. 491-492.

Some time since M. Wöhler stated that when cyanogen acted upon liquid ammonia, amongst other products were oxalic acid and a white crystalline substance; the latter appeared to be formed also whenever cyanic acid was combined with ammonia by double decomposition: it is obtained most readily when cyanate of silver is decomposed by muriate of ammonia, or cyanate of lead by pure ammonia: in the latter way a quantity was prepared for experiment, and appeared as colourless, transparent, four-sided rectangular crystals. Nothing but oxide of lead and the particular substance is formed in this process . . .

When nitric acid is added to it, brilliant scaly crystals were formed, which, when purified by crystallisation, were very acid; these being neutralised gave nothing but nitrates and the peculiar matter in the state it originally possessed. This peculiar action with nitric acid, induced a comparison of it with urea obtained from urine, when the latter body was found to be identical with the peculiar crystalline substance or cyanate of ammonia in all the properties attributed to the former body of Proust, Prout, and others. Mr. Wöhler remarks a circumstance, in addition to those which have been pointed out relative to urea, namely, that when, either it or the artificial compound is decomposed by heat, besides a large quantity of carbonate of ammonia, there is produced, towards the end of the operation, the odour of cyanic acid, resembling that of acetic acid, exactly as in the distillation of the cyanate of mercury or uric acid, or urate of mercury.

If urea is formed by the union of cyanic acid and ammonia, then the composition of the former ought to agree with that of the latter, as M. Wöhler had formerly given it, supposing the cyanate to contain one atom of water, as all the hydrated ammoniacal salts do. The cyanate of ammonia is composed of 56.92 cyanic acid, 28.14 ammonia, and 14.74 of water; so that the ultimate composition of this salt and of urea, as analysed by Dr. Prout, are as follows:

	Cyanate of Ammonia.		Urea (after Prout).	
Azote	4 atoms	46.78	4 atoms	46.650
Carbon	2 ,,	20.19	2 ,,	19.975
Hydrogen	8 ,,	6.59	8 ,,	6.670
Oxygen	2 ,,	26.24	2 ,,	26.650
Total		99.80		99.945

When cyanic acid is decomposed by oxide of copper and heat, it yields two volumes of carbonic acid gas, and one volume of azote: when the cyanate of ammonia is decomposed in the same way, equal volumes of these gasses are obtained: the same ratio of equal volumes was obtained by Prout from urea when decomposed in the same manner.

98

WILLIAM BEAUMONT (1785-1853), UNITED STATES.

Beaumont was a surgeon of the U.S. army. He experimented for many years on his Indian servant, who had a gastric fistule, on the digestion in the human stomach. He published his *Experiments and Observations on the Gastric Juices and the Physiology of Digestion* in 1833. p. 9, 125.

St. Martin, a French Canadian half-bred Indian, 18 years old was of good constitution, robust and healthy. In the service of the of the American Fur Company he was accidentally wounded into his chest by the discharge of a musket on 6.6.1822. I saw him 25 minutes after the accident and found a portion of the lung, protruding, lacerated and burned, and immediately below this another protrusion, a portion of the stomach, lacerated through all its coats, and pouring out the food he had taken for breakfast, through an orifice large enough to admit the fore-finger . . .

At 12 M. on 1.8.1825 I introduced through the perforation, into the stomach, the following articles of diet, suspended by a silk string, and fastened at proper distances to pass in without pain a piece of high seasoned beef; a piece of raw, salted, fat pork; one of raw, salted, lean beef, another of boiled salted beef; a piece of stale bread, and a bunch of raw, sliced cabbage; each piece weighing about 2 drachms. The lad continued his usual employment about the house. At 1.00 P.M. withdrew and examined them. Found the cabbage and bread about half digested, the pieces of meat unchanged. Returned them into the stomach. At 2.00 P.M. withdrew them again. Found the cabbage, bread, pork and boiled beef all cleanly digested and gone from the string, the other pieces of meat but very little affected. Returned them again. At 3.00 P.M. (written: 2.00 P.M.!) examined again. The seasoned beef was partly digested; the raw beef was slightly macerated on the surface, but its general texture was firm and entire. The smell and taste of the fluids of the stomach were slightly rancid. The boy complained of some pain. The experiment interrupted . . . Calomel pills dropped into the stomach had the same effect as when administered by the mouth . . . I think, and subsequent experiments have confirmed the opinion, that fat meats are less easily digested than lean. Generally speaking, the looser the texture and the more tender the fibre of animal food, the easier it is of digestion.

99

CHARLES LYELL (1797-1875), EDINBURGH.

Lyell demonstrated in his *Principles of Geology* (1883. 3 vols. Here Vol. III, chapt. 1, p. 3 ff) that the factors acting at present on the surface of the earth are able to produce all the changes observed in the geological past, provided they are acting over long periods.

In our historical sketch of the progress of geology the reader has seen that a controversy was maintained for more than a century respecting the origin of fossil shells and bones: were they organic or inorganic substances? That the latter opinion should for a long time have prevailed, and that these bodies should have been supposed to be fashioned into their present form by a plastic virtue, or some other mysterious agency, may appear absurd; but it was, perhaps, as reasonable a conjecture as could be expected from those who did not appeal, in the first instance, to the analogy of the living creation, as affording the only source of authentic information. It was only by an accurate examination of living testacea, and by a comparison of the osteology of existing vertebrates with the remains found entombed in ancient strata that this favourite dogma was exploded, and all were, at length, persuaded that these substances were exclusively of organic origin . . .

When the organic origin of fossil shells had been conceded, their occurrence in strata forming some of the loftiest mountains in the world was admitted as a proof of a great alteration of the relative level of sea and land, and doubts were then entertained whether this change might be accounted for by the partial drying up of the ocean, or by the elevation of the solid land . . . The question was agitated whether any changes in the level of sea and land had occurred during the historical period, and by patient research it was soon discovered that considerable tracts of land had been permanently elevated and depressed, while the level of the ocean remained unaltered . . .

We hear of sudden and violent revolutions of the globe, of the instantaneous elevation of mountain chains, of paroxysms of volcanic energy, declining according to some, and according to others increasing in violence, from the earliest to the latest ages. We are also told of general catastrophes and a succession of deluges, of the alternation of periods of repose and disorder, of the refrigeration of the globe, of the sudden annihilation of whole races of animals and plants, and other hypotheses, in which we see the ancient spirit of speculation revived, and a desire manifested to cut, rather than patiently to untie, the Gordian knot.

In our attempt to unravel these difficult questions we shall adopt a different course, restricting ourselves to the known or possible operations of existing causes; feeling assured that we have not yet exhausted the resources which the study of the present course of nature may provide, and therefore that we are not authorised, in the infancy of our science, to recur to extraordinary agents. We shall adhere to this plan, because history informs us that this method has always put geologists on the road that leads to truth: suggesting views which, although imperfect at first, have been found capable of improvement, until at last adopted by universal consent . . . We shall appeal to the best authorities in conchology and comparative anatomy in proof of many positions which, but for the labours of naturalists devoted to these departments, would have demanded long digressions. When we find it asserted, for example, that the bones of a fossil animal at Oeningen were those of man, and the fact adduced as proof of the deluge, we are now able at once to dismiss the argument as negatory and to affirm the skeleton to be that of a reptile, on the authority of an able anatomist; and when we find among ancient writers the opinion of the gigantic stature of the human race in times of old, grounded on the magnitude of certain fossil teeth and bones, we are able to affirm these remains belong to the elephant and rhinoceros, on the same authority.

But since, in our attempt to solve geological problems, we shall be called upon to refer to the operation of aqueous and igneous causes, the geographical distribution of animals and plants, the real existence of species, their successive extinction, and so forth, we were under the necessity of collecting together a variety of facts and of entering into long trains of reasoning, which could only be accomplished in preliminary treatises. These topics we regard as constituting the alphabet and grammar of geology; not that we expect from such studies to obtain a key to the interpretation of all geological phenomena, but because they form the groundwork from which we must rise to the contemplation of more general questions relating to the complicated results to which, in an indefinite lapse of ages, the existing causes of change may give rise.

100

ROBERT BROWN (1773-1858), LONDON.

This important botanical morphologist, phytogeographer, and taxonomist gave the first notion that nuclei are present in almost every cell. Brown also coined the term *nucleus* (1833). He is equally well known as discoverer of the so-called molecular movement of Brown, which he first observed in pollen in a watery medium under the microscope. (From: *Observations on the Organs and Mode of Fecundation in Orchidaceae and Asclepidiaceae.* Trans. Linn. Soc., London, 1833. (Here from *The Miscellaneous Botanical Works of R. Brown.* London, 1866. Vol. I, pp. 511-513).

" In each cell of the epidermis of a great plant of this family a single circular areola, generally somewhat more opaque than the membrane of the cell, is observable. This areole, which is more or less distinctly granular, is slightly convex, yet actually covered by the outer lamina of the cell. There is no regularity as to its place in the cell; it is not infrequently, however, central or nearly so. Only one areole belongs to each cell; it is in the common epidermis cells, but also in the stomata . . . *This areole, or nucleus of the cell*, as perhaps it may be termed, is not confined to the epidermis, being also found not only in the pubescence of the surface, but in many cases in the parenchyme or internal cells of the tissue, especially when these are free from the deposition of granular matter. In the compressed epidermis cells the nucleus is in a corresponding degree flattened; but in the internal tissue, it is often nearly spherical, and more or less firmly adhering to one of the walls, and projecting into the cavity of the cell . . . The nucleus may even be supposed to exist in the pollen . . . This nucleus of the cell is not confined to Orchideae, but is equally manifest in many other Monocotylodonous families, and even have I found them in a few cases in the epidermis of Dicotyledonous plants . . ."

101

HENRI JOACHIM DUTROCHET (1776-1847), BRUXELLES.

Important plant physiologist, who discovered that plants respire like animals. (From *Mémoires pour servir à l'histoire anatomique et physiologiques des végétaux.* 1837).

P. 185. It results from these observations that the plants respire like the insects, i.e. by introducing oxygen into their pneumatic organs which are spread over all parts of their body, and the subsequent assimilation of this oxygen is their respiration.

But there is this difference between plants and animals: the latter take their respiratory oxygen entirely from the environment, while the plants produce by day this respiratory oxygen, and produce it above their needs, releasing the excess into the atmosphere. By night, however, these same plants absorb the atmospheric oxygen as the animals do. This is the subsidiary and imperfect type of their respiration, which alone cannot maintain their life for long. The normal respiration of the green plants is oxygen production in light and its introduction into the pneumatic organs . . . P. 186. Analogy suggests that the oxygen is by no means produced on the surface of the cryptogamic plants which have no stomata, but that it leaves their interior by unknown apertures, which is a topic for future research . . . All the observations produced in this memoir prove definitely a new fact in physiology, namely that respiration is a function essentially identical in plants and animals, differing only in accessory features. One is especially astonished at the likeness of plant and insect respiration. In both the respirable air is distributed throughout all organs in pneumatic tubes: the insect tracheae are often exactly like the pneumatic tubes of the plants . . . And finally the elongate, two-lipped stomata which shut and open on the epidermis are very similar to the tracheal apertures of insects, as described by Réaumur.

102

JOSEPH ANDREAS NICOLAUS FRANZ UNGER (1800-1870), WIEN.

We bring here an important chapter on the presence of sperms in cryptogames. (From *Acta Caes. Leopold Carol.* 1812. 1837. pp. 681-704).

Since I published, three years ago, a short note on the anthers of mosses and their contents, that is beings in plants similar to animal spermatozoa, this discovery in a reproductive organ of a plant of animal beings, so long known in animals as sperms, might have provoked interest and re-examination. Yet to my knowledge nobody has so far made an examination of this peculiar and important phenomenon . . .

The interior of the anther of *Sphagnum capillifolium* and of the other mosses is filled with a thick, slimy fluid. In summer, when these anthers have already reached their final size, only single vesicles will be discovered in this liquid; but with the approach of autumn, it is full of an abundance of animal beings. If a developed anther is carefully pierced under water with a needle, immediately the contents disperse through the surrounding water . . . and you will observe in the outcoming fluid an abundance of very small bodies, the movement of which increases with the thinning of the very viscous fluid of the anthers . . . you may even observe a spontaneous swimming of these animal beings . . .

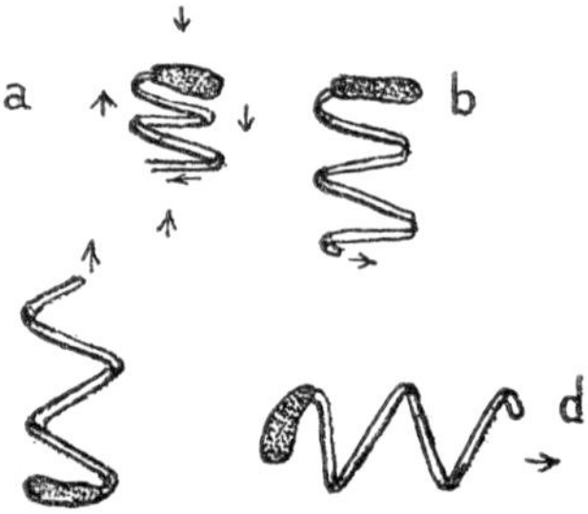

Fig. 33. The spermatozoa of *Sphagnum capillifolium* described as *Spirillum bryozoon n. sp.*, showing various of its motions. From Unger, 1837, pl. 53, fig. 2 a-d.

These animalcules are composed of a thick body and a thin filiform process. The former is 0.0025—0.0020 ''' (0.08 inches) long, apple-green, with no further recognisable organisation. It, and its filiform process are incapable of any visible lengthening or shortening. The process is colourless, very thin and only recognisable in very large magnifications. It is wound from its base into a spiral which grows looser towards the tip, which spiral may expand and contract, yet not be rolled, and hence always shows a certain stiffness. It is about four and a half times the body, giving the animalcule a total length of 0.01 ''' . . . I abstain now from calling this process a tail, as the locomotion does not proceed in the direction of the body, but in that of the spiral, so that the point of the process in the preceding part, to which the body of the spiral follows. As we know many Infusoria possess long mobile snouts (Rüssel), and in the Cercaria which have some resemblance to animalcules, also the tail-end pulls the body after it.

The most primitive movement is rolling around its own axis without locomotion, and this occurs in the thick slime. Locomotion is similar to this primitive motion of the spiral around its axis, but in locomotion the animalcules turn always following the spiral of the snout and swim straight or undulating, 1-3 full rotations being observed in one second. The point of the process is then always in trembling motion . . .

In regard to size and shape our animalcules still agree fairly well with the spermatozoa of animals, also with regard to their total length . . . All these moments point to the affinity of these moss-animalcules with the *sperm animals* of animals. Only the stiffness of the spiral curvature of the snout and the kind of locomotion differ . . . In the tailed spermatozoa always the body leads in locomotion . . . Their position in the system is still uncertain. Yet a majority of reasons has induced me to group them with the genus *Spirillum* (Ehrenberg) and to describe it as *Spirillum bryozoon n. sp.* This is supported by the inclination to class the male spermatozoa with the Vibrios, with which they have more in common than with the Trematoda, with which Ehrenberg (1835) united them.

103

JOHANN SCHLEIDEN (1804-1881), JENA.

Schleiden formulated the cellular theory of plants in his *Beitraeg-zur Phytogenesis.* Arch. Anat., Physiol. u. Wiss. Med. 1838. (Translated by H. Smith (1847) into English: (*a*) = p. 231; (*b*) = p. 231; (*c*) = p. 242.).

(*a*) The general and basic law of human reason, its irrepressible aim to unity of knowledge, has been of dominant importance since early in the history of biology when man began to search for a deeper analysis of the plant and animal kingdoms. Most attempts of this kind must, however, be regarded as failures . . . Recently we have learned that the lower plants all consist of one cell, while the higher ones are composed of (many) individual cells.

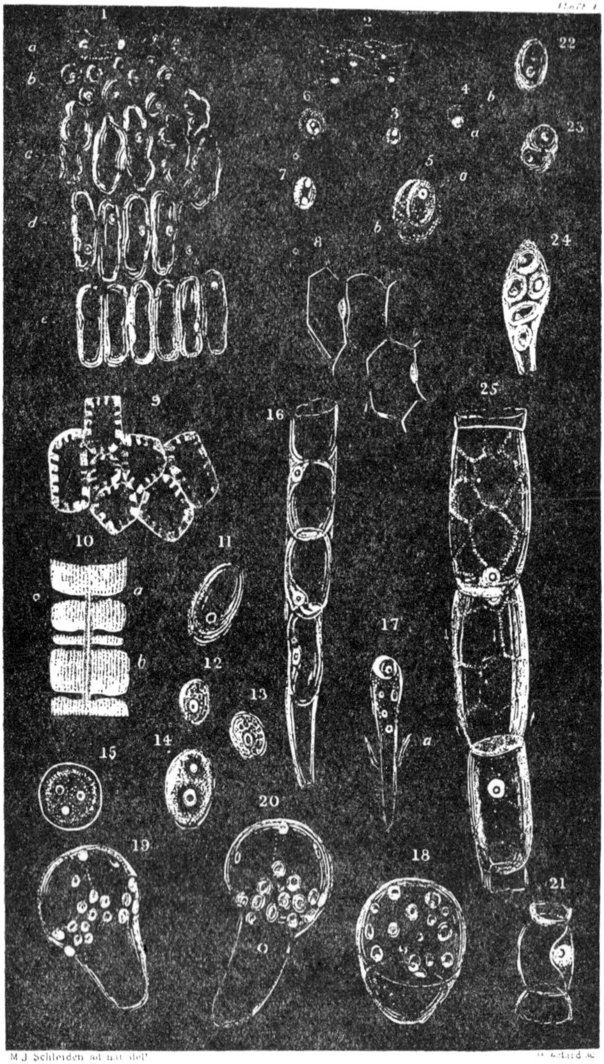

Fig. 34. Cell structure and cell generation in plants. From M. J. Schleiden. 1838. 1, a-e, Cellular tissue from the embryo-sac of *Chaaedorea schiedeana* in the act of formation. 2-10: details from 1; 11: Sporule of *Rhizina* with the cytoblast; 12-15: cytoblasts from *Pimelea drupacea*; 16-25: Cell-formation in embryonal ends of pollen-tubes, sporules, terminal shoots, etc.

Each cell leads a double life, an independent one as a cell-individual, another as a dependent integral part of the plant. It is obvious that the understanding of the life process of the individual cells is the primary, indispensable base for the understanding of plant—as well as of comparative—physiology. Hence the importance of the question: What is the origin of this peculiar small organism, the Cell? Only the far-reaching importance of this subject impels me to publish at this time my preliminary observations on this subject. No real progress in knowledge can come from mere hypotheses concerning nature, especially when facts which may guide us are entirely wanting. I may be exempted from giving a historical introduction since to the best of my knowledge no direct observations have ever been made on the origin of the plant cell . . .

(*b*) Now to the matter itself. New organization of tissue can be observed at two points on the plant in the easiest and safest manner, namely, in the two cavities enclosed in a single membrane comprising the large cell, later containing the albumen of the sperm, and at the end of the pollen tube.

At these two places, small slime bodies within the gum soon originate and cause the hitherto clear and homogeneous gum-solution to become opaque and to increase the quantity of its granulations. Then, a few larger and sharper granules (= nuclei) stand out in this mass, and soon afterwards the cytoblasts (read: cytoplasm) emerge and arrange themselves like a granular coagulation around the nuclei. The cytoblasts grow considerably in this free state, thus I observed an increase in *Fritillaria pyrenaica* from 0.0032 to 0.027 cm. in diameter. As soon as the cytoblasts have reached their full size a transparent vesicle rises upon their surface. This is the young cell which begins as a very flat globular segment; the cytoblast as the plane and the young cell on the convex side resemble a watch glass sitting upon the watch. The cell in the neutral medium is recognizable almost solely by the fact that the space between the convex swelling and the cytoblast is clear as water and translucent, probably filled with watery liquid. This space is limited by the slime granules which are pressed against its circumference (f.1 and 8). While these young cells are isolated, the slime granules can easily be separated by shaking them on a slide. The vesicles slowly extend and grow to be of greater consistency (f. 1b). The wall is now formed —a part of the cytoblast which still forms part of the gelatinous cell wall. In this stage the entire cell grows beyond the margin of the cytoblast, increases rapidly in size and the cytoblast remains only as a (small) part included within one of the cell walls. At the same time the young cell often shows very irregular extrusions (f. 1c) which is evidence that the growth is not effected from one point only. With further growth, however, the circumference becomes more regular, obviously due to internal pressure (f. 1b-e). The cytoblast is still included in the cell wall, on which place it remains throughout life. Only in those cells which are determined for higher development is it dissolved either in its place or expelled into the cell cavity as an organ without (further) use and reabsorbed . . .

(*c*) It is an absolute law that every cell takes its origin as a very small vesicle and grow only slowly to its defined size. The process of cell formation which I have just described *in extenso* is that process which I was able to follow in most of the plants which I have studied. Yet many modifications of this development can be observed. In some plants the observation is difficult in parts or in all cells. Nevertheless, the general law remains incontestable since analogy requires it, and since we fully understand the causes which sometimes prevent complete observation.

104

THEODOR SCHWANN (1810-1882), BERLIN.

Schwann extended the cellular theory from the plants to the animals in his *Mikroskopsiche Untersuchungen ueber die Uebereinstimmung in der Structur und dem Wachstum der Tiere und Pflanzen* (1839). Translated by H. Smith (1847) into English, pp. 161-166).

The Cell Theory. When organic nature, animals and plants, is regarded as a Whole, in contradistinction to the inorganic kingdom, we do not find that all organisms and their separate organs are compact masses, but that they are composed of innumerable small particles of a definite form. These elementary particles, however, are subject to the most extraordinary diversity of figure, especially in animals; in plants they are, for the most part or exclusively, cells. This variety in the elementary parts seemed to hold some relation to their more diversified physiological function in animals, so that it might be established as a principle, that every diversity in the physiological significance of an organ requires a difference in its elementary particles; and, on the contrary, the similarity of two elementary particles seemed to justify the conclusion that they were physiologically similar It was natural that among the very different forms presented by the elementary particles, there should be some more or less alike, and that they may be divided, according to their similarity of figure, into fibres, which compose the great mass of the body of animals, into cells, tubes, globules, etc. The division was, of course, only one of natural history, not expressive of any physiological idea, and just as a primitive muscular fibre, for example, might seem to differ from one of areolar tissue, or all fibres from cells, so would there be in like manner a difference, however gradually marked between the different kinds of cells. It seemed as if the organism arranged the molecules in the definite forms, exhibited by its different elementary particles, in the way required by its physiological function. It might be expected that there would be a definite mode of development for each separate kind of elementary structure, and that it would be similar in those structures which were physiologically identical, and such a mode of development was, indeed, already more or less perfectly known with regard to muscular fibres, blood corpuscles, the ovum, and epithelium-cells. The only process common to all of them, however, seemed to be the expansion of their elementary particles after they had once assumed their proper form. The manner in which their different elementary particles were first formed appeared to vary very much. In muscular fibres they were globules, which were placed together in rows, and coalesced to form a fibre, whose growth proceeded in the direction of its length. In the blood-corpuscles it was a globule, around which a vesicle was formed, and continued to grow; in the case of the ovum, it was a globule around which a vesicle was developed and continued to grow, and around this again a second vesicle was formed . . .

The object, then, of the present investigation was to show that the mode in which the molecules composing the elementary particles of organisms are combined does not vary according to the physiological signification of those particles, but that they are everywhere arranged according to the same laws . . .

In order to establish this point it was necessary to trace the progress of development in two given elementary parts, physio-

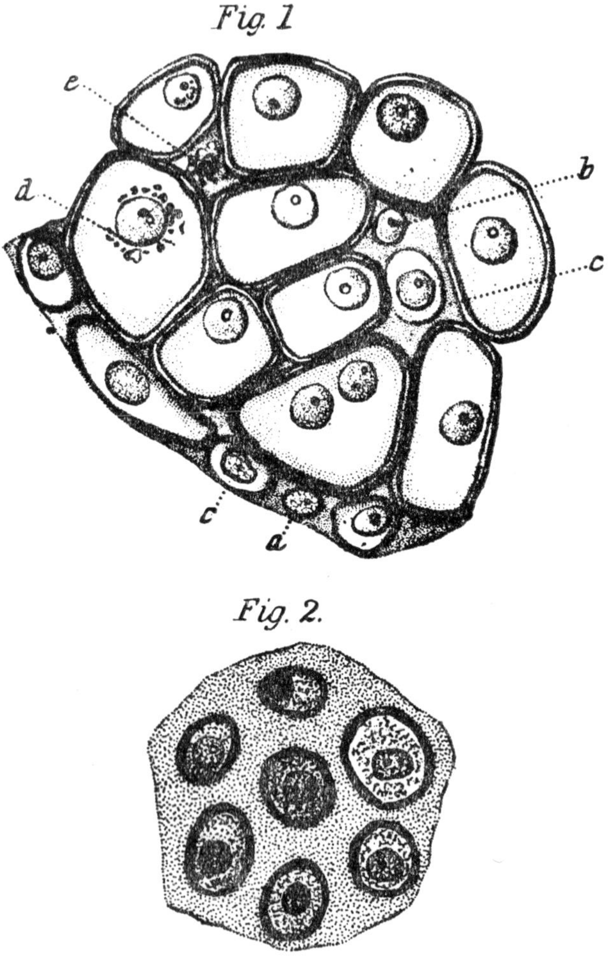

Fig. 35. Cells form all animal tissues. From Schwann. 1839. pl. III, 1-2. Cartilage formation in the early stages of frog (1) and of pig (2).

logically disimilar, and to compare them with one another. If these not only completely agreed in growth, but in their mode of generation also, the principle was established that elementary parts, quite distinct in a physiological sense, may be developed according to the same laws. This was the theme of the first section of this work. The course of development of the cells of cartilage and of the cells of the chorda dorsalis was compared with that of vegetable cells. Were the cells of plants developed merely as infinitely minute vesicles which progressively expand, were the circumstances of their development less characteristic than those pointed out by Schleiden, a comparison, in the sense here required, would scarcely have been possible. We endeavoured to prove in the first section that the complicated process of development in the cells of plants recurs in those of cartilage and of the chorda dorsalis. We remarked the similarity in the formation of the cell-nucleus, and of its nucleolus in all its modifications, with the nucleus of vegetable cells, and the pre-existence of all the cell-nucleus and the development of the cell around it, the similar situation of the nucleus in relation to the cell, the growth of the cells, and the thickening of their wall during growth, the formation of cells within cells, and the transformation of the cell-contents just as in the cells of plants. Here, then, was a complete accordance in every known stage in the progress of development of two elementary parts which are quite distinct, in a physiological sense, and it was established that the principle of development in two such parts may be the same, and so far as could be ascertained in the cases here compared, it is really the same.

But regarding the subject from this point of view we are compelled to prove the universality of this principle of development, and such was the object of the second section . . . There may still be some unknown difference in the laws of the formation of the parts just compared, even though they agree in many points. But, on the contrary, the greater the number of physiologically different elementary parts, which, so far as can be known, originate in a similar manner, and the greater the difference of these parts in form and physiological signification, while they agree in the perceptible phenomena of their mode of formation, the more safely may we assume that all elementary parts have one and the same fundamental principle of development. It was, in fact, shown that the elementary parts of most tissues, when traced backwards from their state of complete development to their primary condition are only developments of cells, which so far as our observations, still incomplete, extend, seemed to be formed in a similar manner to the cells compared in the first section . . . The elementary parts of all tissues are formed of cells in an analogous, though very diversified manner, so that it may be asserted, that there is one universal principle of development for the elementary parts of organisms, however different, and that this principle is the formation of cells.

The same process of development and transformation of cells within a structureless substance is repeated in the formation of all the organs of an organism, as well as in the formation of new organisms; and the fundamental phenomenon attending the exertion of productive power in organic nature is accordingly as follows: A structureless substance is present in the first instance, which lies either around or in the interior of cells already existing; and cells are formed in it in accordance with certain laws, which cells become developed in various ways into the elementary parts of organisms.

This one general principle for the formation of all organic productions may be comprised under the term cell-theory.

105

JUSTUS VON LIEBIG (1803-1873), GIESSEN AND MUENCHEN.

Liebig was one of the outstanding pioneers in the study of chemistry in its application to agriculture, especially the nutrition of plants and animals. His best known book in this respect is *Die Chemie in ihrer Anwendung auf Agricultur und Physiologie* (1840). We quote it here from the English translation by L. Playfair which was prepared from the manuscript and appeared contemporaneously with the German edition (Organic Chemistry in its applications to Agriculture and Physiology).

(*a*) The exhaustion of the soil under agriculture and its prevention. (English translation, pp. 148-151).

(*b*) The importance of phosphoric acid for plant- and animal-nutrition. (English translation, pp. 153-154).

(*c*) The different food requirements of various plants and their consequence for crop-rotation. (English translation, pp. 156-157).

(*a*) Air, water, and the change of temperature prepare the different species of rocks for yielding to plants the alkalies which they contain. A soil which has been exposed for centuries to all the influences which effect the disintegration of rocks, but from which the alkalies have not been removed, will be able to afford the means of nourishment to those vegetables which require alkalies for their growth during many years; but it must gradually become exhausted unless those alkalies which have been removed are again replaced; a period, therefore, will arrive, when it will be necessary to expose it, from time to time, to a further disintegration, in order to obtain a new supply of soluble alkalies, for small as is the quantity of alkali which plants require, it is nevertheless quite indispensable for their perfect development. But when one or more years have elapsed without any alkalies having been extracted from the soil, a new harvest may be expected.

The first colonists of Virginia found a country, the soil of which was similar to that mentioned above; harvests of wheat and tobacco were obtained for a century from one and the same field without the aid of manure, but now whole districts are converted into unfruitful pasture land, which without manure produces neither wheat nor tobacco. From every acre of this land, there were removed in the space of one hundred years 1,200 lbs. of alkalies in leaves, grain, and straw; it became unfruitful therefore, because it was deprived of every particle of alkali, which had been reduced to a soluble state, and because that which was rendered soluble again in the space of one year, was not sufficient to satisfy the demands of the plants. Almost all the cultivated land in Europe is in this condition; fallow is the term applied to land left at rest for further disintegration. It is the greatest possible mistake to suppose that the temporary diminution of fertility in a soil is owing to the loss of humus; it is the mere consequence of the exhaustion of the alkalies.

Let us consider the condition of the country around Naples, which is famed for its fruitful cornland . . . for thousands of years . . . The method of culture in that district completely explains the permanent fertility. It appears very bad in the eyes of our agriculturists, but there it is the best plan which could be adopted. A field is cultivated once every three years, and is in the interval allowed to serve as a sparing pasture for cattle. The soil experiences no change in the two years during which it there lies fallow, further than that it is exposed to the influence of the weather, by which a fresh portion of the alkalies contained in it are again set free or rendered soluble. The animals fed on these fields yield nothing to these soils which they did not formerly possess. The weeds upon which they live spring from the soil, and that which they return to it as excrement, must always be less than that which they extract. The field, therefore, can have gained nothing from the mere feeding of cattle upon them; on the contrary, the soil must have lost some of its constituents.

Experience has shown in agriculture, that wheat should not be cultivated after wheat on the same soil, for it belongs with tobacco to the plants which exhaust a soil. But if the humus of the soil gives it the power of producing corn, how happens it that wheat does not thrive in many parts of Brazil, where the soils are particularly rich in this substance, or in our own climate, in soils formed of mouldered wood; that its stalk under the circumstances attains no strength, and drops prematurely? The cause is this, that the strength of the stalk is due to silicate of potash, and that the corn requires phosphate of magnesia, neither of which substances a soil of humus can afford, since it does not contain them; the plant may indeed, under such circumstances become an herb, but will not bear fruit.

Again, how does it happen that wheat does not flourish on a sandy soil, and that a calcareous soil is also unsuitable for its growth, unless it be not mixed with a considerably quantity of clay? It is because these soils do not contain alkalies in sufficient quantity, the growth of wheat being arrested by this circumstance, even should all other substances be present in abundance.

It is not mere accident that only trees of the fir-tribe grow on sandstone and limestone of the Carpathian mountains and the Jura, whilst we find on soil of gneiss, mica-slate, and granite in Bavaria, of clinckstone on the Rhone, of basalt in the Vogelsberge, and of clay-slate on the Rhine and Eifel, the finest forests of other trees which cannot be produced on the sandy or calcarous soils upon which the pines thrive. It is explained by the fact that trees, the leaves of which are renewed annually, require for their leaves six to ten times more alkalies than the fir-trees or pine. . . .

(*b*) Potash is not the only substance necessary for the existence of most plants, indeed it has been already shown that the potash may be replaced, in many cases, by soda, magnesia, or lime; but other substances, besides alkalies, are required to sustain the life of plants.

Phosphoric acid has been found in the ashes of all plants hitherto examined, and always in combination with alkalies or alkaline earths. Most seeds contain certain quantities of phosphates. In the seeds of different kinds of corn, particularly, there is abundance of phosphate of magnesia. Plants obtain their phosphoric acid from the soil. It is a constituent of all land capable of cultivation, and even the heath at Lueneburg contains it in appreciable quantity. Phosphoric acid has been detected, also, in all mineral waters in which its presence has been tested, and in those in which it has been found, it has not been sought for . . .

The soil in which plants grow furnishes them with phosphoric acid, and they, in turn yield it to animals, to be used in the formation of their bones, and of those constituents of the brain which contain phosphorus. Much more phosphorus is thus afforded to the body than it requires, when flesh, bread, fruit, and husks of grain are used for food, and this excess in them is eliminated in the urine and the solid excrements . . .

(*c*) The fallow-time is that period of culture, during which land is exposed to a progressive disintegration by means of the influence of the atmosphere, for the purpose of sending a certain quantity of alkalies, capable of being appropriated by plants. . . . For the purpose of agriculture, it is quite indifferent whether the land is covered with weeds, or with a plant which does not abstract the potash inclosed in it. Now, many plants in the family Leguminosae are remarkable on account of the small quantity of alkalies or salts in general, which they contain; the *Vicia faba*, for example, contains no free alkalies, and not one per cent. of the phosphates of lime and magnesia (Einhorn). The bean of *Phaseolus vulgaris* contains only traces of salts (Braconnot) . . . Buck-wheat dried in the sun yields only 0.681 per cent. of ashes, of which 0.09 parts are soluble salt (Zenneck). The plants belong to those which are termed fallow-crops, and the cause wherefore they do not exercise any injurious influence on corn which is cultivated immediately after them is, that they do not extract the alkalies of the soil, and only a very small quantity of phosphates.

It is evident that two plants growing beside each other will mutually injure one another, if they withdraw the same food from the soil . . . Plants will, on the contrary, thrive beside each other, either when the substance necessary for their growth which they extract from the soil are of different kinds, or when they themselves are not both in the same stages of development at the same time.

On a soil, for example, which contains potash, both wheat and tobacco may be reared in succession, because the latter plant does not require phosphates, salts which are invariably present in wheat, but requires only alkalies, and food containing nitrogen.

According to the analysis of Posselt and Reimann, 10.000 parts of the leaves of the tobacco-plant contain 16 parts of phosphate of lime, 8.8 parts of silica, and no magnesia; whilst an equal quantity of wheat-straw contains 47.3 parts, and the same quantity of grain of wheat 99.45 parts of phosphates.

Now, if we suppose that the grain of wheat is equal to half the weight of its straw, then the quantity of phosphates extracted from a soil by the same weights of wheat and tobacco must be as 97.7 : 16. This difference is very considerable, the roots of tobacco, as those of wheat, extract the phosphates contained in the soil, but they restore them again, because they are not essentially necessary to the development of the plant.

106

JEAN-BAPTISTE BOUSSINGAULT (1802-1887), PARIS.

The great agricultural chemist Boussingault was an important contemporarean of Liebig. He published in 1843/44 in two volumes his *Economie rurale considérée dans ses rapports avec la chimie, la physique et la météorologie.* Here (from vol. I, pp. 74-82) follows a paragraph demonstrating the assimilation of atmospheric nitrogen by leguminous plants (English translation in C. A. Browne, *A Source Book of Agricultural Chemistry.* Waltham, 1944. pp. 243-244).

I had necessarily to follow a method of inquiry different from any that had yet been taken . . . I called in the aid of elementary analysis, with a view of comparing the seed's composition with that of the crop produced therefrom at the sole cost of water and air. By proceeding in this way I believed a solution, even if not a complete one, be possible: I took as a soil baked clay, or silicous sand, freed of organic matter by proper calcination. In this soil, moistened with distilled water, were sown the seeds whose weight had been determined. In a series of preliminary analyses I determined the moisture which seeds of the same kind, origin and time of sampling lost by drying first in an oven and then in an oil bath at 110°. The porcelain dishes, which contained the sandy soil, were placed in a glass house. During the whole time of growth, the windows were hermetically closed, but their situation permitted free access of sunshine during the whole day. To harvest the crop, the dishes were first dried at a gentle heat. The roots of the plants then came out readily; to free them completely from adhering sand, they were moved about in distilled water but never crushed for fear of losing some of their juice . . . The harvested plant was then dried in the oven so that it could be powdered; final desiccation was accomplished in an oil bath *in vacuo*. A determination by combustion of the weight of ash, enabled a calculation to be made of the weight of crop free from all saline and earthy matter. Elementary analysis then indicated the composition of the crop. All that was now left was to compare this with the composition of the seed in order to know the proportion and nature of the elements which had been assimilated during growth.

(The following experimental results were thus obtained; all in grams:)

Experiments showing assimilation of atmospheric nitrogen by clover and peas:

Crop.	*Weight ash-free substance.*	*Carbon.*	*Hydrogen.*	*Oxygen.*	*Nitrogen.*
Red clover, seed	1.586	0.806	0.095	0.571	0.114
Ditto, crop, 3 months ...	4.106	2.082	0.271	1.597	0.156
	2.520	1.276	0.176	1.026	0.042
Peas, seed	1.072	0.515	0.069	0.442	0.046
Ditto, crop, 3 months ...	4.441	2.376	0.284	1.680	0.101
	3.369	1.861	0.215	1.238	0.055
Wheat, seed	1.644	0.767	0.095	0.725	0.057
Ditto, crop, 2 months ...	3.022	1.456	0.173	1.333	0.060
	1.378	0.689	0.078	0.608	0.003

(Similar results were obtained with oats and the following conclusions drawn):

" (1) Clover and peas, grown in a soil with absolutely no fertilizer, gained in addition to carbon, hydrogen and oxygen an appreciable quantity of nitrogen.

(2) Wheat and oats, grown under the same conditions, also took carbon, hydrogen and oxygen from the air and water, but analyses could not detect any gain in nitrogen after the growth of these cereals.

The purpose of this method of investigation was simply to establish the fact of nitrogen assimilation by certain plants, without entering into the question of the means by which this was effected, and upon this point I can only offer conjectures."

107

JULIUS ROBERT MAYER (1814-1878), HEILBRONN.

In his *Die organische Bewegung in ihrem Zusammenhange mit dem Stoffwechsel* (1845) applies Mayer the (yet undiscovered) law of the conservation of energy to animals and plants. English translation from F. S. Taylor, *Science past and present.* 1945. pp. 244-246.

The second question refers to the cause of the chemical tension (i.e. chemical energy of our days) produced in the plant. This tension is a physical (read: natural) force. It is equivalent to the heat obtained from the combustion of the plant. Does this force, then, come from the vital processes, and without the expenditure of some other form of force? The creation of a physical force, of itself hardly thinkable, seems all the more paradoxical when we consider that it is only by the help of the sun's rays that the plants perform their work. By the assumption of such a hypothetical action of the 'vital force,' all further investigation is cut off, and the application of the methods of exact science to the phenomena of vitality is rendered impossible. Those who hold a notion so opposed to the spirit of science would be thereby carried into the chaos of unbridled phantasy. I therefore hope that I may reckon on the reader's assent when I state as an axiomatic truth, that:

During vital processes a conversion only of matter, as well as of force, occurs, and that creation of either the one or the other never takes place.

The physical force collected by plants becomes the property of another class of creatures—of animals. The living animal consumes combustible substances belonging to the vegetable world and causes them to reunite with the oxygen of the Atmosphere. Parallel to this process runs the work done by animals . . . In the animal body chemical forces are perpetually expended. Ternary and quaternary (read: complex organic compounds with 3 or 4 elements) compounds undergo, during the life of the animal, the most important changes, and are, for the most part given off in the form of binary compounds (such as carbon dioxide and water), as burnt substances. The magnitude of these forces, with reference to the heat developed in these processes, is by no means determined with sufficient accuracy (we are in the very beginnings of thermochemistry!). But here where our object is simply the establishment of a principle it will be sufficient to take into account the heat of combustion of the pure carbon . . .

(Follows calculation).

If the animal organism applied the disposable, combustible material solely to the performance of work, the quantities of carbon just calculated would suffice for the times mentioned. In reality, however, besides the production of mechanical effects there is in the animal body a continuous generation of heat. The chemical force contained in the food and inspired oxygen is therefore the source of two other forms of power, namely mechanical motion and heat; and the sum of these physical forces produced by an animal is the equivalent of the contemporaneous chemical processes. Let the quantity of mechanical work performed by an animal in a given time be collected and converted by friction or some other means into heat; add to this the heat generated immediately in the animal body at the same time, we have then the exact quantity of heat corresponding to the chemical processes that have taken place.

In the active animal the chemical changes are much greater than in the resting one. Let the amount of the chemical processes accomplished in a certain time in the resting animal be x, and in the active one be: $x + y$. If during activity the same quantity of heat were generated as during rest, the additional chemical force y would correspond to the work performed. In general, however, more heat is produced in the active organism than in the resting one. During work, therefore, we shall have x + a portion of y heat, the residue of y being converted into mechanical effect. The maximum mechanical effect produced by a working mammal hardly amounts to one fifth of the force derivable from the total quantity of carbon consumed. The remaining four-fifths are devoted to the generation of heat.

108

CHARLES DARWIN (1809-1882), DOWN, ENGLAND.

Darwin gained fame by his *Origin of Species* (1859), which gave a solid basis to the theory of transformation of species and of evolution. He describes as mechanism of evolution the struggle of existence which brings about natural selection.

(*a*) Animals of the Galapagos Islands compared with those of the neighbouring continent of South America. From the *Journal of Researches into the Natural History and Geology of the countries visited during the voyage of H.M.S. "Beagle" round the world.* 1839/45. Chapter XVII.

(*b*) The struggle for existence. *Origin of Species.* From Chapter III.

(*c*) On natural selection. *Origin of species.* From Chapter IV.

(*a*) If this character were owing merely to immigrants from America, there would be little remarkable in it; but we see that a vast majority of all land animals, and that more than half of the flowering plants of the Galapagos Islands are aboriginal productions. It was most striking to be surrounded by new birds, new reptiles, new shells, new insects, new plants, and yet by innumerable trifling details of structure, and even by the tones of voice and plumage of the birds, to have the temperate plains of Patagonia, or the hot dry deserts of Northern Chile, vividly brought before my eyes. Why, on these small points of land, which within a late geological period must have been covered by the ocean, which are formed of basaltic lava, and therefore differ in geological character from the American continent, and which are placed under a peculiar climate—why were their aboriginal inhabitants, associated, I may add, in different proportions both in kind and number from those on the continent, and therefore acting on each other in a different manner—why were they created on American types of organisation? It is probable that the islands of the Cape de Verd group resemble, in all their physical conditions, far more closely the Galapagos Islands than these latter physically resemble the coast of America; yet the aboriginal inhabitants of the two groups are totally unlike; those of the Cape de Verd Islands bearing the impress of Africa, as the inhabitants of the Galapagos Archipelago are stamped with that of America.

I have not as yet noticed by far the most remarkable feature in the natural history of this archipelago: it is, that the different islands to a considerable extent are inhabited by a different set of beings. My attention was first called to this fact by the Vice-Governor, Mr. Lawson, declaring that the tortoises differed from the different islands, and that he could with certainty tell from which island any one was brought. I did not for some time pay sufficient attention to this statement, and I had already partially mingled together the collections from two of the islands. I never dreamed that islands, about fifty or sixty miles apart, and most of them in sight of each other, formed of precisely the same rocks, placed under a quite similar climate, rising to a nearly equal height, would have been differently tenanted; but we shall soon see that this is the case. It is the fate of most voyagers, no sooner to discover what is most interesting in any locality, than they are hurried from it; but I ought, perhaps, to be thankful that I obtained sufficient materials to establish this most remarkable fact in the distribution of organic beings . . .

The distribution of the tenants of this archipelago would not be nearly so wonderful, if, for instance, one island had a mocking-thrush, and a second island some other quite distinct genus; if one island had its genus of lizard, and a second island another distinct genus, or none whatever; or if the different islands were inhabited, not by representative species of the same genera of plants, but by totally different genera, as does to a certain extent hold good; for, to give one instance, a large berry-bearing tree at James Island had no representative species in Charles Island. But it is the circumstance, that several of the islands possess their own species of the tortoise, mocking-thrush, finches, and numerous plants, these species having the same general habits, occupying analogous situations, and obviously filling the same place in the natural economy of this archipelago, that strikes me with wonder. It may be suspected that some of these representative species, at least in the case of the tortoise and of some of the birds, may hereafter prove to be only well-marked races; but this would be of equally great interest to the philosophical naturalist. I have said that most of the islands are in sight of each other . . . I must repeat, that neither the nature of the soil, nor height of the land, nor the climate, nor the general character of the associated beings, and therefore their action one on another, can differ much in the different islands. If there be any sensible difference in their climates, it must be between the windward group (namely Charles and Chatham Islands), and that to leeward; but there seems to be no corresponding difference in the production of these two halves of the archipelago.

(*b*) The term 'struggle for existence' used in a large sense. I should premise that I use this term in a large and metaphorical sense including dependence of one being on another, and including (which is more important) not only the life of the individual, but success in leaving progeny. Two canine animals, in a time of dearth, may be truly said to struggle with each other which shall get food and live. But a plant on the edge of a desert is said to struggle for life against the drought, though more properly it should be said to be dependent on the moisture. A plant which annually produces a thousand seeds, of which only one of the average comes to maturity, may be more truly said to struggle with the plants of the same and other kinds which already clothe the ground. The mistletoe is dependent on the apple and a few other trees, but can only in a far-fetched sense be said to struggle with these trees, for, if too many of these parasites grow on the same tree, it languishes and dies. But several seedling mistletoes, growing close together on the same branch, may more truly be said to struggle with each other. As the mistletoe is disseminated by birds, its existence depends on them; and it may metaphorically be said to struggle with other fruit-bearing plants, in tempting the birds to devour and thus disseminate its seeds. In these several senses, which pass into each other, I use for convenience sake the general term of 'struggle for existence.'

(*c*) Summary of chapter IV. If under changing conditions of life organic beings present individual differences in almost every part of their structure, and this cannot be disputed; if there be, owing to their geometrical rate of increase, a severe struggle for life at some age, season, or year, and this certainly cannot be disputed; then, considering the infinite complexity of the relations of all organic beings to each other and to their conditions of life, causing an infinite diversity in structure, constitution, and habits, to be advantageous to them, it would be a most extraordinary fact if no variations had ever occurred useful to each being's own welfare, in the same manner as so many variations have occurred useful to man. But if variations useful to any organic being ever do occur, assuredly individuals thus characterised will have the best chance of being preserved in the struggle for life; and from the strong principle of inheritance, these will tend to produce offspring similarly characterised. This principle of preservation, or the survival of the fittest, I have called 'natural selection.' It leads to the improvement of each creature in relation to its organic and inorganic conditions of life; and consequently, in most cases, to what must be regarded as an advance in organisation. Nevertheless, low and simple forms will long endure if well fitted for their simple conditions of life.

Natural selection, on the principle of qualities being inherited at corresponding ages, can modify the egg, seed, or young, as easily as the adult. Amongst many animals, sexual selection will have given its aid to ordinary selection, by assuring to the most vigorous and best adapted males the greatest number of offspring. Sexual selection will also give characters useful to the males alone, in their struggles or rivalry with other males; and these characters will be transmitted to one sex or to both sexes, according to the form of inheritance which prevails.

Whether natural selection has really thus acted in adapting the various forms of life to their several conditions and stations, must be judged by the general tenor and balance of evidence given in the following chapters. But we have already seen how it entails extinction; and how largely extinction has acted in world's history, geology plainly declares. Natural selection, also, leads to divergence of character; for the more organic beings diverge in structure, habits, and constitution, by so much the more can a large number be supported on the area—of which we see proof by looking to the inhabitants of any small spot, and to the productions naturalised in foreign lands. Therefore, during the modification of the descendants of any one species, and during the incessant struggle of all species to increase in numbers, the more diversified the descendants become, the better will be their chance of success in the battle for life. Thus the small differences distinguishing varieties of the same species, steadily tend to increase, till they equal the greater differences between species of the same genus, or even of distinct genera.

We have seen that it is the common, the widely-diffused and widely-ranging species, belonging to the larger genera within each class, which vary most; and these tend to transmit to their modified offspring that superiority which now makes them dominant in their own countries. Natural selection, as has just been remarked, leads to divergence of character and to much extinction of the less improved and intermediate forms of life. On these principles, the nature of the affinities, and the generally well-defined distinctions between the innumerable organic beings in each class throughout the world may be explained. It is a truly wonderful fact—the wonder of which we are apt to overlook from familiarity—that all animals and all plants throughout all time and space should be related to each other in groups, subordinate to groups, in the manner which we everywhere behold—namely, varieties of the same species most closely related, species of the same genus less closely and unequally related, forming sections and sub-genera, species of distinct genera much less closely related, and genera related in different degrees, forming sub-families, families, orders, sub-classes and classes. The several subordinate groups in any class cannot be ranked in a single file, but seen clustered round points, and these round other points, and so on in almost endless cycles. If species had been independently created, no explanation would have been possible of this kind of classification; but it is explained through inheritance and the complex action of natural selection, entailing extinction and divergence of character, as we have seen illustrated in the diagram.

The affinities of all the beings of the same class have sometimes been represented by a great tree. I believe this simile largely speaks the truth. The green and budding twigs may represent existing species; and those produced during former years may represent the long succession of extinct species. At each period of growth all the growing twigs have tried to branch out on all sides, and to overtop and kill the surrounding twigs and branches, in the same manner as species and groups of species have at all times overmastered other species in the great battle for life. The limbs divided into great branches, and these into lesser and lesser branches, were themselves once, when the tree was young, budding twigs; and this connection of the former and present buds by ramifying branches may well represent the classification of all extinct and living species in groups subordinate to groups. Of the many twigs which flourished when the tree was a mere bush, only two or three, now grown into great branches, yet survive and bear the other branches; so with the species which lived during long-past geological periods, very few have left living and modified descendants. From the first growth of the tree, many a limb and branch has decayed and dropped off; and these fallen branches of various sizes may represent those whole orders, families, and genera which have now no living representatives, and which are known to us only in a fossil state. As we here and there see a thin straggling branch springing from a fork low down in a tree, and which by some chance has been favoured and is still alive on its summit, so we occasionally see an animal like the *Ornithorhynchus* or *Lepidosiren*, which in some small degree connects by its affinities two large branches of life, and which has apparently been saved from fatal competition by having inhabited a protected station. As buds give rise by growth to fresh buds, and these, if vigorous, branch out and overtop on all sides many a feebler branch, so by generation I believe it has been with the great Tree of Life, which fills with its dead and broken branches the crust of the earth, and covers the surface with its ever-branching and beautiful ramifications.

109

WILHELM FRIEDRICH BENEDICT HOFMEISTER (1824-1877), HEIDELBERG.

Hofmeister completed the discovery of the comparative life cycle, including the alternation of generations by an asexual and a sexual generation of all groups of cryptogames. From *Vergleichende Untersuchungen ueber Keimung, Entfaltung und Fruchtbildung hoeherer Cryptogamen* (1851). From the conclusions. English translation by F. Currey in the Ray Society in 1862, pp. 434-439.

The comparison of the development of the mosses and liverworts on the one hand, with that of the ferns, Equisetaceae,

Rhizocarpeae and Lycopodiaceae on the other, discloses the most complete uniformity between the fruit formation on the one hand and the embryo-formation on the other. The structure of the archegonium of the mosses—the organ within which the fruit-rudiment is formed—is exactly similar to that of the archegonium of the vascular cryptogams, the latter being that part of the pro-thallium in the interior of which the embryo of the frond-bearing plant originates. In both the large groups of the higher cryptogams there is a cell which originates freely in the large central cell of the archegonium, by the repeated division of which (free) cell, the fruit of the moss and the frond-bearing plant of the fern are produced. In both, the divisions of this cell are suppressed and the archegonium miscarries, unless, at the time of the opening of the top of the latter, spermatozoa find their way to it.

Mosses and ferns, therefore, exhibit remarkable instances of a regular alternation of two generations very different in their organisation. The first generation—that from the spore—is destined to produce the different sexual organs, by the co-operation of which the multiplication of the primary mother-cell of the second generation, which exists in the central cell of the female organ, is brought about. By this multiplication a cellular body is produced which in the mosses forms the rudiment of the fruit, and in the vascular cryptogams, the embryo. The object of the second generation is to form numerous free reproductive cells—the spores—by the germination of which the first generation is reproduced. The leafy plant in the mosses answers therefore to the prothallium of the vascular cryptogams; the fruit in the mosses answers to the fern in the common sense of the word, with its fronds and sporangia. The pro-embryo, that is to say the confervoid process produced by the germinating spore of most of the mosses and many of the liverworts, cannot be looked upon as a special generation any more than the similar organ (the suspensor) in phanerogams. It is to be remembered that when new individuals are produced from single cells of the leaf of a moss, and also during the development of the gemmae of many mosses, the formation of the rudiment of the first leafy axis is preceded by the formation of a similar confervoid pro-embryo. This holds good as well in the mosses as those liverworts which possess a pro-embryo . . .

The vegetative life of the mosses is confined exclusively to the first, and the fructification to the second generation. The leafy stem alone sends forth roots; the spore-forming generation draws its nourishment from the first generation. The life of the fruit is usually much shorter than that of the leaf-bearing plant. In the vascular cryptogams this state of circumstances is reversed. It is true that the prothallia send out capillary roots; this is always the case in the Polypodiaceae and Equisetaceae, and frequently in the Rhizocarpeae and Selignaellae. But the prothallium lives a much shorter time than the leaf-bearing plant, which latter in most cases does not produce fruit for several years. The contrast, however, is not so marked as it appears at first sight. The apparently unlimited life of the leaf-bearing moss depends merely upon continual renovation . . .

In more than one respect the formation of the embryo of the Coniferae is intermediate between the higher cryptogams and the phanerogams . . . Two of the phenomena which led me to compare the embryo sac of the Coniferae with the large spores of the higher cryptogams, is common to the embryo sac of phanerogams, viz., the origin of the ovule from an axis cell, and the want of connexion with the adjoining cellular tissue . . . The Coniferae are closely allied to the phanerogams in the fact that their pollen-grains develop tubes. The phanerogams, therefore, form the upper terminal link of a series, the member of which are the Coniferae and Cycadeae, the vascular cryptogams, the Muscineae, and the Characeae. These members exhibit a continually more extensive and more independent vegetative existence in proportion to the gradually descending rank of the generation preceding impregnation, which generation is developed from reproductive cells cast off from the organism itself . . .

110

CARL WILHELM VON NAEGELI (1817-1891), MUNICH.

Of the physiological experiments of Naegeli we render here some concerning endosmosis and exosmosis in plants. (From *Ueber Endosmose und Exosmose in Pflanzenzelle.* Zürich, 1855. p. 21-35).

In osmotic studies, of course, only the increase or decrease of the cellular liquid and the acceptance or non-acceptance of microscopically visible substances can be observed, as it will probably for ever be impossible to think about a quantitative analysis . . . I chose sugar-solutions of 2.5, 5, 10, 15, 20 and 25%.

If an isolated cell is brought into a sugar solution we see complicated processes with a very simple explanation. First begins a mutual streaming of attracted particles between the sugar solution and the liquid which penetrates the cell membrane. Thereby the osmotic equilibrium between the membrane and the primordial utricle is disturbed, followed by an exchange between both . . . We can assume that membrane and utricle imbibe already at the first contact approximately as much sugar solution as they lose water, and that afterwards they only

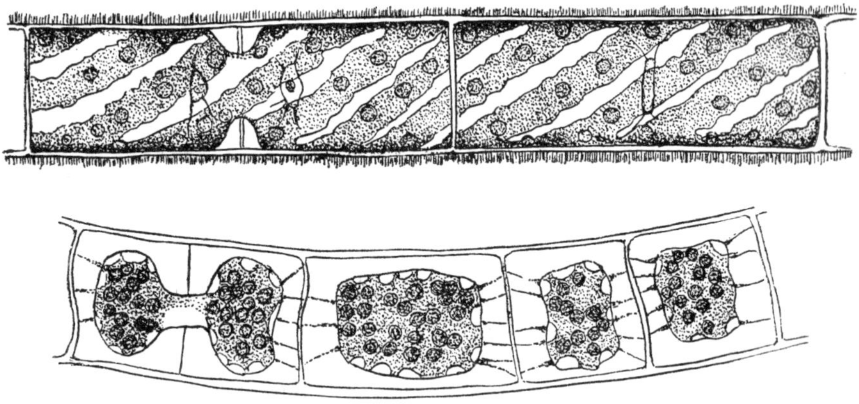

Fig. 36. Some cells of *Nitella* in exosmosis and endosmosis. From von Naegeli 1855, fig. III, 1 and 5.

effect the exchange of the cellular liquid and the environing thicker solution, without undergoing themselves any notable (further) changes. When a plant cell is put into a sugar solution, exosmosis is stronger than endosmosis and induces a corresponding diminution of the cellular liquid. This exerts a negative pressure upon the primordial utricle and thereby upon the membrane, and causes both to contract . . . Yet the latter is stiff, little elastic and can change its surface only to a very limited extent, at most in a ratio of 3 : 2 (in pollen) . . . Yet the diminution of the cellular liquid and therewith that of the hydrostatic pressure in the interior continues. The membrane may diminish further its surface by the formation of small folds . . . Very often, however, the membrane is stiff enough to resist the greater outer pressure, yet just as much sugar solution must pass through the membrane as cellular liquid goes out. And the primordial utricle is under no condition able to resist this unequal pressure, hence it contracts, in separating itself from the membrane . . . , as long as the negative pressure is greater than the adhesion of the utricle to the membrane. As there exists some adhesion of the utricle to the membrane, it usually depends upon the concentration of the sugar solution, whether they separate or not, which is different in different plant cells . . .

The osmotic processes in plants are most complicated. Not only the chemical and physical qualities of both liquids (cell content and water or the cell content of two (neighbouring) cells) are of importance, but also the general factors, as temperature, etc., and in addition the chemical and physical properties of membrane and primordial utricle, such as their thickness and number of layers, their density and water content, their cohesion, elasticity and extensibility, their aggregate condition or the arrangement of their smallest particles and their content of foreign matter. Yet also nowhere in the plant can the exchange follow the osmotic force, resulting from the liquids and the separating vesicles, because the liquids are enclosed in inextensible vessels, but both osmotic streamings, occurring in approximately equal strength, effect first the changes of the pressure, which correspond to the attractive affinities of the liquids. For this reason the cellular liquid in water plants is always exposed to a greater pressure than the surrounding water: e.g., if one breaks a thread of an algua without damaging any cell, the terminal surfaces which were straight before, grow convex . . .

Since the discovery of end- and exosmosis one has often given an easy explanation of the metabolism and the sap streaming in the plant. It was assumed, that the differences of concentration in the lower and upper parts of the plants and the transpiration of the leaves draw upwards crude nutritive sap imbibed by the root. This is, of course, important, and it can be demonstrated artifically by bringing one end of one of the long tubular cells of a *Nitella* into water, the other one into a sugar solution. Under the microscope a disturbance of the rotation of the cell content can be observed and a stream from the end in water to that in the sugar solution, which takes with it all movable particles and carries them to the latter end. But the facts known show that transpiration and osmotic attraction together are unable to raise a liquid to the height which occurs in the plant and that both these forces are unable to explain the direction of the sap stream of the plant. We have to assume additional forces, which have their seat perhaps in the longitudinal, yet more probably in the transversal walls and which strengthen the osmosis from below to above. (Follows experimental analysis of concentration of sap in lower and upper parts of the plant, the effect of transpiration, the effect of transverse partitions in a tube, the effect of capillarity, etc., which are all insufficient).

111

ALFRED RUSSEL WALLACE (1823-1913), ENGLAND.

This famous collector of animals, mainly in the Malayan archipelagos and in Brazil, came independently from Darwin to the theory of natural selection as the mechanism of evolution. Both published the outlines of their theories together in two papers in the Proceedings of the Linnean Society in 1858. Wallace became also the most important founder of animal geography, well prepared by his many and long travels. We give here the principles on which zoological regions should be formed from his *The geographical distribution of animals.* (London, 1876, p. 53-55).

Principles on which Zoological Regions should be formed. It will be evident in the first place that nothing like a perfect zoological division of the earth is possible. The causes that have led to the present distribution of animal life are so varied, their action and reaction have been so complex, that anomalies and irregularities are sure to exist which will mar the symmetry of any rigid system. On two main points every system yet proposed, or that probably can be proposed, is open to objection; they are,—firstly, that the several regions are not of equal rank ; secondly, that they are not equally applicable to all classes of animals. As to the first objection, it will be found impossible to form any three or more regions, each of which differs from the rest in an equal degree or in the same manner. One will surpass all others in the possession of peculiar families; another will have many characteristic genera;

while a third will be mainly distinguished by negative characters. There will also be found many intermediate districts, which possess some of the characteristics of two well-marked regions, with a few special features of their own, or perhaps with none; and it will be a difficult question to decide in all cases which region should possess this doubtful territory, or whether it should be formed into a primary region itself. Again, two regions which have now well-marked points of difference, may be shown to have been much more alike at a comparatively recent geological epoch; and this, it may be said, proves their fundamental unity and that they ought to form but one primary region. To obviate some of these difficulties a binary or dichotomous division is sometimes proposed; that portion of the earth which differs most from the rest being cut off as a region equal in rank to all that remains, which is subjected again and again to the same process.

To decide these various points it seems advisable that convenience, intelligibility, and custom, should largely guide us. The first essential is a broadly marked and easily remembered set of regions; which correspond, as nearly as truth to nature will allow, with the distribution of the most important groups of animals. What these groups are we shall presently explain. In determining the number, extent, and boundaries of these regions, we must be guided by a variety of indications, since the application of fixed rules is impossible. They should evidently be of a moderate number, corresponding as far as practicable with the great natural divisions of the globe marked out by nature, and which have always been recognized by geographers. There should be some approximation to equality of size, since there is reason to believe that a tolerably extensive area has been an essential condition for the development of most animal forms; and it is found that, other things being equal, the numbers, variety and importance of the forms of animal and vegetable life, do bear some approximate relation to extent of area. Although the possession of peculiar families or genera is the main character of a primary zoological region, yet the negative character of the absence of certain families or genera is of equal importance, when this absence does not manifestly depend on unsuitability to the support of the group, and especially when there is now no physical barrier preventing their entrance. This will become evident when we consider that the importance of the possession of a group by one region depends on its absence from the adjoining regions; and if there is now no barrier to its entrance, we may be sure that there has once been one; and that the possession of the area by a distinct and well balanced set of organisms, which must have been slowly developed and adjusted, is the living barrier that now keeps out intruders.

112

MARCELIN PIERRE EUGENE BERTHELOT (1827-1907) PARIS.

It was Berthelot and not Woehler who, in his *Chimie Organique fondée sur la synthèse* (1860, vol. II, pp. 806-808), established that organic matter can be composed from inorganic compounds without any participation of any *vis vitalis.* His slogan was: "Chemistry is one," which means that no basic difference exists between inorganic and organic chemistry.

Thus the barrier is finally demolished, which for so many years has been erected between organic and inorganic chemistry. Until now, all efforts to synthesise organic matter in general from its elements, discovered by analysis and to reproduce artificially the infinite variety of its structure and transformations have remained fruitless. In order to understand the full difficulty of such a task, we need to remember that organic compounds are met with only in living beings, that they represent the combinations of a few elements only, according to fixed proportions for every substance, and show nevertheless an almost infinite variety in number and properties . . . To explain our impotence recourse was taken to a special vital force, alone able to build up the organic substances . . . It was said: "It is this mysterious force which determines exclusively the chemical phenomena observed in organisms; it acts by laws which differ essentially from those which rule the movement of a matter capable only of rest and motion . . .

Two things have been confused with one another in this assertion of our absolute impotence in the production of organic matter: the formation of chemical substances which compose the organism, and the formation of the organs themselves. The latter is no problem of chemistry . . . But what chemistry cannot do in the building (the organisms), it can do in making substances contained in the organism . . . The synthesis (of these substances) and the recording of changes in their weights within the organism, this constitutes a vast and beautiful field of study which can be cultivated only with the aid of synthetic chemistry.

This new general point of view is developed in the present book: it is devoted to the study of methods by which synthesis can be performed without the aid of forces particular to living nature. We have proven that chemical affinities, heat, light, and electricity are sufficient to determine the aggregations of the elements and to form organic compounds. We make use of these forces as we choose, according to known laws; in our hands these forces produce combinations infinite in number and variety. Thus, already today we produce many natural substances and have the legitimate hope that we will be able to produce all others. By this synthesis and by imitation of the mechanisms present in plants and animals, we can establish, contrary to earlier belief, that the chemical processes of life proceed only by the play of physical and mechanic forces. In both cases the molecular forces which act are the same, as their effects are the same. Organic chemistry will continue its march towards synthesis, until it has conquered all its domains and has defined its limits, as inorganic chemistry is already able to do today.

113

RUDOLPH LEUCKART (1823-1898), LEIPZIG.

Leuckart belongs to a generation which did much for the elucidation of the life-cycles of parasitic worms. We render here from *Die menschlichen Parasiten* (1863, vol. I, pp. 227-284) part of the studies of the cycle of *Taenia solium*, the common bladder-worm of the pig.

It belongs to the lasting merits of Küchenmeister (1852) to have recognized the relations between *Taenia solium* and the common bladder-worm of the pig, *Cysticercus cellulosae*, and thereby to have opened up a new orientation in the life history of this parasite. He deduced the identity of both forms from the full agreement in the morphology of the head and of the hooks in both forms . . .

The first experiment was made by van Beneden (1854) who fed a mature tape-worm of this species to a piglet and found it four and a half months later full of bladder-worms. Yet only the experiments of Haubner (1855), who fed proglottids or greater fractions of this tape-worm to five piglets from a farm which was free from any tape- and bladder-worms in its pigsty. Three of these piglets showed afterwards many bladder-worms of this species, the two others, none.

The (final) experimental proof, namely the transformation of pig meat with *Cysticercus cellulosae* within the human being into *Taenia solium* was first attempted by Küchenmeister (1855) who infested a criminal three days before his execution by 75 bladder-worms. Two days later he found ten young tape-worms, 3 to 4 mm. long, which by the shape of their abdomen showed that they had just hatched from the bladder-worm . . .

In continuation of these experiments, I (Leuckart) found a well educated young man of 30 years, of healthy constitution, whom I informed about our knowledge concerning the life-history of this tape-worm, who volunteered to undergo a suitable experiment. He swallowed, in my presence, four cysticercae in milk, after I had slit open their tail vesicles. Three and a half months later, this young man observed some proglottids in his excrements, such as he had never seen before. One month later I purged him by some dosages of Cousso from two tape-worms of this species, which were two meters long. . . .

Thus, the last doubt must disappear. We are fully justified to assume that the hook-bearing *Taenia solium* is the transformation of *Cysticercus cellulosae*. To doubt this in our days means to close the eyes before the truth. . . .

Jews, who do not consume the meat of pigs, are, of course, free of *Taenia solium*, while they are not less infested with *T. saginata* (of cattle) than the Christians among whom they live.

114

CLAUDE BERNARD (1813-1878), PARIS.

Bernard was one of the great physiological experimenters and thinkers of the 19th century. We bring here the following selections:

(*a*) The Action of Curare. From *Introduction à la Médecine Expérimentale.* 1865. English translation 1927. pp. 157-158.

(*b*) Production of Sugar (glycogenesis) in the animal. *Ibidem.* pp. 163-165.

(*c*) On the *milieu fixe interne.* From *Leçons sur les phénomènes de la vie commun aux animaux et aux végétaux.* Paris, 1879. English translation by T. S. Hall.

(*a*) *The Action of Curare.* In 1845, Monsieur Pelouze gave me a toxic substance, called curare, which had been brought to him from America. We then knew nothing about the physiological action of this substance. From old observations and from the interesting accounts of A. von Humboldt and of Roulin and Boussingault, we knew only that the preparation of this substance was complex and difficult, and that it very speedily kills an animal if introduced under the skin. But from the earlier observations I could get no idea of the mechanism of death by curare; to get such an idea I had to make fresh observations as to the organic disturbances to which this poison may lead. I therefore made experiments to see things about which I had absolutely no preconceived idea. First, I put curare under the skin of a frog: it died after a few minutes; I opened it at once, and in this physiological autopsy I studied in succession what had become of the known physiological properties of its various tissues. I say physiological autopsy purposely, because no others are really instructive. The disappearance of physiological properties is what explains death, and not anatomical changes. Indeed, in the present state of science, we see physiological properties disappear in any number of cases without being able to show, by our present means of observation, any corresponding anatomical change; such, for example, is the case with curare. Meantime, we shall find examples, on the contrary, in which physiological properties persist, in spite of the very marked anatomical changes with which the functions are by no means incompatible. Now in my frog poisoned with curare, the heart maintained its movements, the blood was apparently no more changed in physiological properties than the muscles, which kept their normal contractility. But while the nervous system had kept its normal anatomical appearance, the properties of the nerves had nevertheless completely disappeared. There were no movements, either voluntary or reflex, and when the motor nerves were stimulated directly, they no longer caused any contraction in the muscles. To learn whether there was anything accidental or mistaken in this first observation, I repeated it several times and verified it in various ways; for when we wish to reason experimentally, the first thing necessary is to be a good observer and to make quite certain that the starting point of our reasoning is not a mistake in observation. In mammals and in birds, I found the same phenomena as in frogs, and disappearance of the physiological properties of the motor nervous system became my constant fact. Starting from this well established fact, I could then carry analysis of the phenomena further and determine the mechanism of death from curare. I still proceeded by reasonings analogous to those quoted in the earlier example, and, from idea and experiment to experiment, I progressed to more and more definite facts. I finally reached this general proposition, that *Curare causes death by destroying all the motor nerves, without affecting the sensory nerves.*

(*b*) *Production of Sugar* (*Glycogenesis*) *in the animal.* In 1843, in one of my first pieces of work, I undertook to study what becomes of different alimentary substances in nutrition. I began with sugar, a definite substance that is easier than any other to recognize and follow in the body economy. With this in view, I injected solutions of cane sugar into the blood of animals, and I noted that even when injected in weak doses the sugar passed into the urine. I recognized later that, by changing or transforming sugar, the gastric juice made it capable of assimilation, i.e., of destruction in the blood.

Thereupon I wished to learn in what organ the nutritive sugar disappeared, and I conceived the hypothesis that sugar introduced into the blood through nutrition might be destroyed in the lungs or in the general capillaries. The theory, indeed which then prevailed and which naturally was my proper starting point, assumed that the sugar present in animals came exclusively from foods, and that it was destroyed in animal organisms by the phenomena of combustion, i.e. of respiration. Thus sugar had gained the name of *respiratory nutriment.* But I was immediately led to see that the theory about the origin of sugar in animals, which served as a starting point, was false. As a result of the experiments which I shall describe further on, I was not indeed led to find an organ for destroying sugar, but, on the contrary, I discovered an organ for making it, and I found that all animal blood contains sugar even when they do not eat it . . . I therefore abandoned my hypothesis on the spot, so as to pursue the unexpected result which has since become the fertile origin of a new path for investigation and a mine of discoveries that is not yet exhausted.

In these researches I followed the principles of the experimental method that we have established, i.e. that, in presence of a well-noted, new fact which contradicts a theory, instead of keeping the theory and abandoning the fact, I should keep and study the fact, and I hastened to give up the theory, thus conforming to our precept given before . . . According to the established theory, supported by the most illustrious chemists of our day, animals were incapable of producing sugars in their organism. If I had believed in this theory absolutely, I should have to conclude that my experiment was vitiated by some inaccuracy; and less wary experimenters than I might have condemned it at once, and might not have tarried longer at an observation which could be theoretically suspected of including sources of error, since it showed sugar in the blood of animals on a diet that lacked starchy or sugary materials. But instead of being concerned about the theory, I concerned myself only with the fact whose reality I was trying to establish. By new experiments . . . I was led to find that the liver is the organ in which animal sugar is formed in certain given circumstances, to spread later into the whole blood supply and into the tissues and fluids.

Animal glycogenesis is now an acquired fact for science . . . but we have not yet fixed on a plausible theory accounting for the phenomenon . . .

To sum up, theories are only hypotheses, verified by more or less numerous facts. Those verified by the most facts are the best, but even then they are never final . . . It is true that an hypothesis based on a theory produced the experiment; but as soon as the results of the experiment appeared, theory and hypothesis had to disappear, for the experimental facts were now just an observation, to be made without any preconceived ideas.

In sciences as complex and as little developed as physiology, the great principle is therefore to give little heed to hypotheses or theories and always to keep an eye alert to observe everything that appears in every aspect of an experiment. An apparently accidental and inexplicable circumstance may occasion the discovery of an important new fact, as we shall see in the continuation of the example just noted . . .

The chemists usually see in secretion a direct doubling of the blood, a separation of immediate elements from that liquid, like the relation between the blood which enters the gland and that which leaves it, increased by the product of the secretion . . . With such ideas on secretion I began my research on the sugar secretion of the liver. But soon I had to recognize that they were not in agreement with the facts.

I found that the liver of an animal contains sugar. But another analysis, 24 hours later, shows an increased sugar content. We also know that if the spinal cord of a rabbit is cut at a certain height, the sugar disappears from its liver, but a substance enters the liver which can be transformed into sugar. These facts show clearly that the sugar which is produced in the liver is there not secreted directly by the blood which passes through the organ. Also the other experiences which I reported before forced us to abandon the theory of secretion as a doubling of blood. . . .

Many varied experiments brought me to the conclusion, that the sugar formation in the liver, like all other secretions, are not a direct phenomenon. The production of sugar is the result of a series of organic transformations and the secretion must give necessarily something else than sugar, i.e. a substance which changes into sugar. Secretion is thus a double process: a vital and a chemical one. The vital phenomenon stops its activity with death, the chemical one continues its work after death. Life produces in the liver a matter which is changed by a transformation into sugar. This transformation may continue after the animal's death in a liver from which by washing the blood is removed, whilst the glycogenous substance remains.

(*c*) Constant, or free, life is in addition to the oscillating life of the cold-blooded animals and the latent life of spores, the third form of existence and pertains only to the animals with the highest organization. In them life is never suspended. It pursues a constant course, apparently indifferent to changes of the cosmic environment and of those in material environmental conditions. Organs, mechanisms, tissue function in a stable manner, without those considerable variations of the animals with variable (oscillating) life. This comes about because the internal environment which surrounds organs, tissues, and tissue elements does not change; atmospheric variations are checked (buffered) by it, so that the physical conditions of the environment are, in the higher animal, constant. It is enveloped in an internal environment which acts for it as an atmosphere of its own in the midst of an ever changing outer cosmic environment. The higher organism has, in effect, been placed in a hot house. Here it is beyond the reach of the perpetual changes of the cosmic environment. It is not bound up in them; it is free and independent.

I believe I was the first to insist upon this idea that there are for the animal really two environments: an external environment in which the organism is situated, and an internal environment in which the tissue elements live. . . . The invariability of the internal environment is the essential condition of free independent life: the mechanism which permits this constancy is precisely that which insures the maintenance in the internal environment of all conditions necessary to the life of the elements . . . Far from being indifferent to the external world, the higher animal is, on the contrary, narrowly and wisely attuned to it in such a way that, from the continual and delicate compensation, established as if by the most sensitive balance, equilibrium results . . . These conditions , the same as needed for simple organisms, are needed for the life of the elements, which must be mobilized and kept constant in the internal environment already known to us: water, oxygen, heat, chemical substances, or reserves. The nervous system is called upon to preserve in them harmony.

Conclusion. We have successively examined the three general forms in which life appears: latent life, oscillatory life, constant life, in order to see whether in any of them we would find an interior vital principle capable of causing manifestations of life, independent of the exterior physico-chemical conditions. In conclusion, we see that in latent life the being is dominated by exterior physico-chemical conditions to such a point that every vital manifestation can be stopped. In oscillating life, although the living creature is not absolutely submitted to these conditions, yet it remains so bound up in them that it undergoes all their variations. In constant life, the animal appears to be free, and the vital manifestations seem to be effective and controlled by an internal vital principle, entirely free from the influence of external physico-chemical conditions. This appearance is an illusion. Quite to the contrary, it is exactly in the mechanism of constant or free life that these narrow relationships are particularly evident.

We cannot, therefore, admit in living organisms a free vital principle struggling against the influence of physical conditions. The opposite has been proved, and thus all of the contrary conceptions of the vitalists are seen to be overthrown.

115

JOHAN GREGOR MENDEL (1822-1884), BRNO.

Mendel, a monk belonging to the Augustinian order, studied for many years quantitatively the heredity of two alternative characters in peas. His conclusions, now regarded as basic in modern genetics as the Mendelian rules, were neglected, until, in 1900, three students rediscovered them independently one from the other: De Vries in Amsterdam, Correns in Berlin, and Czermak in Prague. The *Versuche ueber Pflanzenhybriden* (Verhandl. Naturforsch. Ver. Bruenn. 1868. pp. 3-47) were translated into English as an appendix by Bateson in his *Heredity* (1925. pp. 313-353).

P. 313: Experience of artificial fertilisation, such as is effected with ornamental plants in order to obtain new variations in colour, has led to the experiments which will here be discussed. The striking regularity with which the same hybrid forms always reappeared whenever fertilisation took place between the same species induced further experiments to be undertaken, the object of which was to follow up the development of the hybrids in their progeny. . . .

P. 314: The experimental plants must necessarily—

1. Possess constant differentiating characters.
2. The hybrids of such plants must during the flowering period, be protected from the influence of all foreign pollen, or be easily capable of such protection. . . .

P. 317: The characters (of peas) which were selected for experiments relate to the difference:

1. In the form of the ripe seeds (*roundish* or angular);
2. In the colour of the seed albumen (*yellow* or green);
3. In the colour of the seed coat (*grey* or white);
4. In the form of the ripe pods (*simply inflated* or deeply constricted;
5. In the colour of the unripe pods (*green* or yellow)

6. In the position of the flowers (*axial* or terminal);
7. In the length of the stem (not of the floral axis! *long* or short).

P. 319: *The forms of the hybrids.* Experiments which in previous years were made with ornamental plants have already afforded evidence that the hybrids, as a rule, are not exactly intermediate between the parent species. With some of the more striking characters, those for instance which relate to the form and size of the leaves, the pubescence of the several parts, etc., the intermediate, indeed, is nearly always to be seen; in other cases, however, one of the two parental characters is so predominant that it is difficult, or quite impossible, to detect the other in the hybrid . . . Henceforth those characters which are transmitted entire in the hybridisation . . . are termed the *dominant*, and those which become latent in the process *recessive*.

P. 321: *The generation from the hybrids* (F2). In this generation there reappear, together with the dominant characters, also the recessive ones with their peculiarities fully developed, and this occurs in the definitely expressed average proportion of 3 : 1, so that among four plants of this generation three display the dominant character and one the recessive. This relates, without exception to all the characters which were investigated in the experiments . . . Transitional forms were not observed in any experiment.

P. 324: *The second generation (bred) from the hybrids* (F3). Those forms which in the first generation (F2) exhibit the recessive character do not further vary in the second generation (F3) as regards this character; they remain constant in their offspring.

It is otherwise with those which possess the dominant character in the first generation (bred from the hybrids). Of these *two*-thirds yield offspring which display the dominant and recessive characters in the proportion of 3 : 1, and thereby show exactly the same ratio as the hybrid forms, while only *one*-third remains with the dominant character constant.

P. 325: It is now clear that the hybrids from seeds having . . . one or other of the two differentiating characters, and of these one half develop again in hybrid form, while the other half yield plants which remain constant and receive the dominant or recessive characters (respectively) in equal numbers.

P. 325: *The subsequent generations (bred) from the hybrids.* The proportion in which the descendants of the hybrids develop and split up in the first and second generations presumably hold good for all subsequent progeny . . . The offspring of the hybrids separated in each generation in the ration of 2 : 1 : 1 into hybrids and constant forms. . . .

P. 326: *The offspring of hybrids in which several differentiating characters are associated.* In the experiments above described plants were used which differed only in one essential character. The next task consisted in ascertaining whether the law of development discovered in these applied to each pair of differentiating characters when several diverse characters are united in the hybrid by crossing. As regards the form of the hybrids in these cases, the experiments showed throughout that this invariably more nearly approaches to that one of the two parental plants which possesses the greater number of dominant characters. If, for instance, the seed plant has a short stem, terminal white flowers, and simply inflated pods; the pollen plant, on the other hand, a long stem, violet-red flowers distributed along the stem, and constricted pods; the hybrid resembles the seed parent only in the form of the pod; in the other characters it agrees with the pollen parent. Should one of the two parental types possess only dominant characters, then the hybrid is scarcely or not at all distinguishable from it. . . .

P. 331: There is no doubt that for the whole of the characters involved in the experiments, the principle applies that the offspring of the hybrids in which several essentially different characters are combined, exhibit the terms of a series of combinations, in which the developmental series for each pair of differentiating characters are united. It is demonstrated at the same time that the selection of each pair of different characters in hybrid union is independent of the other differences in the two original parental stocks . . . Repeated crossing gives the practical proof that the constant characters which appear in the several variations of a group of plants may be obtained in all the combinations which are possible according to the (mathematical) laws of combination by means of repeated artificial fertilisation. . . .

P. 337: Experimentally, therefore, the theory is confirmed that the pea hybrids from egg and pollen cells which, in their constitution represent in equal numbers all constant forms which result from the combination of the characters united in fertilisation . . .

P. 344: It is more than probable that as regards the variability of cultivated plants there exists a factor which so far has received little attention. Various experiments force us to the conclusion that cultivated plants, with few exceptions, are members of various hybrid series, whose further development in conformity with law is varied and interrupted by frequent crossings *inter se*. The circumstance must not be overlooked that cultivated plants are mostly grown in great numbers and close together, affording the most favourable conditions for reciprocal fertilisation between the varieties present and the species itself. The probability of this is supported by the fact that among the great array of variable forms solitary examples are always found, which in one character or another remain constant, if only foreign influence be carefully excluded. . . . Whoever studies the colouration on which results, in ornamental plants, from similar fertilization, can hardly escape the conviction that here also the development follows a definite law, which possibly finds its expression in the combination of several independent colour characters.

116

EDUARD STRASSBURGER (1844-1912), BONN.

Strassburger was the first botanist to discover mitotic cell-division and its sexual partner meiosis. From: *Ueber Befruchtung und Zellteilung*. Jenaische Zeitschr. Naturwiss. 11 (N.S. 4), 1877. pp. 435-550, pl. 27-36.

P. 517: The nuclear spindle (pl. 33, fig. 48) of *Nothoscordum fragrana* is built from thick fibres which converge towards the poles and meet in the equatorial plain. The entire nuclear substance has been used in their formation. These spindles are at the opposite extreme from those cases where the nuclear fibres are almost invisible, but where the nuclear plate is strongly developed. When the nuclear spindle of *Nothocordum* is dividing, both halves separate in the equatorial plain. Now the fibres or rodlets of that side to which both nuclear halves are turning, diverge somewhat in the shape of a fan; at the same time the two poles are flattening and the rodlets begin to fuse there (f. 49). Later on the free ends of the rodlets curve inwards (f. 50) and begin to fuse also. The two new nuclei have now acquired a completely closed, smooth contour (f. 51). Between these, a cell plate (f. 52, 53) appears in the usual way of plant cells. Earlier (f. 50) or later (f. 53) the nuclei become homogenous, but may still retain signs of their composition.

I also met with in *Nothocordum* a distinct nuclear condition which precedes its differentiation into a spindle. In the nuclear substance of a nucleus of the embryo-sac (f. 46) a differentiation had started, so that denser nuclei of irregular contour filled the

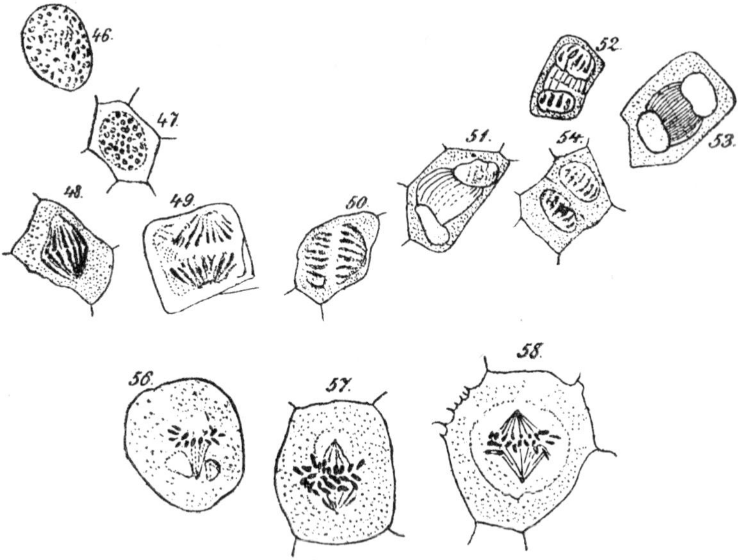

Fig. 37. Mitotic cell division in *Nothocordum*. From Strassburger (*Jena Zeitschr. Naturwiss.*, Vol. 11, 1877, pl. 3, fig. 46-53). Mitotic cell divisions in frog endothel (after W. Maysel, *ibidem*, fig. 56-58).

sac fairly equally spaced. Also the nucleolus showed the beginning of a similar differentiation; it seems to indicate the separation of the dense nuclear substance from the watery nuclear sap. In another nucleus (f. 47) the nucleolus had already disappeared. . . . The comparison of many figures in dividing animal cells shows considerable agreement with these observed on *Nothocordum*.

117

CHARLES WYVILLE THOMSON (1830-1882), EDINBURGH.

Thomson was much interested in the geography and fauna of *The Depths of the Sea* (1873). He participated in the famous 'Challenger' expedition (1873-1876). From his preliminary account on the results of this expedition in *The Atlantic* (London, 1877. Vol. II, pp. 352-354) we have selected the concluding remarks on the fauna of the depth, the abyssal.

The first general survey of the deep-sea collections, undertaken with a knowledge of the circumstances under which the specimens were procured, justify us, I believe, in arriving at the following conclusions:

(1) Animal life is present on the bottom of the ocean at all depths.

(2) Animal life is not nearly so abundant at extreme, as it is at moderate depths; but, as well developed members of all the marine invertebrate classes occur at all depths, this appears to depend more upon certain causes affecting the composition of the bottom deposits, and of the bottom water involving the supply of oxygen, and of carbonate of lime, phosphate of lime, and other materials necessary for their development, than upon any of the conditions immediately connected with depth.

(3) There is every reason to believe that the fauna of deep water is confined principally to two belts, one at and near the surface, and the other on and near the bottom; leaving an intermediate zone in which the larger animal forms, vertebrate and invertebrate, are nearly or entirely absent.

(4) Although all the principal marine invertebrate groups are represented in the abyssal fauna, the relative proportion in which they occur is peculiar. Thus Mollusca in all their classes, Brachyourous Crustacea, and Annelida, are on the whole scarce; while Echinodermata and Porifera greatly preponderate.

(5) Depths beyond 500 fathoms are inhabited throughout the world by a fauna which presents generally the same features throughout; deep-sea genera have usually a cosmopolitan extension, while species are either universally distributed, or, if they differ in remote localities, they are markedly

representatives, that is to say, they bear to one another a close genetic relation.

(6) The abyssal fauna is certainly more nearly related than the fauna of shallow water to the faunae of the tertiary and secondary periods, although this relation is not so close as we were at first inclined to expect, and only a comparatively small number of types supposed to have become extinct, have yet been discovered.

(7) The most characteristic abyssal forms, and those which are most nearly related to extinct types, seem to occur in greatest abundance and of largest size in the southern ocean; and the general character of the faunae of the Atlantic and of the Pacific gives the impression that the migration of species has taken place in a northerly direction, that is to say, in a direction corresponding with the movement of the cold under-current.

(8) The general character of the abyssal fauna resembles most that of the shallower water of high northern and southern latitudes, no doubt because the conditions of temperature, on which the distribution of animals mainly depends, are nearly similar.

118

PATRICK MANSON (1844-1922), LONDON.

This famous specialist of tropical diseases discovered, amongst others, that some *Filaria* worms which cause disease in human blood, are transmitted to man by mosquitoes. From: *On the development of* Filaria sanguinis hominis, *and on the Mosquito considered as a nurse.* J. Linn. Soc. London, Zool. 14. 1878. p. 304 ff.

I have calculated that in the blood of certain dogs and men there exist at any given moment more than two millions of embryos. Now the individuals of such a swarm could never attain anything approaching the size of the mature worm without certainly involving the death of the host. The death of the host would imply the death of the parasite before a second generation of *Filariae* could be born, and this, of course, entails the extermination of the species; for in such an arrangement reproduction would be equivalent to the death of both parent and offspring, an anomaly impossible in nature.

PLATE XVIII

LEVANTINE PHYSICIANS ARE SEEN TO EXTRACT CAREFULLY THE GUINEA-WORM FROM THE LEGS OF INFESTED PEOPLE. FROM G. VELSCHIUS, *Exercitationes de Vena Medinensi*, AUGSBURG, 1674.

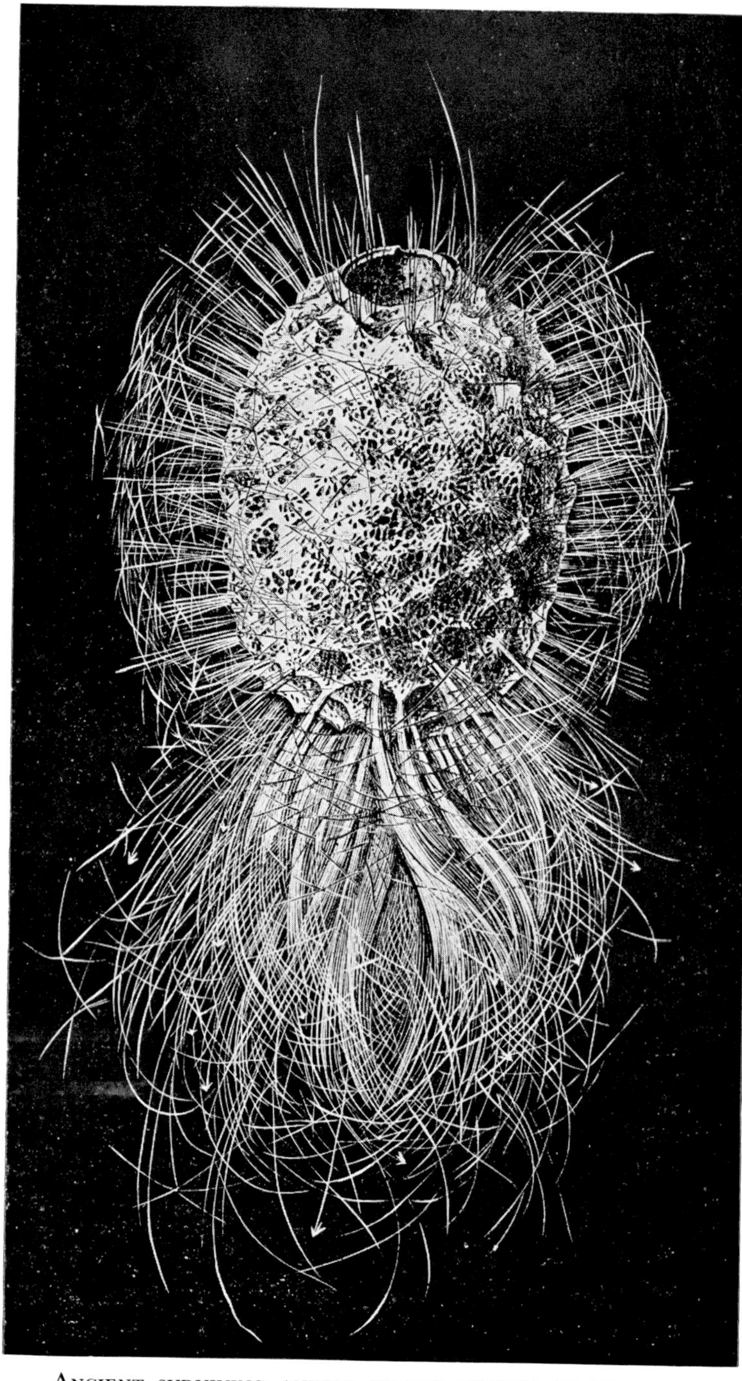

ANCIENT SURVIVING ANIMAL IN THE DEPTHS OF THE SEA, THE *Rossella velata*. FROM C. W. THOMSON, *The Depths of the Sea*). 1874.

The Embryo must escape from the original Host. The embryo, in order to continue its development and keep its species from extermination, must escape from the first host in some way. After accomplishing this it either lives an independent existence for a time, during which it is provided with organs for growth not possessed by it hitherto (as in most threadworms); or it is swallowed by another animal which treats it as a nursling for such time as is necessary to fit it with an alimentary system (some tapeworms).

Where embryo *Filariae* are not in great abundance in the blood, we may infer that there are only one or two parent worms; they often disappear completely for a time, to reappear after the lapse of a few days or weeks. From this circumstance I infer: first, that reproduction is of an intermitting and not of a continuous character; and secondly, that the embryos, after a certain time, are either disintegrated in the blood or voided in the excretions. The latter does occur in the urine and in the tears. In this way they may have an opportunity of continuing development either free (as in many threadworms) in the media into which the excretions are voided, or in the body of another animal which has intentionally or accidentally fed on these (in many tapeworms). Man, in his turn, may then swallow this hypothetic animal containing the embryo suitably perfected, and so complete the circle. This is the history of the Entozoa; but I have evidence to adduce that, if it be one way in which *F. sanguis hominis* is nursed, it is not the only way, and therefore probably not the way at all.

The Mosquito found to be the Nurse. It occurred to me that, as the first step in the history of the haematozoon was in the blood, the next might happen in an animal who fed on that fluid. To test this idea I procured mosquitos that had fed on patient Hinlo's blood, and, examining the expressed contents of their abdomens from day to day with the microscope, I found that my idea was correct, and that the haematozoon which entered the mosquito as a simple structureless animal, left it, after passing through a series of highly interesting metamorphoses, much increased in size, possessing an alimentary canal, and being otherwise suited for an independent existence.

History of the Mosquito after feeding on Human Blood. I have never, in many hundreds of specimens, met with a male insect charged with blood . . . After a mosquito has filled itself with blood (in about two minutes), it is evidently much embarrassed by the weight of its distended abdomen, so that it no longer can wheel about in the air. It accordingly attaches itself to some surface, if possible near stagnant water, where it remains in a comparatively torpid condition, digesting the blood, excreting yellow gamboge-looking faeces, and maturing its ova. In from 3 to 5 days these processes are ended, and the insect now betakes itself to the water, where the eggs are deposited, and on the surface of which they float. The eggs soon hatch, and the embryo emerges and escapes into the water. In the mosquito's abdomen can be distinguished two ovisacs, intestine and oesophagus and the stomach. If the blood contained in the dilated stomach is examined soon after ingestion, the blood corpuscles are seen quite distinct in outline, and behaving very much as when drawn in the ordinary way; but changes rapidly occur. First, the corpuscles lose their distinctness in outline, crystals of haematin appear, and before the eggs are deposited all colouring matter disappears; by the time the eggs are deposited the stomach is quite empty but for the embryo Filariae it may contain.

How to procure Mosquitos containing embryo Filariae. I persuaded a Chinaman, in whose blood I had already ascertained that Filariae abounded, to sleep in a mosquito-house, in a room where mosquitos were plentiful. After he had gone to bed a light was placed beside him, and the door of the mosquito-house kept open for half an hour. In this way many mosquitos entered the 'house'; the light was then put out, and the door closed. Next morning the walls of the ' house ' were covered with an abundant supply of insects with abdomens thoroughly distended . . . The abdomen when torn off, is placed on a glass slide, and a small cylinder rolled over it from the anus towards the severed thoracic attachment. In this way the contents are safely and efficiently expressed. If the contents are white and dry a little water should be added.

Large proportion of Filariae ingested by the Mosquito. The blood in the stomach of a mosquito that has fed on a Filaria-infested man usually contains a much larger proportion of Filariae than does an equal quantity of blood obtained from the same man by pricking the finger . . . From this it would appear that the mosquito has the faculty of selecting the embryo Filariae; and in this strange circumstance we have an additional reason for concluding that this insect is the natural nurse of the parasite.

All Embryos do not attain Maturity. By far the greater number die and are disintegrated, or are expelled in the faeces undeveloped. After a few days, when the stomach is quite empty from food and an embryo could not easily be overlooked, only 2 to 6 are found in the same stage of development.

The Metamorphosis of the Embryo. (The further morphological development of the Filaria in the mosquito is described in detail). These papillae are the boring-apparatus to be used in penetrating the tissues of man and escaping from the mosquito. At this (final stage in the mosquito) it becomes endowed with marvellous power and activity. It rushes about the (microscopic) field, forcing obstacles aside, moving indifferently at either end, and appears quite at home in the water. This formidable-looking animal is undoubtedly the *Filaria sanguis hominis* equipped for independent life and ready to quit its nurse the mosquito.

Future History of the Filaria. There can be little doubt as to the subsequent history of the Filaria, or that, escaping into water in which the mosquito died, it is through the medium of this fluid brought into contact with the tissues of man, and that, either piercing the integuments, or, what is more probable, being swallowed, it works its way through the alimentary canal to its final resting place. Arrived there, its development is perfected, fecundation is effected, and finally the embryo Filariae we meet with in the blood are discharged in successive swarms and in countless numbers. In this way the genetic cycle is completed.

119

EDOUARD VAN BENEDEN (1845-1910), LIÈGE.

Belgian zoologist, who described the cytology of cell-division and of fecundation in the egg of *Ascaris megalocephala.* From: *Recherches sur la Maturation de l'oeuf, la Fécondation et la Division Cellulaire.* (1883. Gand, Paris, Leipzig, p. 403: Résumé).

I hope I have been able to further our knowledge of the phenomena connected with fertilization, in demonstrating:

(1) That not only the chromatic (part of the nucleus) of the zoosperm, but also the achromatic layer which surrounds it, participates in the formation of the male pronucleus.

(2) That the germinative vesicle gives the female pronucleus not only chromatic, elements but also an achromatic body.

(3) That the two pronuclei may, without fusing together, acquire in their progressive maturation the constitution of ordinary nuclei.

(4) That in *Ascaris* no unique nucleus is produced by the two pronuclei. The essential function of fecundation is hence not the fusion of two nuclei, but the formation of these elements in the female gonocyte. One of these nuclei is derived from the egg, the other from the zoosperm. The nuclear elements extended as polar bodies are replaced by the male pronucleus, and once the two half-nuclei, one male, the other female, are constituted, fecundation is accomplished.

(5) In consequence of a series of transformations of the nuclear structure of every pronucleus, transformations analogous to those of the normal cell division, every pronucleus gives rise to two chromatic loops.

(6) These four chromatic loops take part in the formation of the chromatic star, but they remain distinct. Each of them divides itself longitudinally in two secondary twin loops.

(7) The nuclei of the two first blastomeres receive each one half of every primary loop. i.e. four secondary ones, two of which are male, two female.

I do not consider fecundation to be a generation, but an exchange, not the origin of a new cell-individual. One may not say with O. Hertwig that fecundation is the fusion of two nuclei, one male, one female. This does not exist in *Ascaris* . . . Fecundation is the necessary condition for the continuance of life. All other parts of the individual are mortal. By fecundation the parents escape death.

120

LOUIS PASTEUR (1822-1895), PARIS.

Of this great reformer of modern experimental science we mention only the discovery of a remedy against the pebrine of silk-worms which menaced to destroy the flowering silk-industry of S. France; the theory of fermentation by one-cellular organisms; the sterility of boiled water, when no living particles can drop into it; the discovery of many bacteria as causes of infectious diseases; the control of a virus-disease, rabies, by immunisation, etc.

(*a*) On the control of rabies by inoculation with a weakened virus. From: *Méthode pour prévenir la rage après morsure.* C.R. Acad. Sci. Paris, 1885. Vol. 101, pp. 765. English translation by F. R. Moulton and J. J. Schifferes, The Autobiography of Science. New York, 1945. Pp. 431-433.

(*b*) Fermentation and putrefaction are brought about not by the air, but by the particles suspended therein. From: *Les corpuscules organiques de l'atmosphère.* Ann. Chemie et Physique (3) 64. 1862. pp. 66-68. English translation: F. S. H. Taylor, Science past and present. London, 1949. New edition. pp. 206/7.

(*c*) The Pebrine Disease of Silkworms. (Etude sur la Maladie des Vers à Soie. Paris, 1880).

(*d*) The Anthrax experiment. (See: R. Vallery-Radot, La Vie de Pasteur. Paris, 1900. p. 371 ff.). (*c*) and (*d*) from R. Gregory, Discovery. London, Macmillan, 1921. p. 274 f, p. 210 f.

(*a*) After making almost innumerable experiments, I have discovered a prophylactic method which is practical and prompt, and which has already afforded me sufficiently numerous results in dogs, certain and successful enough, to warrant my having confidence in its general applicability to all animals, and even to man himself. This method depends essentially on the following facts:

The inoculation of the infected spinal cord of a dog suffering from ordinary rabies under the *dura mater* of a rabbit always produces rabies after a period of incubation having a mean duration of about fifteen days. If, by the above method of inoculation, the virus of the first rabbit is passed into a second, and that of a second into a third, and so on, in series, a more and more striking tendency is soon manifested towards a diminution of the duration of the incubation period of rabies in the rabbits successively inoculated.

The virus of rabies at a constant degree of virulence is contained in the spinal cords of these rabbits throughout their whole length. If portions a few centimeters long are removed from these spinal cords, with every possible precaution to preserve their purity, and are then suspended in dry air, the virulence slowly disappears, until at last it entirely vanishes. The time within which this extinction of virulence is brought about varies a little with the thickness of the pieces of spinal cord, but more with the external temperature. The lower the temperature the longer is the virulence preserved. These results form the central scientific point in the method.

These facts being established, a dog may be rendered refractory to rabies in a relatively short time in the following way: Every day pieces of fresh infective spinal cord from a rabbit which has died of rabies, developed after an incubation period of seven days, are suspended in a series of flasks, the air in which is kept dry by placing fragments of potash at the bottom of the flask. Every day also a dog is inoculated under the skin with a Pravaz' syringe full of sterilized broth, in which a small fragment of one of the spinal cords has been broken up, commencing with a spinal cord far enough removed in order of time from the day of the operation to render it certain that the cord was not at all virulent (as ascertained by previous experiments). On the following days the same operation is performed with more recent cords, each operation being separated from the last by an interval of two days, until at last a very virulent cord, which has only been in the flask for two days, is used. The dog has now been rendered refractory to rabies. It may be inoculated with the virus of rabies under the skin, or even after trephining, on the surface of the brain, without any subsequent development of rabies.

Never once having failed when using this method, I had in my possession fifty dogs, of all ages and of every breed, refractory to rabies, when three persons from Alsace unexpectedly presented themselves at my laboratory, on Monday the sixth of last July: Th. Vone bitten in the arm on July 4 by his own dog which had gone mad; J. Meister, aged nine years, bitten the same morning by the same dog many times on hands, legs and thighs, some wounds so deep as to render walking difficult, and cauterised twelve hours later with phenic acid. The dog killed by its master was certainly rabid. Meister had been pulled out from under it covered with foam and blood. As Vone had only contusions on the arm, and his shirt was not pierced by the dog's fangs, I told him to return to Alsace without fear, but I retained J. Meister and his mother.

From the weekly meeting of the Académie des Sciences on July 6 I took Prof. Vulpian and Grancher to inspect his fourteen wounds. Both were of opinion that owing to the severity of the bites, Meister was almost certain to take rabies. I then informed them of my new results on rabies. As the death of this child appeared to be inevitable, I decided, not without lively and severe anxiety, to try upon it the method which I had found constantly successful with dogs . . .

Consequently, on July 6, at eight o'clock in the evening, sixty hours after the bites, young Meister was—in the presence of Vulpian and Grancher—inoculated under a fold of skin raised in the right hypochondrium, with half a Pravaz' syringe full of the spinal cord of a rabbit which had died of rabies on June 21. It had been preserved since then, that is to say fifteen days, in a flask of dry air. In the following days fresh inoculations were made. I thus made thirteen inoculations, and prolonged the treatment to ten days. I shall say later on that a smaller number of inoculations would have been sufficient. . . . On the last days I had Meister inoculated with the most virulent virus of rabies, that namely of the dog, reinforced by passing a great number of times from rabbit to rabbit, a virus which produces rabies after seven days' inoculation in these animals, after eight or ten days in dogs. . . .

Meister, therefore, has escaped not only the rabies, which would have been caused by the bites he received, but also the rabies with which I have inoculated him in order to test the immunity produced by the treatment, a rabies more virulent than ordinary canine rabies. The final inoculation with very virulent virus has this further advantage, that it puts an end to the apprehensions which arise as to the consequences of the bites. If rabies could occur it would declare itself more quickly after a more virulent virus than after the virus of the bites. Since the middle of August I have looked forward with confidence to the future good health of Meister. At present three months and three weeks have elapsed since the accident; his state of health leaves nothing to be desired.

(*b*) *Experiments on Spontaneous Generation.* I believe I have rigorously established in the preceding chapters that the organised production of infusions that have previously been heated have no other origin than the solid particles which the air always carries and constantly lets fall on all objects. If there can still remain

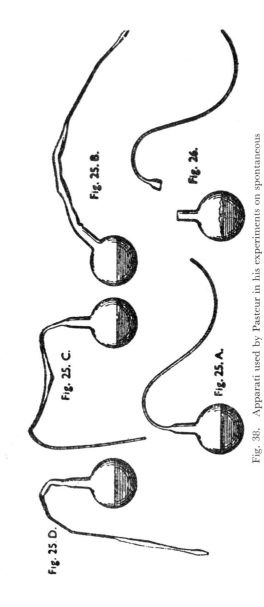

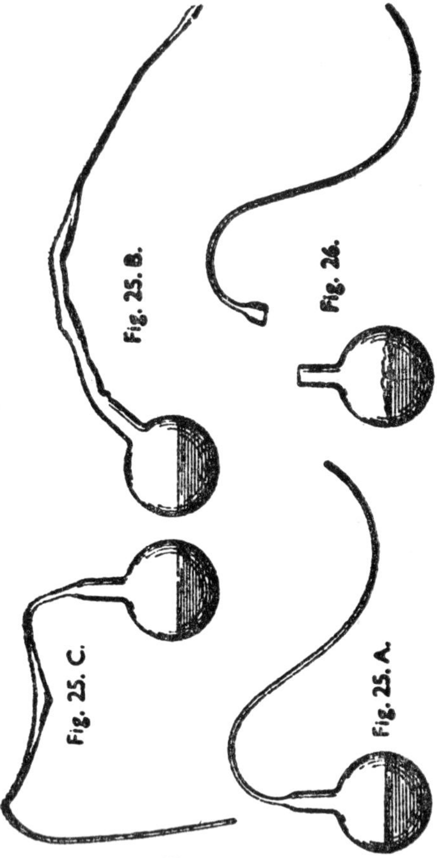

Fig. 38. Apparati used by Pasteur in his experiments on spontaneous generation. (After Taylor, *Science past and present*).

in the mind of the reader even the least doubt in this regard, it will be removed by the experiments of which I am about to speak.

I place in a globular glass flask one of the following liquids, all of which are very easily altered by contact with ordinary air, such as watery extract of yeast, sweetened watery extract of yeast, urine, beetroot-juice, watery extract of pepper; then I draw out, at the blowpipe, the neck of the flask so as to give it various curvatures as indicated (Pl. I, f. 25 A-D, Fig. 45 in Taylor). I then bring the liquid to the boil for several minutes until the steam issues freely from the open end of the drawn-out neck, using no other precaution. I then let the flask cool. It is a remarkable fact, calculated to astonish anybody who is used to the delicacy of experiments relative to the so-called spontaneous generations, that the liquid in the flask remains indefinitely without alteration. One can handle the flask without apprehension, move it from one place to another, allow it to undergo all the temperature-changes of the seasons, and the liquid in it shows not the slightest alteration and keeps its smell and taste: it is an excellent *conserve d'Appert* (a food preserve). There will be no change in its nature other than, in certain cases, a direct oxidation, purely chemical, of the matter. But we have seen by the analyses which I have published in this memoir, how this action of oxygen is limited, and that in no case are any organised productions developed in the liquids.

It appears that the ordinary air, returning with force at the first moment, must reach the flask quite unaltered. That is true, but it encounters a liquid that is still near its boiling-point. The re-entry of the air thereafter occurs more slowly, and by the time that the liquid is so far cooled as no longer to be able to deprive germs of their vitality, the re-entry of the air becomes so slow that it leaves behind in the damp curves of the neck, all the dust particles capable of acting on the infusions and bringing about the production of organisms. At least, I see no other possible explanation of these curious experiments. If, after one or many months' stay in the incubator, the neck of the flask be detached by a nick with a file, without touching the flask in any other way (fig. 26) after 24, 36, or 48 hours, moulds and infusoria begin to show themselves absolutely in the ordinary way or just as if one had sown dust-particles from the air in the flask. . . .

(*c*) *The Pebrine.* In 1865, Pasteur was urged by Dumas to undertake the investigation of a disease of silkworms, called

Pebrine, which had for several years been ruinously prevalent in the south of France. The loss was estimated at four millions sterling; and the disease had spread to many other countries from which silkworm's eggs had been brought to France. Pasteur had no intimate knowledge of silkworms, and he hesitated to take up the study of the causes of the epidemic. Writing to Dumas he said:

" Your proposition throws me into a great perplexity; it is indeed most flattering, and the object is a high one; but it troubles and embarrasses me! Remember, if you please, that I have never touched a silkworm. If I had some of your knowledge on the subject I should not hesitate; it may even come within the range of my present studies. However, the recollection of your many kindnesses to me would leave me bitter regrets if I were to decline your pressing invitation. Do as you like with me.

Pasteur."

Many silkworm cultivators expressed regret that the Government should choose a ' mere chemist ' for the investigation of the disease instead of a zoologist or silkworm cultivator. Pasteur only said, " have patience "; and he began his attack of the problem with the precision and acuteness of observation characteristic of him. He suspected that certain ' corpuscles ' found in the bodies of diseased silkworms and in the moths and their eggs were disease-producing organisms, and directed his chief studies to them, while making also a careful investigation of the whole disease. He proved that the disease was not only contagious, but hereditary—diseased moths laying diseased eggs from which came diseased silkworms which died young or were useless for the production of silk. Thus the disease passed on by inheritance (read: by transmission!) from year to year. The germs in eggs laid by diseased moths survived; but those left on leaves, or in the dust, or in the bodies of dead moths, soon perished. Only in the diseased and living eggs was the contagion maintained. Sir James Paget continues:

" These things were proved by repeated experiments, and by observations by Pasteur in his own breeding-chamber; and they made him believe that the disease might be put an end to by the destruction of all diseased eggs. To this end he invented the plan which has been universally adopted, and has restored a source of wealth to the silk districts. Each female moth, when ready to lay eggs, is placed on a separate piece of linen on which it may lay them all. After it has laid them, and has died, it is dried and then pounded in water, and the water is examined microscopically. If ' corpuscles ' are found in it, the whole of the eggs of this moth, and the linen on which they were laid, are burnt; if no ' corpuscles ' are found, the eggs are kept, to be, in due time, hatched, and they yield healthy silkworms."

(*d*) *The anthrax experiment.* Pasteur proved that the power of conveying anthrax was due to a certain bacillus, and to it alone; and, later, he suggested that, by vaccination of animals with the virus, it would be possible to give them the disease in a mild form, and to secure their immunity from fatal attacks. His conclusions met with the usual reception of ridicule from the practical men concerned with the care of cattle; and the veterinary profession proposed an experiment to test them. Pasteur accepted the challenge, and the conditions of the battle between knowledge and incredulity were drawn up. Sixty sheep were put at his disposal; twenty-five were vaccinated, by two inoculations, with the attenuated virus of anthrax. Some days later these, and twenty-five others, were inoculated with some very virulent cultures of the anthrax bacillus. Ten sheep underwent no treatment at all.

" The twenty-five unvaccinated sheep will all perish," wrote Pasteur; " the twenty-five vaccinated sheep will survive." The result turned out exactly as he had predicted. The sheep which had been vaccinated with a mild form of anthrax, in order to make them more capable of resisting the later inoculation, survived, while those which had not thus been rendered immune, perished. The conclusion of the experiment is one of the most dramatic incidents in the history of science. It was arranged that believers and unbelievers in Pasteur's views should meet on June 2, 1881, at the farmyard where the sheep had been placed, there to celebrate a victory or proclaim a failure.

R. Vallery-Radot describes this event in his *Life of Pasteur*, as follows:

" When Pasteur arrived at two o'clock in the afternoon, at the farmyard of Pouilly le Fort, accompanied by his young collaborators, a murmur of applause arose, which soon became loud acclamation, bursting from all lips. Delegates from the Agricultural Society of Melun, from medical societies, veterinary societies, from the Central Council of Hygiene of Seine et Marne, journalists; small farmers, who had been divided in their minds by laudatory or injurious newspaper articles—all were there. The carcasses of twenty-two unvaccinated sheep were lying side by side; two others were breathing their last; the last survivors of the sacrificed lot showed all the characteristic symptoms of splenic fever (anthrax). All the vaccinated sheep were in perfect health . . . The one remaining unvaccinated sheep died that same night."

121

PAUL EHRLICH (1854-1915), FRANKFURT A. M.

This founder of modern chemo-therapy was also the first to use vital dyes for the study of physiological problems. From: *Das Sauerstoft-Beduerfnis des Organismus. Eine farbenanalytische Studie.* Berlin, 1885. pp. 114-117.

The oxygen-zones of the protoplasm should, I think, be separated into three. The first comprehends all those places of highest oxygen affinity, which remain sated throughout the normal cell activity. When the cell is forced to exist under want of oxygen, they form a reserve of oxygen for the protoplasm. The second zone includes all those regions of oxygen which are in action during normal cell activity, sometimes oxydizing, sometimes reducing. The third zone includes those regions which remain without oxygen even during the normal cell activity and which, hence, have a continuous power of attraction for the blood-oxygen. It follows from these conclusions that the active protoplasm shows a Janus-face, oxydising certain compounds through its oxygen-sated zones, while reducing others with the aid of the unsated zones . . .

I have described in the experimental part how very quickly the organs dyed with indophenolblue turn white under the influence of oxygen-lack and of function; under these latter conditions the the oxygen satiation of the protoplasm decreases and its reductive power increases. . . .

(Our analysis shows) that the protoplasm itself is indeed the region within which the granules of indophenol are deposited during the experiment (i.e. following its injection). We are fully justified in assuming that the behaviour of the intracellular indophenol is a direct quantitative measure for the changing power of reduction of the protoplasm.

122

WILHELM ROUX (1850-1924), HALLE.

Founder of experimental embryology (*Entwicklungs-Mechanik*). From: *Ueber die kuenstliche Hervorbringung halber Embryonen durch Zerstoerung einer der beiden ersten Furchungszellen.* Virchow's Archiv. 114. 1888.

P. 518 ff. *Results:* After the destruction of one of the two first blastomeres the other one may develop in the normal way into an entirely normal half embryo. Such lateral or anterior half-embryos were obtained by us with the corresponding transition steps of semi-neurula, semi-blastula and semi-gastrula. Also three-quarter embryos with absence of one lateral head—half were obtained by piercing the egg after the second segmentation. This leads to the thesis; The development of the gastrula and of the embryo directly developed from it is, for the moment of the segmentation of the egg into four blastomeres, a mosaic process of at least four vertical, essentially independent parts.

From the teratology of the anentoblastic individual we learned that the ecto- and meso-derm may differentiate even in absence of the intestinal entoblast, their specific differentiations, even if the entire shape of the embryo becomes abnormal in consequence of that absence. Also, at *both* sides one half of the chorda dorsalis lateralis is formed.

That blastomere which was robbed by the segmentation of its developmental faculty can slowly be revived. This reorganisation occurs partly by immigration of a greater number of cell nuclei (with protoplasma?) from the normally developed egg-half, by spreading of the immigrant nuclei within the entire yolk-mass, as far as it is not already furnished with nuclei of its own, and by later multiplication of both these groups of nuclei. This nucleization of the operated blastomere is followed later by cellulisation, as around every nucleus the yolk forms a cell . . .

This reorganisation of the operated egg-half is later followed by a *post-generation* which may lead to complete restitution of the missing lateral or posteral half of the embryo. This post-generation is not made in the same way as the normal development of the primarily formed half, it is merely a normal, belated development. In post-generation the germ-layers in the newly formed half do not originate as in primary development in dependent primary formation of the germ layers, but is produced by that half of the already developed germ-layers present in the remaining half. And this only from regions where the primary germ-layers of the embryo were already separated in such a way that every germ-layer touches the undeveloped egg-half with a free margin, with an interrupted surface, as in an artificial wound. Therefore, in the segmentation of the lateral half-formations no real gastrulation occurs. The post-generative development of the newly formed cells is a progressive differentiation of the 'resting' cell material. The various differentiations needed for the development of a germ-layer spread with unequal speed into the still indifferent cell-material.

As the various yolk-materials and the nuclei of the operated egg-half have *no typical arrangement*, but their mutual position is determined by chance, we cannot assume that the *typical* spreading and the typical result of post-generation is determined by a typical arrangement of special substances, gifted with self-differentiation.

We conclude, that certain differentiative stimuli go out from the already differentiated material to the neighbouring still undifferentiated cell-material. Whilst our observations have shown that the primary or direct development of the first blastomeres is a 'self-differentiation' of these cells and their derivated (offspring) cells, the regenerating parts of the egg are only capable of 'dependent differentiation' . . .

123

JACQUES LOEB (1859-1924), CHICAGO AND BERKELEY

Leading biologist, much involved in the problems of artificial development (i.e. without fecundation), physiology of proteins. One of his earliest discoveries were the taxes of animals. From: *Der Heliotropismus der Thiere und seine Uebereinstimmung mit dem Heliotropismus der Pflanzen.* Wuerzburg, 1890. The résumé of this booklet runs as follows:

Résumé:

I. The dependence of animal locomotion upon light is in all respects the same as the dependence of the movements of plants caused by the same stimulation: The direction of the median plane, i.e., that of the locomotion of animals, coincides with the direction of the light ray.

The rays of greater refraction (staerker brechbaren) of the visible solar spectrum are stronger in their effect than those of lesser refraction, as is the case in the orientation of plants. The light acts, if at constant intensity, permanently as heliotropic stimulation on animals as well as on plants. The intensity of the light is important for the heliotropism of the animals because heliotropic locomotion occurs only above a certain light intensity; with increasing light intensity the orientation of animals in the direction of the rays grows more accurate. In winged insects (ants, butterflies, aphids) direct sunlight induces flight, whilst in the light of the sky the heliotropic reactions are usually confined to running. Yet positively heliotropic animals sometimes move to the source of light, when this leads them from places of higher light intensity to those of lower light intensity; and negatively heliotropic animals move in the opposite sense, even if thereby they come from places of lower to those of higher light intensity. The heliotropic movements are observed only within certain limits of temperature. Within this range there is one temperature at which the orientative movements of animals towards light is speediest and the most precise, as is the case with plants.

II. The orientation of animals towards a source of light depends also on the shape of the body of the animals, as the light-orientation of plants depends upon their shape. Symmetrical points of the body surface of animals in dorsoventral orientation have the same intensity of sensitivity. The heliotropic sensitivity of the oral pole in animals is different in intensity from that of the aboral pole, that of the oral pole being usually greater. The sensitivity of the ventral surface differs in intensity from that of the dorsal surface. These three circumstances together have the effect that the dorsoventral orientation bring their median plane into the directions of the rays and move from or towards the source of light. Eyeless animals, such as the maggots of *Musca vomitoria* (read: *Calliphora erythrocephala*) react in this respect just as animals with eyes.

III. The heliotropic sensitivity of an animal manifests itself often only in certain periods of its life, as in ants during the nuptial flight, in plant-lice, when they are winged. In the maggots of *Calliphora*, the negative heliotropism is most pronounced in the full-grown maggot, while the ventral surface is most energetically turned towards the source of light immediately after its hatching. In very many animals the heliotropism is in the larval and in the adult condition of opposite sign. Both moths and butterflies are positively heliotropic, behaving as other positively heliotropic animals. Yet the period of sleep of moths is during the day, and hence their positive heliotropism manifests itself usually only during the night.

IV. The intensity of heliotropic sensitivity of many animals coincides with their sexual activity. Apart from the nuptial flight of ants, the fact is pertinent that in ants and butterflies the males are more sensitive to light than the females.

V. The behaviour of an animal depends upon the sum of its various kinds of sensitivity. Thus, it may occur that the caterpillar of *Cossus cossus* and the *Cuma rathkii*, which live in darkness, are strongly positive heliotropic, without this sensitivity having any use for them, with many heliotropic plant roots.

VI. A certain type of thigmotaxis is very widespread among animals, and little attention has been paid to it so far, and it may easily be mistaken for a negative heliotropism. Many animals are forced to orientate their body in a certain way towards the surface of other solid bodies, or to bring their body as far as possible on all sides into contact with other solid bodies (stereotropismus). Certain animals occupy in hollow cubes only the concave corners and edges (earworm, ants, *Amphipyra*, maggots of *Calliphora*, *etc.*), whilst other animals occupy with equal constancy the convex edges and corners of bodies (e.g. the caterpillars of *Porthesia chrysorrhoea*).

VII. A dark source of heat has some influence on the animal's orientation, yet will generally not be able to dictate a rectilinear direction of its locomotion to or from the source. Thus it comes to pass that animals which move away from the source of heat, may be forced by the direction of the light rays to move from diffuse daylight into direct sunlight and to remain continuously exposed to the direct sunlight, in spite of the fact that in doing so they die. The effect of a dark heat-source may best be compared to that of a weak source of light, which still suffices to prevent, the locomotion of a negatively heliotropic animal towards the light, but is insufficient to force the animal to move straight in the direction of its rays.

A general conclusion is, that in animals which have nerves the movements of orientation concerning light are in all details the same in the same external conditions, and depend in the same way on their external body-shape as is the case in plants. Hence, these heliotropic phenomena cannot be a specific property of the central nervous system, as is still assumed in nerve-physiology, in calling the attraction of a moth to light as an instinctive effect or a reflex.

On negative heliotropism of fly maggots. P. 74/75. Some experiments: I placed fly-eggs on sooty glass, to await the hatching of the maggots. As these remove the soot along their route, they registered themselves the path they followed from the egg. The glass sheets were upon a horizontal plate in a room which received light from one side only. The traces lead almost without exception towards the (dark) room side of the plate. In a few exceptional cases the traces which first led towards the window, turned soon and also continued to the room side. Their direction is not following the dictates of a mystic ' hereditary instinct,' but only the direction of the light rays, just as it is gravity which determines the orientation of butterflies hatching from their pupal skin.

If diffuse light came from two windows which were at an angle of 90°, the traces of the maggots run approximately in the diagonal between the two planes In an absolutely dark room, the traces were arranged concentrically around the eggs, the animals showing an equal dispersal, but in contrast to the long path of the maggots retreating from light, their locomotion was of short duration, and they did not leave the glass sheet. Constant light-intensity acts, as in positively heliotropic animals, as a continuous stimulus, which forces the animal to steady movement towards or away from the source of light, until other sensory stimuli participate to modify or to counterbalance the stimulation by light.

124

ILJA (ELIAS) PAUL METSCHNIKOFF (1845-1916), ODESSA AND PARIS.

Metschnikoff is the discoverer of phagocytosis. He described this phenomenon in his *Leçons sur la pathologie comparée de l'enflammation.* Paris, 1892. English translation from T. S. Hall, *A source book in animal biology.* (New York, 1951. pp. 544-547).

I was resting from the shock of events which provoked my resignation from the university (of Odessa) and indulging enthusiastically in researches in the splendid setting of the straits of Messina. One day when the whole family had gone to a circus to see some performing apes, I remained alone with my microscope, observing the life in the mobile cells of a transparent starfish larva, when a new thought suddenly flashed across my brain. It struck me that similar cells might serve in the defense of the organism against intruders. Feeling that there was something in this of surpassing interest, I felt so excited that I began striding up and down the room and even went to the seashore to collect my thoughts. I said to myself that, if my supposition was true, a splinter introduced into the body of a starfish larva, devoid of blood vessels or of a nervous system, should soon be surrounded by mobile cells as is to be observed in a man who runs a splinter into his finger. This was no sooner said than done. There was a small garden to our dwelling. I fetched from it a few rose thorns and introduced them at once under the skin of some beautiful starfish larvae as transparent as water. I was too excited to sleep that night in the expectation of the result of my experiment, and very early the next morning I ascertained that it had fully succeeded. That experiment formed the basis of the phagocyte theory, to the development of which I devoted the next twenty-five years of my life.

The essential factor in the inflammatory reaction is an endeavour on the part of the protoplasm to digest the harmful object. This digestive action, in which the whole or almost the whole organism of the Protozoa takes part, is undertaken by the entire plasmodic mass of the Myxomycetes, while, from the sponges upwards, it is confined to the mesoderm. In those cases where the victory remains with the invaded organism, the phagocytic cells of this layer assemble, englobe and destroy the injurious agent. This phagocytic reaction, in the lower scale of animal life, is slow owing to the progression of these cells towards the injurious body being dependent solely on their amoeboid movements; but as soon as a circulatory or vascular system makes its appearance in the course of evolution, it becomes much more rapid. By means of the blood-current the organism can at any given moment send along to the threatened spot a considerable number of leucocytes to avert the evil. When the circulation is partially carried on by a lacunar system there is nothing to intercept the movement of the leucocytes towards the seat of the injury. But when these cells are enclosed within the vessels, they are obliged to adapt themselves specially to fulfil their object, which they do by passing through the vascular wall.

If we accept this conclusion that inflammation in the higher animals is a salutary reaction of the organism and that diapedesis and its accompaniments form part of this reaction, several details of inflammatory phenomena will appear clear to us. For instance the lobed and polymorphous shape of the nucleus of the pus-corpuscles has been remarked. This particular shape is peculiar to the polynuclear leucocytes, which frequent the vast majority (75%) of the total number of white cells. As it was noticed that a quantity of pus-corpuscles died in the exudation, this fact became associated with the curious form of the nucleus; it was said, and is still maintained, that the polynuclear leucocytes are cells predestined to perish and incapable of any considerable activity. On the contrary these leucocytes are precisely the most active cells in the organism. The shape of their nucleus may be more adequately explained as a special adaptation for passing through the vessel-wall. If the process of diapedesis be watched, the difficulty experienced by the nucleus in getting through will at once be noticed. Directly this has occurred, the rest of the protoplasm follows rapidly. It is obvious that a nucleus divided into several lobes can pass through the wall more easily than one not so separated. Hence in pus the polynuclear leucocytes are more numerous than the mononuclear leucocytes, and hence the lobed shape of the nucleus is found only in the leucocytes adapted for diapedesis and does not occur among the invertebrates (except in a few Cephalopoda . . .).

To sum up: Inflammation generally must be regarded as a phagocytic reaction on the part of the organism against irritants. This reaction is carried out by the mobile phagocytes sometimes alone, sometimes with the aid of the vascular phagocytes or of the nervous system. The theory here indicated might be termed the biological or comparative theory of inflammation, since it is founded on a comparative study of the pathological phenomena presented by living cells.

125

EMIL VON BEHRING (1854-1917), MARBURG.

Founder of immunology, especially merited for the prevention of diphtheria and tetanus. Here we have selected (p. 141) from his *Die Geschichte der Diphtherie. Mit besonderer Beruecksichtigung der Immunitaetstheorie.* (1893).

While Jenner, Pasteur and Koch worked with immunisation of man and animals which were infected by a light form of a disease (rabies, small-pox, tuberculosis) and produced a light form of a dangerous disease in healthy men which became thus immune against any later severe infestation of that disease, Behring utilized the changes in the blood serum after a disease had passed an individual to immunise by its serum other individuals against this disease, initially against diphtheria and tetanus.

For some infections we could find in the cell-free body liquids very specific differences in the same individual before, during and after the infection. These specific differences in the different conditions of the body liquids of the same individual are as follows:

1. The body liquids of the healthy individual, are not producing a disease when they are transmitted to individuals of the same or related species.

2. The body liquids of a diseased individual can transmit the same disease upon other individuals, even if the presence of living agents of the disease are excluded with certainty.

3. The body liquids, especially the blood of the individual which is healthy after the disease, can influence healthy individuals in such a way that an infection does not produce a disease,—we may say, that they have become *immune*. The same blood, after it has been cleaned carefully of all cellular elements, will also heal individuals when they become infected with the infectious agents of the same disease.

This has been shown for diphtheria and tetanus.

126

ANDREAS FRANZ WILHELM SCHIMPER (1856-1901), BASEL.

Schimper founded plant ecology. His *Pflanzengeographie auf physiologischer Grundlage*, (1898), was translated into English by W. R. Fischer in 1903. (Here quoted from German, pp. 42, 45, 53, 658, 681; English translation pp. 36, 38, 47, 626, 647).

(*a*) On temperature limits.
(*b*) The *Artemisia*-steppe of Central Asia (after Krassnov).
(*c*) The Desert of Chili (after Niederlein).

(*a*) *Heat.* Every plant can exist only between two extremes of temperature which may be more or less distant, namely the upper and lower temperature threshold. If one of these thresholds is overstepped for a shorter or longer time, this induces death, sooner of later, but at the latest after two or three days. These thresholds are different for every species of plants, but the individuals of the same species, as far as they have developed under approximately the same external conditions, have the same threshold. The absolute limits of plant life are not identical with the limit of every individual function. Every function has its own limits and reaches at a certain degree its temperature optimum. Thus we have to consider three cardinal points in temperature relations. In the same way as the thresholds are characteristic for every function of plant life is also their optimum. It is specific and usually the higher, the higher the minimum threshold of a given plant is. These data form the only basis for the research of temperature influences upon the mode of life and distribution of plant species . . .

Death from cold is, without any doubt, in very many cases a result of the want of water and not of the low temperature. . . .

For the hitherto discussed functions and aggregates of functions the optima lie at high temperatures. We know, however, certain physiological processes for which not only the optima, but also the upper thresholds are so low that, as a rule, they can take place only from late autumn to early spring. This is, of course, the case only in plants of middle and higher altitudes, while tropical plants require high thresholds. . . .

(*b*) *The Artemisia-steppe of Central Asia* (after Krassnov): The main characteristics of this steppe-formation of *Artemisia* is the low growth of its plants, in their separated growth, leaving wide stretches of bare soil between these bunches of growth, and, above all, in the strong prevalence of greyish-green herbs, which are hairy, strongly growing under the heavy sun radiation, and are rich in etheric and aromatic oils. The change from one species to another is in these steppes very rapid and common. When new forms appear almost no trace remain of the wilted predecessors. The steppe is really never completely dried up, even if she often gives this impression. This deceptive appearance is due to the fact that usually very few species with bright flowers are in blossom, except in early spring, when tender, juicy species of Ranunculaceae, Cruciferae, Papaveraceae and Liliaceae, and, among the grasses, *Poa bulbosa* prevail. These are later replaced by *Achillea gerberi* and by the great mass of grasses with rolled and stiff leaves. With the increasing heat and dryness follow: *Alhagi camelorum*, *Xanthium spinosum*, *Ceratocarpus arenarius* and *Eryngium campestre*, which are all extremely thorny plants, whose tender spring leaves are replaced in the dry season by projecting thorns. Towards the end of summer *Artemisia frigida* and *A. maritima* strongly dominate together with salt-plants, whose roots enter five metres deep into the soil and still gain some moisture, when all their neighbours die from drought.

(*c*) *The desert of Chili* (after Niederlein). The vegetation of this desert is perfectly adapted to the terrestrial and physical features of the environment. Stiff are the few grasses; woody the perennial herbs, usually viscous or hairy; squarrose, scrubby, and thorny and—owing to their very few leaves—apparently dead are the shrubs which reach a height of one to three metres. The entire aspect is, with few exceptions, a mixture of tall and short oval, globose, elongated or otherwise shaped dark, greyish-green or yellowish-green, loose (and then shadeless) or thickly grown masses of thorny, woody or brushlike plants on a hard, grey or reddish soil, formed of pebbles, gravel or sand, or on dunes, but eventually entirely hidden within the shifting sand driven by the many storms. The vegetation is here dense, there with great stretches of bare soil. Here isolated individuals, there groups of individuals, often struggling one with another for existence, according to the meagre soil and the level of the ground water. One kind of shrub grows vigorously upward, yet with the appearance of stunted existence; another creeps; a third is pressed against the ground; a fourth contracted into a dwarfed condition; a fifth grows in cushions, and soon most of the shrubs are densely, shortly, vertically and often crookedly branched, thorny, shaggy, or otherwise malformed. The older branches are often dead. The strongly corky bark is black-brown, greyish-green or yellowish-green, usually rough. In some plants it also exudes wax, in others resin or gum. The leaves are, as a rule, tiny and caducous, sometimes scale-like (as in *Fabiana hieronymi*), fleshy as in the salt-shrub, in other bushes leathery, or hard and thorny, in still others they are transformed entirely into thorns or prismatic needles. In *Mimosa* and others the leaves reappear periodically, but are entirely absent in *Monthea aphylla*, *Cassia aphylla*, and some other shrubs. Only a few of the flowers excel by beauty, scent or size.

127

HUGO DE VRIES (1848-1935), AMSTERDAM.

De Vries is the founder of the mutation theory. His first paper in this repect appeared in the C. R. Acad. Sci. Paris. 131. 1900. p. 561, as *Sur la Mutabilité de l'Oenothera lamarckiana.*

Almost always wild species show themselves to our direct observation as unalterable, but highly polymorphic: this means that the seeds of one individual may produce all the forms (of the species), while, in a species with mutability, individuals possessing new characters may appear which in isolated reproduction produce only this newly appeared form.

In cultivated plants the mutability is usually apparent, only it is rather based upon a polymorphism than the result of (real) changes. The truly changing forms of our cultures are almost always based upon hybridization.

It is very rare to find a pure species in mutation. I have obtained such a mutability from *Oenothera lamarckiana*, which I have bred for over twelve years in my experimental garden. It constantly produces new forms. Most of these are unable to develop normally and die before they reach the maturity necessary to the production of seeds; others are completely sterile. Seven forms, however, have produced seeds in numbers large enough to permit of an exact study.

These seven species are *Oe. gigas*, which I have already described (Op. coll. VI, p. 253); *Oe. albida*, with very narrow, whitish leaves, pale yellow flowers and short fruits; *Oe. oblonga* with oblong, petiolate leaves, with short stem, ending in a cluster of flowers smaller than those of the parent plant and with small fruits; *Oe. rubrinervis* with fragile stem due to the imperfect development of the bast-fibres; *Oe. lata* changed into a female by the complete abortion of the pollen (accompanied by an abnormal development of the inner cellular layer of the wall of the anthers) and easily recognized by the broadness of all its organs; *Oe. scintillans* with narrow, dark green, almost shining leaves, with small flowers and fruits; and *Oe. nanella*, a dwarf form only some decimeters high.

Oe. gigas has appeared once only; the other species have more or less regularly appeared in every generation and often in fairly large numbers.

Oe. lamarckiana has been cultivated in its first three generations. from 1886 to 1891 as biennial. The seed-bearers, 6 to 10 in every generation, have flowered every time on a well-isolated square. The following five generations were annuals (1895 to 1899); the seed-bearers followed in parchment bags and were artificially fertilized.

The following genealogical table indicates the number of mutated individuals produced directly by normal *Oe. lamarckiana* seeds:

Generation.		*gigas.*	*albida.*	*oblonga.*	*rubrin.*	*lamarckiana.*	*nanella*	*lata.*	*scintill.*
VIII.	1899		5	1		1700	21	1	
VII.	1898			9		3000	11		
VI.	1897		11	29	3	1800	9	5	1
V.	1896		25	135	20	8000	49	142	6
IV.	1895	1	15	176	8	14000	60	73	1
III.	1890/1				1	10000	3	3	
II.	1888/9					15000	5	5	
I.	1886/7					9			

The notes made every year on this cultivation and on the seeds of a certain number of transformed individuals have led me to the following conclusions:

I. The new species appear abruptly, without any intermediary or preliminary forms; the transformed individual has all the characters of a new type, although it is derived from perfectly normal parents and grandparents.

II. All the seeds of the transformed individuals show the new type, without any return to the characters of *Oe. lamarckiana.* They remain stable after their first appearance. Hence, I am justified in considering them as new species. Yet, *Oe. scintillans* is an exception from this rule; certain individuals reproduce their type only in one third, two thirds or slightly more of their offspring. *Oe. lata* is entirely female and does not reproduce when crossed with the parent species or with other forms; its stability cannot, therefore, be determined. *Oe. nanella* can be regarded as a dwarf variety.

III. The new forms differ in almost every character from the parent species, and thus correspond to the 'small species' of the florists, and not to the varieties of cultivated plants.

IV. The new species appear usually in a fairly large number of individuals, in one and the same generation, as well as in a series of generations. Their number can be estimated to be 1 to 3. This observation seems to confirm the theories of W. B. Scott regarding mutation, derived by him from the continuity of palaeontological series.

V. The characters of the new species show no evident relationship to the ordinary varieties of the parent species. Mutation seems to be independent of "variation."

The new characters appear without any direction, as they are supposed to appear in the Darwinian principle of development. They concern all the organs and transform them in all directions; they are either injurious, or indifferent, or in some cases probably beneficial for the individuals concerned. Most of the forms described above are weaker or more fragile than *Oe. lamarckiana;* only *Oe. gigas* seems to be in all respects more robust. Many forms are sterile; these have not been mentioned in our table.

128

GERRIT GRIJNS (1865-1944), NETHERLANDS.

Grijns, public health officer in Netherlandsch India, studied the vitamin diseases caused by inadequate diet. One of the earliest studies in this field is his paper *Over Polyneuritis gallinarum* (from Geneesk. Tijdschr. Nederlandsch India. 41. 1901. p. 3 ff.).

We had in polyneuritis gallinarum, as described by Eijkman, a disease which, both in its clinical symptoms and in the changes which it produces in the peripheral nervous system, very much resembles beri-beri. The cause of both was hitherto unknown. There was one remarkable difference that, while a very close, direct connection appeared to exist between polyneuritis of fowls and the nature of the food, beri-beri could not be so directly connected with the feeding; indeed except for the results of Vorderman's enquiry there were not many observations which appeared to support this connection. . . .

In my opinion an explanation of the peculiar symptoms which occur in the disease in question can be sought in two directions. Either we can presume a deficiency, a partial starvation, or we can imagine that there is an agent distributed in nature, which exercises a degenerative influence on the nerves and that it depends on the nature of the food, whether the peripheral nervous system has enough power of resistance to get the better of this influence. In the latter case, it is most likely that the harmful agent is a microorganism. . . .

When, therefore, in certain foods the substances indispensable for the nervous system are lacking or are present in insufficient quantity, in the first place any reserve supply, which is present either in the nerve itself or in the blood or in some other organ, will be used up. When this occurs, disturbances will develop, just as in albumen starvation the circulating albumen is used up first and then that of the organs. Therefore, from the time the deficiency begins, a certain time must elapse before symptoms set in in the nervous system; there will be a sort of incubation period. This will be the longer, the smaller the deficiency of the nutritive substances, which are indispensable for the nerve, in the food taken.

If besides these 'protective substances', as I have named the still unknown compounds which I have mentioned now and then without any prejudice, albumen is also withheld, which is the case with an absolute diet, then, as is well known, the albumen indispensable for metabolism is drawn from the organs, especially the muscles. As the muscular substance also possesses protective qualities, it is probable that with the albumen from the muscles enough of the substances are liberated and become available for the nerves, to prevent polyneuritis. This is a simple explanation why, when there is absolute starvation no polyneuritis is observed.

129

CLARENCE ERWIN McCLUNG (1870-1946). CARNEGIE FOUNDATION, U.S.A.

McClung and Montgomery were among the first to discover the so-called X- or sex-determining chromosomes. See his: *The accessory chromosome as sex-determinant.* Biol. Bull. 3. 1902. pp. 43 ff.

Briefly stated, then, my conception of the function exercised by the accessory chromosome is that it is the bearer of those qualities which pertain to the male organism, primary among which is the faculty of producing sex cells that have the form of spermatozoa. I have been led to this belief by the favourable response which the element makes to the theoretical requirements conceivably inherent in any structure which might function as a sex determinant.

These requirements, I should consider, are that: (*a*) The element be chromosomic in character and subject to the laws governing the action of such structures. (*b*) Since it is to determine whether the germ cells are to grow into the passive, yolk-laden ova or into the minute motile spermatozoa, it should be present in all the forming cells until they are definitely established in the cycle of their development. (*c*) As the sexes exist normally in about equal proportions, it should be present in half the mature germ cells of the sex that bears it. (*d*) Such disposition of the element in the two forms of germ cells, paternal and maternal, should be made as to admit of the readiest response to the demands of environment regarding the proportion of the sexes. (*e*) It should show variations in structure in accordance with the variations of sex potentiality observable in different species. (*f*) In parthenogenesis its function would be assumed by the elements of a certain polar body. It is conceivable, in this regard, that another form of polar body might function as the non-determinant bearing germ-cell. . . .

Sex, then, is determined sometimes by the act of fertilization and cannot be subsequently altered. But between this extreme and the other of marked instability there may be found all degrees of response to environment. It must accordingly be granted that there is no hard-and-fast rule about the determination of sex, but that specific conditions have to be taken into account in each case. . . .

Finally, with respect to the evidence to be derived from parthenogenesis, it should be remembered that we are here dealing with a practical suppression of sexuality and it is to be expected that extensive modifications of the ordinary process will follow. If the egg takes upon itself all the functions commonly exercised by it in conjunction with the spermatozoan, it must be that the determination of sex is included. This, in some instances, is a final choice on the part of the ovum and ever afterward one sex only is produced by it; again, however, it maintains a responsive attitude toward environments and gives rise to the sex most needed by the species. It is to be hoped that the very promising field of artificial parthenogenesis will throw much light upon these vexed problems.

130

IVAN PETROVICH PAVLOV (1849-1936), LENINGRAD.

This famous Russian physiologist is most famous by his so-called ‘ conditioned reflexes ’ which are no reflexes, but much more complicated processes connected with learning. His first comprehensive paper appeared in 1897: *Die Arbeit der Verdauungsdruesen*, reprinted in the Comprehensive Work of the Digestive Glands. (1902).

We give here a short, later résumé on the conditioned reflexes (from *Experimental Psychology and Psycho-pathology in Animals*. Int. Congr. Medic., Madrid. 1903).

All the reactions of adaptation depend upon a simple reflex act which has its beginning in certain external conditions, affecting only certain kinds of centripetal nerve endings. From here the excitation runs along a certain nerve path to the centres whence it is conducted to the salivary glands, calling out their specific function. . . .

Here is another series of constantly recurring facts. The object acts upon the salivary glands at a distance not only as a complex of all its properties, but through each of its individual properties. You can bring near the dog your hand having the odour of the meat powder, and that will be enough to produce a flow of saliva. In the same manner the sight of the food from a further distance, and consequently only its optical effect, can also provoke the reaction of the salivary glands. But the combined action of all these properties always gives at once the larger and more significant effect, i.e., the sum of the stimuli acts more strongly than they do separately. . . . If you combine food with an undesired object, or even with the qualities of this object, the reaction is that to an undesired object. . . .

All the above facts lead, on the one hand, to important and interesting conclusions about the processes in the central nervous system, and, on the other hand, to the possibility of a more detailed and successful analysis. Let us consider some of our facts physiologically, beginning with the cardinal ones. If a given object—food or a chemical—is brought in contact with the special oral surface, and stimulates it by virtue of those of its properties upon which the work of the salivary glands is especially directed, then it happens that at the same time other properties of the object, unessential for the activity of these glands, or the whole medium in which the object appears, stimulate simultaneously other sensory body surfaces. Now these latter stimuli become evidently connected with the nervous centre of the salivary glands, whither (to this centre) is conducted through a fixed centripetal nervous path also the stimulation of the essential properties of the object. It can be assumed that in such a case the salivary centre acts in the central nervous system as a point of attraction for the impulses proceeding from the other sensory body surfaces. Thus from the other excited body regions, paths are opened up to the salivary centre. But this connection of the centre with accidental pathways is very unstable and may of itself disappear. In order to preserve the strength of this connection it is necessary to repeat time and again the stimulation through the essential properties of the object simultaneously with the unessential. There is established in this way a temporary relation between the activity of a certain organ and the phenomena of the external world. This temporary relation and its law (reinforcement by repetition and weakening if not repeated) play an important role in the welfare and integrity of the organism; by means of it the fitness of the adaptation between the activity of the organism and the environment becomes more perfect. Both parts of this law are of equal value. If the temporary relations to some object are of great significance for the organism, it is also of the highest importance that these relations should be abandoned as soon as they cease to correspond to reality. Otherwise the relations of the animal, instead of being delicately adapted, would be chaotic.

131

ROSS GRANVILLE HARRISON (1870-19—). U.S.A.

From this great experimental zoologist we bring a chapter on the early discovery of living tissue cultures. From: *Observations on the living developing nerve fibre.* (Anatom. Record, 1907, p. 116 ff.).

The immediate object of the following experiments was to obtain a method by which the end of a growing nerve could be brought under direct observation while alive, in order that a correct conception might be had regarding what takes place as the fibre extends during embryonic development from the nerve centre out to the periphery.

The method employed was to isolate pieces of embryonic tissue, known to give rise to nerve fibres, as for example, the whole or fragments of the medullar tube, or ectoderm from the branchial region, and to observe their further development. The pieces were taken from frog embryos about 3 mm. long at which stage, i.e., shortly after the closure of the medullary folds, there is no visible differentiation of the nerve elements. After carefully dissecting it out, the piece of tissue is removed by a fine pipette to a cover slip upon which is a drop of lymph freshly drawn from one of the lymph-sacs of an adult frog. The lymph clots very quickly, holding the tissue in a fixed position. The cover slip is then inverted over a hollow slide and the rim sealed with paraffin. When reasonable asceptic precautions are taken, tissues will live under these conditions for a week and in some cases specimens have been kept alive for nearly four weeks. Such specimens may be readily observed from day to day under highly magnifying powers.

While the cell aggregates, which make up the different organs and organ complexes of the embryo, do not undergo normal transformation in form, owing, no doubt, in part, to the abnormal conditions of mechanical tension to which they are subjected; nevertheless, the individual tissue elements do differentiate characteristically. Groups of epidermis cells round themselves off into little spheres or stretch out into long bands, their cilia remain active for a week or more and a typical cuticular border develops. Masses of cells taken from the myotomes differentiate into muscle fibres showing fibrillae with typical striations. When portions of myotomes are left attached to a piece of the medullary cord the muscle fibres which develop will, after two or three days, exhibit frequent contractions. In pieces of nervous tissue numerous fibres are formed, though, owing to the fact that they are developed largely within the mass of transplanted tissue itself, their mode of development cannot always be followed. However, in a large number of cases fibres were observed which left the mass of nerve tissue and extended out into the surrounding lymph-clot. It is these structures which concern us at the present time. . . .

These observations show beyond question that the nerve fibre develops by the outflowing of protoplasm from the central cells. This protoplasm retains its amoeboid activity at its distal end, the result being that it is drawn out into a long thread which becomes the axis cylinder. No other cells or living structures take part in this process.

The development of the nerve fibre is thus brought about by means of one of the very primitive properties of living protoplasm, amoeboid movement, which, though probably common to some extent to all the cells of the embryo, is especially accentuated in the nerve cells at this period of development.

The possibility becomes apparent of applying the above method to the study of the influences which act upon a growing nerve. While at present it seems certain that the mere outgrowth of the fibres is largely independent of external stimuli, it is, of course, probable that in the body of the embryo there are many influences which guide the moving end and bring about contact with the proper enstructure. The method here employed may be of value in analyzing these factors.

132

LELAND OSSIAN HOWARD (1857-ca.1954), WASHINGTON.

L. O. Howard was the founder of modern agricultural and medical entomology. He paid great attention to the biological equilibrium in nature, attributing it mainly to the parasites. He expresses his views, characteristic for early animal ecology in a study on *Lymantria dispar* in New England, together with W. F. Fiske (Bull U.S. Bur. Entom. 91. 1911. p. 107 ff.).

To put it dogmatically, each species of insect in a country where the conditions are settled is subjected to a certain fixed average percentage of parasitism which in the vast majority of instances and in connection with numerous other controlling agencies, results in the maintenance of a perfect balance. The insect neither increases to such abundance as to be affected by disease or checked from further multiplication through lack of food, nor does it become extinct, but throughout maintains a degree of abundance in relation to other species existing in the same vicinity, which, when averaged for a long series of years, is constant.

In order that this balance may exist it is necessary that among the factors which work together in restricting the multiplication of the species there shall be at least one, if not more, which is what is here termed facultative, and which, by exerting a restraining influence which is relatively more effective when other conditions favour undue increase, serves to present it.

A very large proportion of the controlling agencies, such as the destruction by (extreme) climatic conditions, is to be classed as catastrophic, since they are wholly independent in their activities upon whether the insect which incidentally suffers is rare or abundant. The average percentage of destruction remains the same, no matter how abundant or how near to extinction the insect may have become.

Destruction through birds and other predators, works in a radically different manner. These predators are not directly affected by the abundance or scarcity of any single item in their varied menu. Like all other creatures they are forced to maintain relatively constant abundance among the other forms of animal and plant life, and since their abundance from year to year is not influenced by the abundance or scarcity of any particular prey they cannot be ranked as elements in the facultative control of such species. On the contrary, they average to destroy a certain gross number of individuals each year, and since their destruction is either constant, or, if variable, is not correlated in its variations to the fluctuations in abundance of the insect preyed upon, it would most probably represent a heavier percentage when that insect was scarce than when it is common. A natural balance can only be maintained through the operation of facultative agencies which effect the destruction of a greater proportionate number of individuals as the insect in question increases in abundance. Of these facultative agencies parasitism appears to be the most subtle in its action. Disease or insufficient food supply does not as a rule become effective until the insect has increased to far beyond its average abundance.

133

THOMAS HUNT MORGAN (1886-1945), NEW YORK AND PASADENA.

This great zoologist is the founder of the modern genetical theory of chromosomes. In his book *The Mechanism of Mendelian Heredity* (New York, 1915, pp. 64-69), published together with his most important pupils Sturtevant, Muller and Bridges, contains the technique by which the mapping of the genes of a chromosome was done.

The fact that crossing over makes less likely another crossing over in a nearby region, or in a sense *interferes* with a second crossing over nearby, is called *interference*. As has been shown, interference decreases with increase of distance.

In the construction of a chromosome map the distance taken as a unit is that within which 1 per cent of crossing over will occur. Thus, yellow and white are placed one unit apart, since there is 1 per cent of crossing over between yellow and white. White and bifid give 5 per cent of crossing over, hence they are placed five units apart; and since yellow and bifid give 6 per cent, bifid must be on the other side of white from yellow. In a similar way the relative positions of the other factors have been plotted, the position of any factor on the map being determined, as far as possible, by the per cent of crossing over between it and the factor nearest to it. In general, it may be said that the number of units of distance on the map between any two factors A and C—, will equal the per cent of crossing over that will actually be observed between them in an experiment involving these two pairs of factors, even although their distance on the map may not have been obtained directly from their linkage with each other, their positions having, instead, been determined by their linkage with other factors. On account of double crossing over, however, this would not be expected to hold for very long distances; and, as has been explained, we do actually find that, if long distances are involved, the distance between A and C determined as on the map, by adding the intermediate distances A—B and B—C, is longer than the distance A—C as directly determined in an experiment involving only these two pairs of factors. It nevertheless remains true that, given the distance between any two factors on the map, the per cent of crossing over between them can always be calculated from this distance (since the amount of discrepancy due to double crossing over also depends on the distance); this shows that the amount of crossing over between them is an expression of their position in a *linear* series. This striking fact, that the mathematical relations between the various linkage values conforms to a linear series, is a strong argument that the factors are actually arranged in line in the chromosomes. If the relations between the various linkage values were not determined by some linear relation of the factors but were of a random sort, these relations could not be calculated from a linear map. . . .

This may be illustrated by an actual case. The first formula shows the composition of a hybrid female which has received from her mother the mutant factors: yellow, white, abnormal, bifid, vermilion, miniature, sable, rudimentary, and forked; and from her father the normal allelomorphs of these factors, together with the dominant mutant factor, bar:

y w a bi v m s r f b′
———————————————
Y W A Bi V M S R F B′

A number of females of that type have been made up by Muller. The next formula shows the kinds of eggs that were produced by one of these females and the numbers of each kind that were produced:

None-crossovers:	y	w	a	bi	v	m	s	r	f	b′	—6
	Y	W	A	Bi	V	M	S	R	F	B′	—8
Single crossovers:	Y	W	a	bi	v	m	s	r	f	b′	—2
	Y	W	A	Bi	v	m	s	r	f	b′	—2
	y	w	a	bi	V	M	S	R	F	B′	—2
	Y	W	A	Bi	V	M	S	r	f	b′	—1
	Y	W	A	Bi	V	m	s	r	f	b′	—1
Double crossover:	y	w	a	bi	v	m	s	R	F	B′	—1
	y	w	a	bi	V	M	S	R	F	b′	—1

Counts of over 600 offspring from females of the same type have given similar results. The characteristic method of interchange here demonstrated may perhaps be better realized by contrasting the combinations just given with the following, which illustrate types of eggs found *not* to be produced by such females:

y W a Bi V m S r f B′
Y W a bi V m s R f B′.

It is not supposed, however, that the per cent of crossing over represents precisely the distance between the factors, for it may be that crossing over is more likely to take place in one region of the chromosome than in another. In that case the distances between factors in this region calculated from the amount of crossing over between them, would be relatively greater than the actual distance. It is supposed, however, that at least the order of the factors in the diagram represents their real order. Sturtevant has found definite factors which alter the amount of crossing over in the chromosomes, and these factors actually do affect the amount of crossing over differently in the different regions . . . It is to be noted, however, that the order of the factors remains unchanged.

Index